From Jungle Rubber to Rubber Agroforestry Systems

History of Rubber Agroforestry Practices in the World

Éric Penot, editor

Éditions Quæ

To cite this book:
Penot É. (ed), 2024. *From Jungle Rubber to Rubber Agroforestry Systems. History of Rubber Agroforestry Practices in the World*. Versailles, éditions Quæ, 240 p.

Cover: Agroforestry system based on rubber, ananas and durian in double spacing design in Myanmar in 2019 during immature period
(© É. Penot, Cirad).

Éditions Quæ
RD 10, 78026 Versailles Cedex
www.quae.com – www.quae-open.com

© Éditions Quæ, 2024

ISBN (print): 978-2-7592-3934-4   ISBN (PDF): 978-2-7592-3935-1
ISBN (eBook): 978-2-7592-3936-8

Table of contents

Acknowledgments

This book is dedicated to the memory of Dr. Afs Budiman, chairman of GAPKINDO from 1993 to 2001. Budiman was a close friend of the author and played a key role in the development of RAS (Rubber Agroforestry Systems) in the 1990s through both his personal involvement and funding of the very first research actions in 1994. He was a strong supporter of RAS as a way of diversifying rubber farmers' income, improving farm resilience and making the Indonesian rubber sector sustainable. He always helped and supported funding for research. The RAS research program (SRAP, Smallholder Rubber Agroforestry Project 1994/2007) and most of the ideas reported in this book were discussed, shared and promoted by Budiman. He passed away in 2009 and the agroforestry research team will always remember his exceptional personality.

We would like to thanks all researchers, students (16), technicians, extension agents, farmers, and any persons involved in agroforestry activities I met in the last 30 years that held us to define originally the SRAP project in Indonesia and later on all the studies implemented in many countries (Indonesia, Thailand, Cambodia, Vietnam, Myanmar, Sri Lanka, Colombia, Brazil and Côte d'Ivoire mainly), to understand the "who and what" of agroforestry practices.

We would like to thank in particular Ms Daphne Goodfellow who implemented the complete English editing of this book.

Foreword

Rubber Agroforestry Systems (RAS) have been developed by local farmers in Southeast Asia (Indonesia, Thailand and Malaysia) as soon as the 1920s initially through the development of jungle rubber. Jungle rubber is a very practical and easy way to develop at very low cost non clonal rubber (seedlings) plantations with forest regrowth. Jungle rubber was the main smallholding rubber cropping system until 1950s. Then, for political reasons, clonal plantations with a better productivity were developed though national planting programs in the 1950s in Malaysia, in the 1960s in Thailand and later on in the 1970s in Indonesia. In the 1990s, jungle has disappeared in Malaysia and Thailand when in Indonesia, jungle rubber was still covering 3 million ha for 70% of the rubber national production. In 2023, most of the jungle rubber has disappeared or is not anymore tapped, replaced by monoclonal plantation.

However, in some countries, some local farmers continue to adopt or develop agroforestry practices, basically associating rubber with various number and types of plants and trees in both immature and mature period, in order to increase global productivity at plot level and to diversify sources of incomes to increase farms' resilience. These countries are India, Sri Lanka, Indonesia and Thailand and to a lesser extend Columbia and Brazil. In other countries, more recent rubber booms or lack of history and knowledge about agroforestry limited rubber development to monoculture with good success as well.

In this book, we try to explain what have been the historical and societal conditions for agroforestry to develop in these countries, in particular Thailand and Indonesia. The interest for local farmers to develop agroforestry systems is still very important: incomes diversity to tackle with low rubber prices and with positive environmental externalities. Long periods of low rubber prices since the 1990s increase interest of many farmers for agroforestry practices.

In 2024, environmental concerns, cropping systems sustainability and more globally positive externalities are largely taken into account not only by farmers but also by governments, research and extension bodies as well as most Non Gouvernmental Associations (NGOs).

There is evidently a future for RAS in the current world with global economic uncertainty. However, this is still relatively difficult for most farmers to develop agroforestry practices in countries with no local knowledge and know-how such as Vietnam, Cambodia, Myanmar, China, as well as central America and Western Africa. We listed all constraints for agroforestry adoption.

This is mainly now a political decision for governments to allocate funds to promote agroforestry where it could be possible and locally adapted. RAS is not the perfect "panacea" for agricultural economic rubber development but it might help in many situations depending on farmers situations and strategies as well as local existing markets for the various associated products such as fruits, timber, food, resins, spices, medicinal plants, rattan and other plants.

The objective of this book is to provide evidence of RAS interest and constraints as well as an analysis of local historical evolution of RAS in order to understand how to develop potentially such systems in other countries. Crop diversification is still a very important component for most farmers strategies in the world, and some crops in monoculture might also be in concurrence with rubber (oil palm, pepper, fruit trees…). In agroforestry systems, the objective is to find complementary between crops within one plot. The book integrates various sources from the author and associated researchers and students, written between 1994 and 2024 that have been updated. All original sources dans dates will be precised.

The introduction presents the rubber world and the definition of the agroforestry concept. Chapter 1 presents the original development of jungle rubber based on the use of seedlings as the main agroforestry system in Southeast Asia and the development in the 1990s of the RAS concept (Rubber Agroforestry Systems) based on the use of clonal planting material. Chapter 2 illustrates the development of RAS in Indonesia and Thailand and the way to develop it through "innovation platforms". Chapter 3 presents the current state of RAS in the world. Chapter 4 displays current expectations of RAS, impacts and contribution to today's main challenges on biodiversity, eco-systemic services, environmental concerns, externalities and impact on farmers' income. The conclusion suggests some potential tracks and perspectives for further agroforestry development in the very next future.

I personally strongly believe that if historically famers develop on their own such adapted agroforestry systems in some countries, there is still a future for these very flexible and locally adapted agroforestry systems in many different situations in the tropical world where rubber is present, depending now on government's willingness to tackle with farmers objectives and global environmental concerns.

Introduction

Éric Penot, Joseph Adelegan, Lekshmi Nair, Hugo Lehoux,
Adrien Perroches, Lucie Poline, Jerôme Sainte-Beuve

▸▸ Rubber in the world

The place of rubber

This section has been originally partly published in a Cirad report for AFD using IRSG data in Penot et al. (2020)[1].

Natural rubber is a key product for the global economy because its elasticity and strength have never been perfectly reproduced in synthetic rubber. Natural rubber is extensively used in the tire industry, whose growth due to increasing transport by car, truck and plane, has a direct impact on the demand for rubber (Sainte-Beuve, 2015). As a result, in one decade, rubber plantations grew by more than 2 million hectares to reach 12 million hectares worldwide (Figure I.1).

World production of natural rubber (2017) reached 13.5 million tons, while synthetic rubber production accounted for 15.06 million tons (IRSG, 2018). The vast majority of natural rubber is produced in Asia (Figure I.2)[2].

The order of the top 10 rubber producing countries has remained virtually unchanged since the 2000s, but in the decade 2007 to 2017, annual world production increased from 10.1 to 13.55 million tons (Figure I.3).

The top 25 producing countries can be classified in five groups based on their annual production (Table I.1). The following pages describe the increase in production in each of these groups.

---

1. Éric Penot, Philippe Thaler, Yann Nouvellon, Bénédicte Chambon, Jérôme Sainte Beuve, 2020. Revue de la littérature sur les Standards de la Filière Hévéa. Rapport AFD. 41 p.
2. Partial source: IRSG data and report by Hugo Lehoux, Adrien Peroches, Lucie Poline, Éric Penot, Jérôme Sainte-Beuve. Rubber in the world. Rubber growing throughout the World. Overview of production dynamics, market and value-chain sustainability challenges. FTA project. Montpellier, 2019.

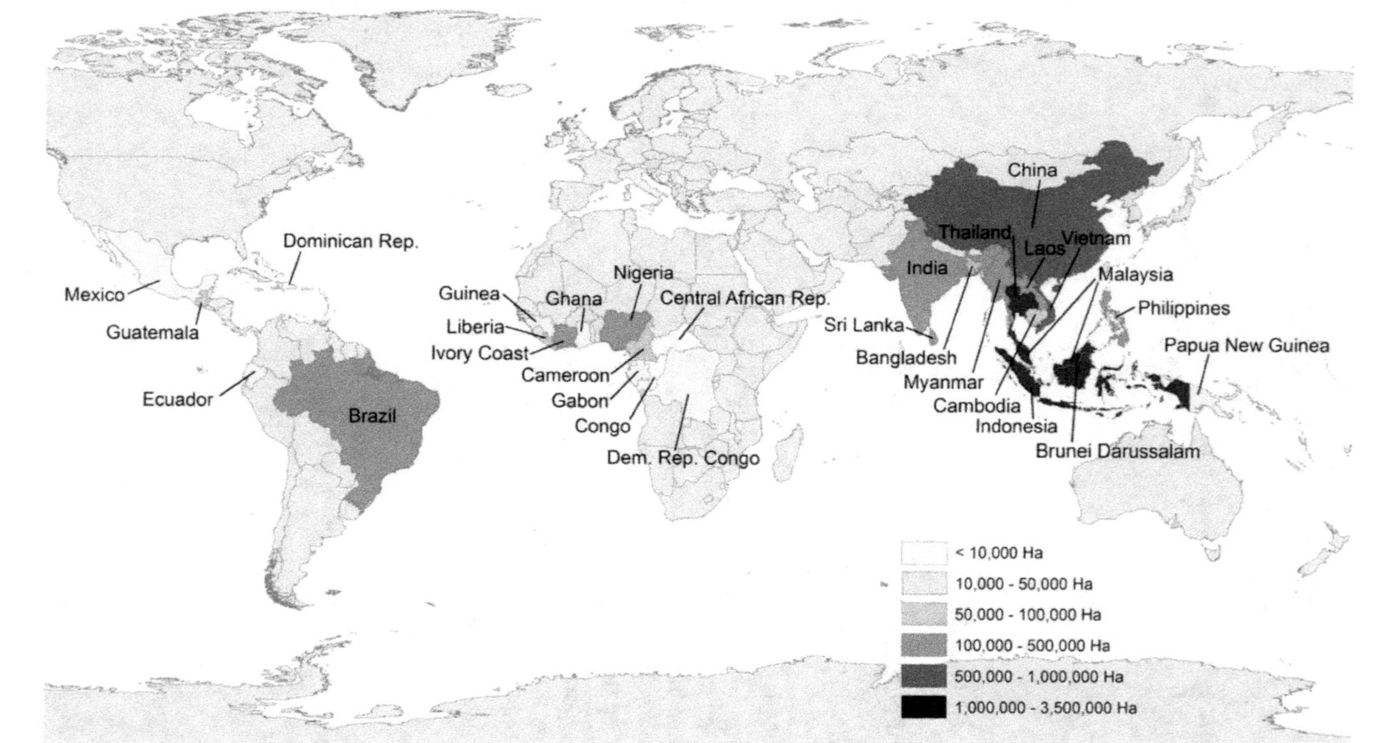

Figure I.1. Surface area of plantations in rubber producing countries (© 2015 The Authors Conservation Letters Published by Wiley Periodicals, Inc.; Warren-Thomas et al., 2015)

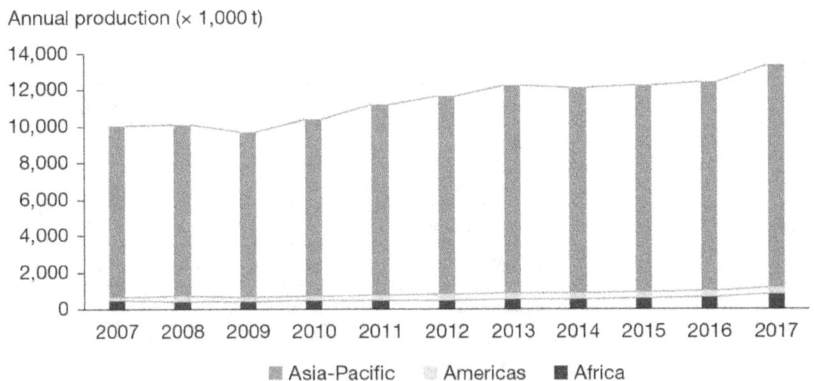

Figure I.2. Annual rubber production per continent between 2007 and 2017 (IRSG, 2018)

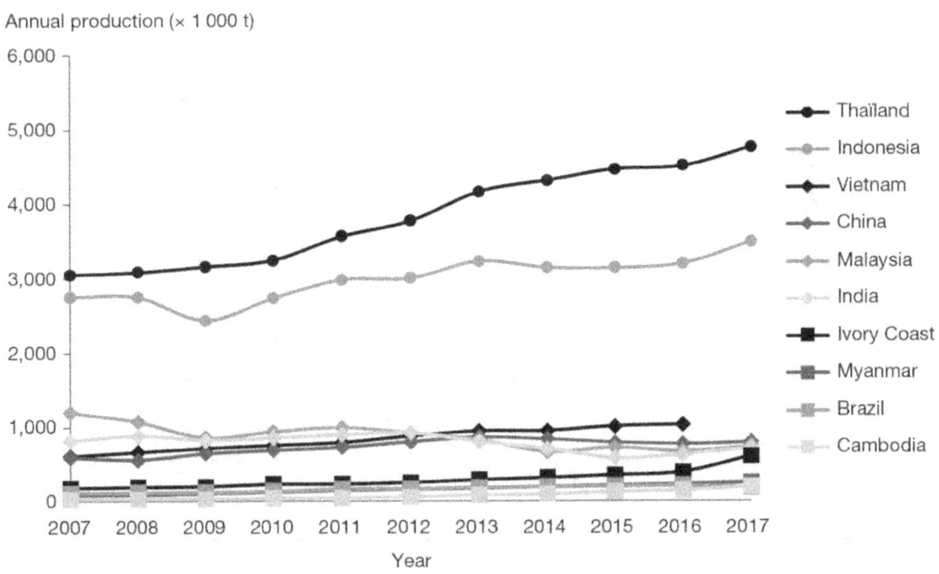

Figure I.3. Annual production of the 10 biggest rubber producing countries (IRSG, 2018)

Table I.1. Countries grouped based on their annual production

Category	Country	Production in 2018 (x 1,000 tons)
Group A 3 to 5 million tons	Thailand	4,755.0
	Indonesia	3,499.0
Group B 0.5 to 1 million tons	Vietnam	1,032.0
	China	798.0
	Malaysia	741.0
	India	713.0
	Côte d'Ivoire	604.0
Group C 100 to 250,000 tons	Myanmar	249.0
	Brazil	184.0
	Cambodia	193.3
	Philippines	1022
	Guatemala	100.2
	Sri Lanka	83.1
	Laos	78.3
Group D 40 to 60,000 tons	Liberia	63.0
	Nigeria	53.0
	Cameroon	53.0
	Ghana	37.0
Group E Less than 20,000 tons	Gabon	21.2
	Bangladesh	21.0
	Guinea	17.0
	Mexico	18.1
	Republic of Congo	13.2
	Colombia	12.0
	Americas – Other countries	9.2
	Bolivia	5.7
	Papua New-Guinea	5.7
	Africa – Other countries	4.8

The development of rubber plantations. Booms in China, Côte d'Ivoire, Laos, Cambodia and Vietnam

Two countries account for 61% of world production: Thailand and Indonesia. This is also the case for the last countries to have joined or are in the process of joining the leading group (Myanmar, Cambodia, Côte d'Ivoire, Vietnam and Brazil, a special case), which significantly increases the total area planted (Table I.2).

Table I.2. Increasing production by countries with growing global production dynamics (IRSG, 2018)

Country	Rank	2007 (x 1,000 tons)	2017 (x 1,000 tons)	Increase (x 1,000 tons)	Increase (Percentage)
Thailand	1	3,056	4,775	+ 1,719	+56%
Indonesia	2	2,755	3,499	+ 744	+27%
Vietnam	3	606	1,094	+ 488	+80%
Côte d'Ivoire	7	183	604	+ 421	+230%
Myanmar	8	89	249	+ 160	+ 180%
Brazil	9	116	184	+ 68	+ 59%
Cambodia	10	33	193	+ 160	+ 485%

These countries operate under different models. While production in Thailand and Côte d'Ivoire is mainly based on village plantations (family and family business), Cambodia is developing industrial plantations on private concessions. Private investment is all the more important as land prices are low and the country's policy favour the development of large plantations (Chambon et al., 2018).

It is interesting to note that in Cambodia, the development of large industrial plantations follows more or less the same logic as in this group of countries at the beginning of the last century: attractive selling prices for raw materials and labour, public policies favouring domestic and foreign private investment and the desire to develop an agricultural model based on agribusiness (Byerlee, 2014).

In the case of Brazil, production is growing slightly but suffers from severe constraints. The rubber comes from industrial plantations under the pressure of the phytopathogenic fungus *Microcyclus ulei*, which limits its expansion to so-called "escape" zones. An escape zone corresponds to an area whose agro-ecological conditions limit the development of *Microcyclus ulei*. The rest of the rubber comes from tapping in spontaneous forest areas.

Thailand is currently the world's largest producer with 4.77 million tons in 2017, corresponding to 37.1% of global production. Rubber production has continued to increase at an average rate of 4.3% per year for the last 5 years. Thailand is the only country where rubber cultivation has been developed exclusively by family farms. This is largely due to the fact that the country has never been colonised, the Thai state has strongly supported these family farms (Fox and Castella, 2013), and had no policy to encourage private investment and large-scale industrial plantations. Support for family farms in southern Thailand, the traditional cradle of rubber cultivation in the 1950s and 1960s, was also established for political reasons, mainly to counter the communist rebellion (like in Malaysia) and to esure a good source of income for local farmers. Industrial plantations occupy a very small place in Thailand, 3.5% to 7% of the total rubber production area (Chambon et al., 2018; IRSG, 2018). The world's major industrial groups have acquired land in Thailand, but most of it is fragmented, with an average area of 63 ha for this type of industrial plantations. It should be noted that the model of family business plantations is in full development, particularly in the so-called marginal areas (Chambon et al., 2018; Fox and Castella, 2013). Their average surface area ranges between 10 and 300 ha.

Indonesia[3] has been the world's second largest producer for many years, with 3.4 million tons in 2017, or 25% of the world's natural rubber production. Like in many countries in Southeast Asia, rubber tree cultivation was developed under colonisation in the form of "domains". At the beginning of the 20th century, natural rubber prices were very volatile, which pushed tire manufacturers to internalise the production stage. In Indonesia, this was the case of Goodyear (Barlow, 1978). Small family farms very quickly adopted rubber in the 1920s, which reached 85% of the country's total area (Fox and Castella, 2013; IRSG, 2018). In the 1960s, some of the large plantations were nationalised to form the state-owned company PTP (PT Perkebunan Nusantara III). The State introduced concession policies promoting the development of oil palm in the 1990s, thereby contributing to massive deforestation (Byerlee, 2014; Feintrenie *et al.*, 2010).

Over the last decade, natural rubber production has increased at an average rate of 2.4% per year, with a slight slowdown in growth since 2013. This growth is due to the conversion of land formerly dedicated to cocoa, tea and coffee to rubber, but also to oil palm and *Acacia mangium* (Feintrenie *et al.*, 2010). The slowdown is being caused by the reduction in available land and competition with other types of speculation. Yields in Indonesia are reported to be lower than in other producing countries, mainly due to the use of unproductive tree planting material in jungle rubber systems (ANRPC[4]), the ageing of the trees and competition with oil palm when replanting.

Industrial plantations represent 14% of the area planted with rubber trees (Chambon *et al.*, 2018), but are decreasing in favour of village plantations, but also of industrial plantations of *A. mangium* and oil palm. These industrial plantations often belong to the state-owned PTP or are foreign-owned private plantations (owners from China and Singapore). Private plantations can be very large: 35,000 hectares for Michelin and 24,000 hectares for Bridgestone. These plantations mainly follow the hybrid planting model called NES[5], i.e., an industrial plantation feeding a factory, surrounded by family plantations. The possibility to increase the extent of industrial plantations is currently quite limited. Some private plantations may consider taking over concessions that have fallen into the public domain or planting in areas that are still untouched but difficult to access. Access to land is strongly dependent on policies, which are not as favourable to major concessions as in the past. At the moment, the Indonesian government does not wish to open new concessions, but this could change in the future.

GAPKINDO, the association of Indonesian rubber producers, introduced a policy to improve rubber quality in the 1990s. Actions to improve quality are also successfully implemented by private companies themselves (Dao, 2015).

India is the world's sixth largest producer with 713,000 tons produced in 2017, representing 5.3% of global production. Developed under English colonisation, rubber plantations have always been in the hands of smallholders, with production for the domestic market. These are intensive, small-scale family farm systems, resulting in very high land productivity (Viswanathan and Shivakoti, 2008); 89% of Indian production

---

3. This section has been originally published in: Éric Penot, Bénédicte Chambon, Jérôme Sainte Beuve, 2023. An analysis and comparaison of the rubber smallholder sector in 5 countries: Cote d'Ivoire, Thailand, Indonesia, Vietnam and Cambodia. FTA/CIFOR final report, Montpellier France.
4. The Association of Natural Rubber Producing Countries (ANRPC). http://www.anrpc.org/
5. PIR/NES: NES = Nucleus Estate Scheme which has its Indonesian equivalent; PIR = Perkebunan Inti Rakhyat.

is provided by family plantations (Fox and Castella, 2013). India wanted to increase its domestic market for natural rubber, in line with its increasing economic growth. It therefore put in place policies to support plantation renewal, including through its rubber production department at the Indian Rubber Office. However, the policies do not seem to have been sufficient to generate a strong trend towards planting or renewal of rubber plantations. The very significant ageing and fragmentation of the plantations has limited the expected growth. Over the past ten years, production has declined at an average annual rate of 1.2%. Currently, prices are subsidised to maintain or even restart plantations, with the aim of producing for the domestic market. It should be noted that there are also plantations in the seven northeastern states, but yields are lower there (ANRPC).

The dynamics and problems are similar in Sri Lanka with an additional constraint due to the prevalence of heavy rains. Production drops drastically when heavy rains fall every day. This has led farmers to implement rain protection practices such as installing a rainguard to protect the notch and the cup. Agroforestry systems in association with tea were developed in the 1990s.

The years following 2010 saw significant rubber booms in some countries: Côte d'Ivoire, Vietnam, Laos, Cambodia, and China, mainly through the development of rubber monoculture. Rubber agroforestry systems (RAS) remain an interesting alternative in some other countries such as Indonesia, Thailand, Sri Lanka, Columbia and India.

▸▸ The concept of agroforestry and agroforestry systems

This section has been published in 1999 as a Cirad working document[6].

The objective of agroforestry can be defined as the reconciliation of two types of land exploitation which have deeply affected the countryside of both tropical and temperate countries in recent centuries: agriculture and forestry. The main feature which characterises agroforestry is the combination, or association, of several annual and perennial plants in the same field.

Agroforestry systems (AFS) are one type of "cropping system", in which the field is homogeneously managed, using a particular technical pathway (or "technological pattern") and a defined plant succession. One may consider AFS as cropping systems, possibly based on one main species. A systemic approach is appropriate to define systems in which labour, inputs, land use and know-how are managed under a particular strategy. Agroforestry strategies can be defined mainly through three features: (i) the minimisation of risk (crop failure), (ii) the optimisation of labour efficiency, different levels of intensification depending on the system, and (iii) the possible use of improved planting material and inputs, according to a strategy that takes land tenure and occupation into account. At the field level, combinations of crops, planted or the result of natural regeneration, lead to interactions between plants: competition and sometimes complementarity.

The traditional definition of ICRAF (International Center for Research in Agroforestry) is the following: *"A collective name for land use systems and practices in which woody perennials are deliberately integrated with crops and/or animals on the same land management unit".*

---

6. Éric Penot, Bernard Malet, 1999. Agroforestry systems: some definitions and contribution to forests dynamics. Cirad, Montpellier.

Agroforestry is generally practiced with the intention of developing a more sustainable form of land use that can improve farm productivity and the welfare of the rural community (Leakey, 1996).

The general definition provided by Somarriba in 1992 seems to us to be a less "reducing" definition: *"Agroforestry involves diverse technical practices that have in common the following: (i) there are at least 2 different plants in biological interaction, (ii) one of the 2 plants is a perennial and (iii) one of the 2 plants is a forage crop, a food crop or a tree crop."*

The definition was revisited by Leakey in 1996: *"Agroforestry should be considered as a dynamic, ecologically based, natural resource management system that, through the integration of trees in farm and rangeland, diversifies and sustains smallholder production for increased social, economic and environmental benefits".*

Another definition was suggested by the *"Laboratoire de Botanique Tropicale"* in Montpellier, France: *"Agroforestry is a land use system, controlled by the local population in which perennial trees are associated with agriculture and/or stock farming on the same piece of land in such a way that the resulting ecosystem tends to mimic the natural forest ecosystem in terms of aerial and soil biomass, vegetation structure and species richness".*

That definition paves the way for complex agroforestry (CAF). The definition of an "agroforest" was made by de Foresta and Michon: *"Agroforests are a particular kind of agroforestry land use, but the word "agroforest" is sometimes understood as the end-result of all agroforestry systems, whatever their structure and composition. For us[7] as for many scientists and laymen, using the word "agroforest" to describe structures that have no forest features, like alley-cropping or trees on contour lines systems, represents a language abuse that only leads to confusion"* (de Foresta and Michon, 1996). This definition is perfect for jungle rubber and complex agroforestry systems.

A typology is therefore necessary to classify agroforestry systems. Many typologies have been defined (King, 1979; Huxley, 1883; Nair, 1985; Macdicken, 1990; Somarriba, 1992; Mary and Besse, 1996; Torquebiau, 1998) and are generally based on their components (crops, trees and livestock) and their combination in space and over time. De Foresta and Michon proposed another classification with two components: simple agroforestry (SAF) systems and complex agroforestry systems (de Foresta and Michon 1965), that perfectly reflect most agroforestry situations: *"Simple agroforestry systems (SAF) refers to associations involving a small number of components arranged in obvious, usually well-ordered patterns: one or a couple of tree species, either as a continuous canopy, in equally distant lines or in edges, and some annual species for ground cover".*

The tree component is generally a crop of major economic importance, coconut, rubber, clove, teak and now oil palm, or plays a qualitative or environmental role, with *Erythrina, Leucaena, Calliandra* planted for fodder or to improve soil fertility. The annual species are usually important economically as intercrops during the immature period, such as paddy, maize, vegetables, forage crops or banana, pineapple, cassava or sugarcane. These simple agroforestry associations represent the classical

---

7. This represents the "Montpellier group *Laboratoire de Botanique Tropicale*" with F. Hallé, J.M. Bombard, F. Mary, G. Michon, H. de Foresta, E. Torquebiau, and other Cirad/ICRAF researchers (E. Penot, F. Besse, etc.).

agroforestry model most favoured in the development programmes of most institutions dealing with agroforestry (Steppler and Nair, 1987; Nair, 1989) as they are simple to promote (shading systems with coffee/cocoa, alley-cropping, hedgerows, improved fallows, etc.). The structure and functioning of these SAF do not resemble a "forest structure" and do not provide the same environmental outputs in the humid tropics.

Complex agroforestry systems are tree-based systems with a forest-like configuration that associate a large number of components, including trees as well as tree-lets, lianas, herbs, crops, and medicinal plants. Agroforests mimic the structure of natural forests, with a complex multi-strata structure and a closed canopy dominated by a few tree species (definition in van Noordwijk et al., 1997). The word "complex agroforestry systems" (CAF) is far more appropriate for agroforestry systems that match this definition such as jungle rubber, RAS (Rubber agroforestry systems, etc.). According to these authors, the CAF concept implies relative continuity in space and over time. Forest biodiversity in agroforests is usually quite high, as most farmers do not systematically eliminate "unused species", thereby allowing the regeneration of numerous forest species. CAF functioning is close to that of natural ecosystems. Complex systems are encountered almost exclusively in agriculture in the humid tropics. Except for home-garden systems, a particular form of CAF association that is relatively well documented worldwide, complex systems are now better recognised after having been ignored for decades. The functional reference to a natural forest ecosystem is one of the main features that distinguish "complex" from "simple" agroforestry systems. CAF are far more relevant for the analysis of forest dynamics as their ecological structure and physiological functioning in the mature period is very similar to that of a forest. However, as perennial cropping systems, CAF are also closer to plantations than to forests in terms of investment, management, economic strategies and outputs.

The dimension of the concepts of simple and complex agroforestry systems goes far beyond this physiognomic description or its intrinsic implications for the respective qualities of both systems. SAF and CAF relate to two different, though potentially complementary, conceptions of land development. One refers to field management: SAF address the integration of trees in agricultural lands. The other refers to resource management: CAF address the integration between forests and agriculture. This difference does not only involve important ecological aspects but has also essential socio-political implications, especially concerning the global role and interest of smallholder farmers in the management of forest lands and resources (de Foresta and Michon, 1995).

In the case of rubber, some AFS with rubber include only one tree, which could be one fruit tree (Thailand) or only one associated species (coffee or cocoa). Historically, jungle rubber was the most developed and famous CAF in southeast Asia and, in particular, in Indonesia. Modern RAS such as CAF also exist with several fruit tree and timber species all mixed together.

Chapter 1

Definition and history of RAS

Éric Penot, Bénédicte Chambon, Pascal Montoro, Wilfried Shueller

▸▸ Rubber in Southeast Asia from 1900 to 2023

The rubber boom and the development of jungle rubber

This section has been originally published in 2004[8].

Rubber *(Hevea brasiliensis)* was introduced in Indonesia from Malaysia by the Dutch at the turn of the 20th century in North Sumatra and was originally cropped in private estates in the form of monoculture in the "estate belt", following the trend observed among English estates in the western part of Malaysia. At that time, the market for natural rubber was booming due to a constant increase in demand and is still sustained in 2024 by a permanent demand for around 14 million tons per year (world consumption in 2022). In the 1910s and 1920s, Sumatra and Kalimantan were sparsely populated, with 1-4 inhabitants per km². Shifting cultivation was the usual practice involving slash and burn of primary forest or old secondary forest[9], one or two years of upland rice cropping followed by a long fallow lasting up to 30/40 years depending on land availability. Land was plentiful and there was no particular pressure to force farmers to change to another system. The system was sustainable as long as the population remained relatively stable, which was not the case in Java. In Sumatra in the 1910s, rubber seeds were collected from estates in the north and then distributed or sold by Chinese traders and missionaries in the south (Riau, Jambi and South Sumatra provinces) creating a tremendous demand for rubber in pioneer zones. In Borneo, the first seedlings were introduced in 1882 (Treemer, 1864, cited in Dove, 1995). Seeds were distributed to the indigenous population in 1908 by the Sarawak government. In Kalimantan, rubber seeds were introduced in 1909 (Uljee, 1925, cited in King et al., 1988) and were spread by Chinese merchants and Catholic missionaries in the Kapuas river basin.

Farmers immediately saw rubber as a new source of income, and in addition, it was easy to integrate in their existing agricultural practices. They began to collect seeds from surrounding estates or existing plantations and started their own rubber plantations. Rubber was cultivated in a very intensive way on the estates using fertilisers and

---

8. Didier Babin (ed), 2004. *Beyond tropical deforestation. From tropical deforestation to forest cover dynamics and forest development*, UNESCO/Cirad, 488 p.

9. At the turn of the 20th century, the peneplains in Sumatra and Kalimantan were still largely covered by primary forest.

continuous weeding that required a high investment in labour and capital. Local farmers, along with spontaneous migrants, some of whom came from the estate sector, adopted their own system, according to their limited cash and labour resources. They planted rubber trees with rice after traditional slash and burn (*ladang*). Rubber trees then grew among the secondary forest. No weeding or inputs were required. This system is called jungle rubber (*hutan karet*). A higher planting density than that of estates is used in order to compensate for the loss of trees due to competition and depredation (between 800 to 2000 seedlings/ha). Eventually, after a longer immature period (8-15 years for jungle rubber compared to 5-7 years on estates), the number of trees that can be tapped (between 350 and 500/ha) is comparable. From the point of view of the estates, whose objective was monoculture, unselected rubber proved to be perfectly suited to this "new environment" (agroforestry), whereas in its original habitat in the forests of South America monoculture is not possible[10].

The emergence of the jungle rubber system

The jungle rubber system has been well described (Gouyon and Penot., 1995; Penot and Ruf, 2001) and from a botanical point of view, is defined as a "complex agroforestry system" (de Foresta and Michon, 1995). Originally this system implied fallow enriched with rubber trees. The lifespan of rubber (35 years) is the same as the traditional fallow period necessary to restore soil fertility and get rid of weeds. The Kantus Dayaks, considered jungle rubber (or rubber gardens) as "managed swidden fallows" (Dove, 1993). "*Swidden cultivators use simple land and labour resources within the swidden system to cultivate rubber*" as explained by Dove (1993). The suitability of rubber seedlings for agroforestry, with no inputs required, only marginal labour requirements at planting, and very limited risks are the factors that triggered the rubber boom. Labour requirements shifted from cyclic (a period of four months a year for upland rice) to permanent for rubber (from 6:00 to 11:00 a.m. every day) although with no competition between the two systems. The afternoons are still available for *ladang* activities (rain fed upland crops such as rice or groundnuts). Rubber proved to be the perfect crop to grow with rice. Beside land, labour is the main available factor of production. The main limiting factors are capital and technical information, but these were not necessary in the initial stages of setting up jungle rubber plantations. Thus, in the original farming system, rubber and *ladang* rice could be grown together satisfactorily. Rubber was never seen as an alternative to rice, although this is becoming less and less true due to intensification and the increasing pressure on land in some provinces as was the case in North and South Sumatra in 2002. In 1997 already and still in 2024, rubber, and in particular clonal rubber, provide income and return on labour far above that of upland rice (return on labour was 4 times that of rice in 1997, idem in 2022).

From a historical point of view, farmers changed to rubber not because they were obliged to (as were many farmers during the "forced crop" period from 1830 to 1870 under the Dutch) or were under pressure to change to another more intensive system

---

10. The rubber tree is a forest species whose original habitat is the Amazon Basin in Brazil. In this respect, farmers in Indonesia gave rubber a second chance to grow in its original environment: the forest. This was possible because there is no leaf blight (*Microcyclus ulei*) in Southeast Asian forests thus enabling wide dispersion of rubber throughout the sub-continent. In Amazonia, rubber trees only survive if they are isolated in forests and cannot be grown in pure plantations.

(as were Javanese farmers during the green revolution) but because it suited the local environment and was sustained by a constant market. It proved to be a safe and easy way to increase farm income without fundamentally disturbing local farming systems, at least in the beginning. Rubber enabled local farmers to improve their standard of living and welfare. At the same time, it enabled increasing numbers of migrants to settle permanently in these areas, either on a spontaneous basis or through transmigration programmes. These changes in both autochthone and allochthone populations triggered a change in population density and eventually led to increased pressure on available resources. The average population density in Sumatra was already at 35 inhabitants per km^2 in 1997 (30/50 in West Kalimantan), 120 in 2024, and land is becoming scarce in some provinces (North and South Sumatra, Jambi, Lampung).

The average area of jungle rubber per family is between 2.7 and 4 ha (Barlow et al., 1986; Gouyon, 1995). Rubber generates 55% for Barlow in 1982 to 80% for Gouyon in 1995 (Penot and Ruf, 2001) of the total farm income. Jungle rubber and shifting cultivation are not at all antinomic, as both systems can co-exist in local farming systems. The notion of "composite system" was developed by Dove (1993): *"there is little analysis of the relationship between the 2 systems (rubber as swidden agriculture with rice) and thus little understanding of why this combination historically proved to be so successful".*

Farmers profit from a no input/no labour rubber cropping system. During the immature stage, planting jungle rubber requires four days of additional work (Levang et al., 1997), which involves a certain amount of income diversification as the jungle rubber system also yields fruits, nuts, timber for housing a well as other products such as rattan and non-timber forest products (NTFP)[11]. The cost advantage of smallholders vs. estates in setting up a rubber plantation has been assessed at 13 to 1 during the colonial era (Dove, 1995), at 6 to 1 vs. estates in 1982, and between 3 to 1 and 11 to 1 vs. government rubber schemes (Barlow et al., 1982, 1989), showing that farmers always had very competitive cost advantages with rubber.

The fact that production per hectare of unselected rubber seedlings is very similar in monoculture and agroforestry systems shows that unselected rubber can compete and maintain its yield in association with a relatively high density of other trees (200 to 300/ha). However, this needs to be verified with improved agroforestry systems using clonal rubber. In the case of jungle rubber, the advantages are quite clear: no establishment costs (the use of unselected seeds with no monetary value, and no fertilisers), low labour investment (only a few additional days to plant the rubber as the land has been already cleared for upland rice) and no maintenance during the immature stage. These three components explain the success of jungle rubber, which became the biggest source of income for most smallholders in inland Sumatra and West Kalimantan. The disadvantages of jungle rubber compared with clonal plantations are the delay in production due to the longer immature stage and relatively low productivity.

The success of jungle rubber and the future of this cropping system

Jungle rubber was in fact very well suited to the situation farmers faced in 1900. Five conditions triggered the replacement of shifting cultivation by a sustainable rubber cropping system that was still being used by more than 1.2 million farmers in Indonesia

---

11. NTFP: medicinal plants, *gaharu*, resins, local vegetables, etc.

in the 1990s: (i) land was plentiful and unspoiled (primary forest; in this respect farmers profited from the "forest rent" theory proposed by Ruf in 1987)[12] and jungle rubber conserves soil fertility and biodiversity, enabling renewal of the system every 30 or 40 years. Land opportunity cost was low, and remains low in remote or pioneer areas on the outskirts of traditional rubber areas; (ii) the particularly satisfactory adaptation of unselected rubber to the forest environment in complex agroforestry systems; (iii) a labour pool was available in Java that enabled land colonisation by both local Dayak or Malayu farmers as well as Javanese transmigrants, who originally came as tappers for estates or sharecroppers for local farmers; (iv) the sustained demand for rubber and the pricing policy was almost always positive for farmers with continuous incentives for the further extension of land and an increase in production. In this respect, Indonesia is still very well placed on the world market with low labour costs and the capability of significantly increasing its production if more farmers change to clonal rubber. Demand is still sustained and will probably continue for the next 20 years as substitution with synthetic rubber is not possible for at least 10% of the total demand for rubber; the demand requires natural rubber with specific characteristics in terms of heat and shock resistance for the tyre industry; (v) no real alternatives were available up to the 1980s.

The ecological advantages of jungle rubber

It is clear that rubber initially triggered deforestation (Prasetyo and Kumazaki, 1995). Timber concessions (see the example of South Sumatra) originally had less impact on forest cover than any other land use. The paradox lies in the fact that in areas where the forest has disappeared, jungle rubber is now the main reservoir of biodiversity (de Foresta and Michon, 1995). In comparison with other land-use systems based on oil palm, coconut, coffee, cocoa or pulp trees, rubber agroforestry systems are among the best adapted to maintaining a certain level of biodiversity. Jungle rubber proved to be better adapted to this "new" environment than estates, especially as yields were comparable, with 500 kg/ha/year of rubber[13] (Djikman, 1951) as long as both farmers and estates used the same unselected rubber planting material, which was the case up to the 1930s.

Conservation of biodiversity is a spin-off of jungle rubber. Plant biodiversity in a mature old jungle rubber system is close to that of primary forest or old secondary forest (de Foresta and Michon, 1992, 1995). Environmental benefits in terms of soil conservation (Sethuraj, 1996) and water management due to its forest-like characteristics are also significant. The biomass of a 33-year-old rubber plantation (444.9 t/ha dry weight) is similar to that of humid tropical evergreen forest in Brazil (473 t/ha, from Jose et al., 1986, cited in Wan Abdul Rahaman Wan Yacoob et al., 1996, or Sivanadyan, 1992) or in Malaysia (475-664 t/ha, from Kato et al., 1978, cited in Wan Abdul Rahaman Wan Yacoob et al., 1996).

According to Sethuraj (1996), the potential photosynthetic capacity of rubber leaves is comparable to or even better than many forest species (about 1,150 g/m^2/year in a well-managed plantation). A total area of 10 million ha under rubber worldwide would fix about 115 million tons of carbon annually (of which 1/3 in Indonesia). Soil fertility is

---

12. See Clarence-Smith and Ruf, 1996.
13. Rubber yield is always presented as Dry Rubber Content (DRC) 100% and not as kg of raw material (rubber sheets or cup lumps) or litres (latex).

maintained or even improved (Sethuraj, 1996; Dijkman, 1951) as rubber increases the nutrient content of the upper layer of soil due to leaf litter (4 to 7 tons/year/ha; Sethuraj, 1996). Of course, removing rubber trees for timber implies considerable nutrient exports that must be replaced through equivalent fertilisation rates at replanting. Soil moisture is very high under rubber, which probably also leads to a faster rate of decomposition and better nutrient turnover. From a nutritional point of view, mature rubber is a self-sustaining ecosystem, unlike oil palm for instance. Nutrient cycling is likely to approach that of forest ecosystems (Shorrocks, 1995, cited in Tillekeratne, 1996). Farmers do not view these benefits as main components of the system but rather as incidental "gifts", comparable to the "gifts" provided by the long-term fallow in the original system ("slash and burn, S&B). These "gifts" may be included in what Ruf called "forest rent" (1994), which provides advantages comparable with planting tree crops in a forestry or agroforestry environment as oppose to degraded land or land which is already cropped.

Historically, forest products formed the basis of commercial exchanges between "farmers-gatherers" and foreign traders like the Chinese as early as the 5th century AD, or the Arabs after the 9th century, for various products including resins, spices, nuts, or latex — gutta-percha[14] — for insulating marine telegraph cables in the 1840s (Dove, 1995). In the 19th century, rubber was also the product extracted from various other plants including creepers. Conserving biodiversity in agroforests is still considered to be a useful by-product by many farmers, in addition to the fact that agroforestry practices enable savings in both labour and inputs. This is particularly true in degraded and depleted areas like the low altitude mountains in West Sumatra in the Pasaman area or land that has been invaded by *Imperata cylindrica* (such as the West Kalimantan plains), as agroforests are a source of seeds for valuable fruit and timber trees. But of all these potentially profitable products, only rubber triggered the very large-scale development of agroforests (of which there were more than 2.5 million ha in Indonesia in 1997). Rubber became a strategic product as early as 1839 with the discovery of vulcanisation by J. Goodyear, and, later on, with the development of the tyre industry, which began in 1888, and accounted for 78% of world consumption of natural rubber in 2020.

Jungle rubber sustains development in pioneer zones

Four factors explain the rapid change in local agrarian systems and the adoption of rubber by more than 1 million farmers in Indonesia[15]. The first reason was the availability of rubber seeds on a large scale and at no cost (from estates and an increasing number of smallholder plantations) and the perfect adaptability of rubber as an enrichment species for fallows. The second reason was the apparently endless amount of available land and the possibility to extend plantations at a very large scale, originally using the river network. Rubber is not perishable, so its transport and sale is problem free. The third reason was the availability of labour from a reservoir of migrants from

---

14. Gutta-percha is a natural latex obtained from trees of the species *Palaquium spp.* Like natural rubber (from *Hevea bresiliensis*) from the rubber tree, gutta-percha is a polyisoprenoid, very rigid and partly crystallized at room temperature, which makes it much less elastic than natural rubber. It was used at the beginning of the 20th century to make golf balls.
15. Contributing to 3 million ha of which 2.5 million ha were under jungle rubber in 2001.

over-populated Java encouraged by estates contracting labour for their plantations and also spontaneous migration, which was later followed by official transmigration programmes[16]. The last reason is the fact that planting rubber represents a land-acquisition process that gives the planter land and tree tenure very similar to that of full ownership, at least under the traditional law ("*adat*" in Indonesian).

These factors relate to pioneer zones. Land and labour being almost inexhaustible (at least so it seemed up to the 1990s), success was guaranteed by a plant that required no capital investment or major labour at planting and could easily be integrated in local farming systems along with agroforestry practices that had already been developed in other agroforestry systems (such as *tembawang*[17] in Kalimantan or durian or durian/cinnamon and damar agroforests in Sumatra).

The three stages of jungle rubber development

Historically, rubber expansion can also be characterised in three stages. The first stage, from 1900 to the 1930s, was the enrichment of fallow with unselected rubber (improved fallow). Rubber was considered as a source of income but for obvious reasons connected with the need for a food supply, priority was still given to rice as the main staple food in shifting cultivation. Farmers rapidly shifted to the agroforestry rubber-cropping system and rubber became their main source of income as a result of the constant improvement due to selected farming practices. The second stage, from 1930s to the 1990s, was the shift from an improved fallow based on rubber to a real rubber-based complex agroforestry system that integrated some cultivation practices: planting in line, selective weeding at specified intervals, selection of associated trees, etc. The third stage, from 1990s to the present day, was when external technical innovations (such as clonal planting material, use of herbicides to control *Imperata cylindrica*, pesticides and fertilisers) were integrated in jungle rubber in order to improve land and labour productivity. In this way, jungle rubber was progressively transformed by some farmers into improved Rubber Agroforestry Systems called RAS.

In the 1980s, only 8% of rubber farmers were affected by government rubber programmes, vs. 16% in 2002 (unknown in 2024). A total of 350,000 ha has been planted or replanted as productive plantations in 1998. Meanwhile, 94% of intensive irrigated rice farms were involved in government programmes during the green revolution (Booth, 1988). Consequently, the diffusion of techniques, skills, and information on improved rubber was perhaps limited, although all farmers knew about and wanted to acquire clonal rubber. The commodity system did not benefit from a "first priority" government development policy, as was the case for rice with the objective of self-sufficiency. One major challenge is to ensure the diffusion of certain technical innovations, in particular clonal planting material, to farmers irrespective of the rubber cultivation system they use (monoculture or agroforestry), which would result in the full recognition of the advantages of agroforestry practices as a true component of cropping systems.

---

16. Some of which focused on tree crops and particularly rubber (NES: Nucleus Estate Scheme programme).
17. Tembawang are local timber/fruit-based agroforestry systems developed by Dayaks people in west Kalimantan and are still very popular in 2024.

During the colonial era, each time a natural resource was the subject of a commercial boom, active restriction measures were taken by the government to control and restrict its exploitation (Dove, 1995). Spices in the 18th century, in particular *jelutong* (from *Dyera* spp.), rubber (through the "international Rubber Regulation Agreement" from 1934 to 1944), as well as other timber such as teak (in the 1820s) are examples of such policies. In 1997, farmers did not have the right to exploit, cut and sell their timber trees if the land was classified as "forest area". Forests officially cover 74% of Indonesia and are under the control of the forestry department. This "tree tenure" policy was clearly restrictive and did not provide any incentive for the improvement or optimisation of timber production in agroforests outside the estate sector. Consequently, timber products from agroforests, and in particular in the jungle rubber system, could not be adequately valorised, underlining the fact that agroforestry practices were not officially considered to be "modern and efficient". By comparison, oil palm monoculture, which uses large quantities of fertilisers and capital, is considered a "modern" tree crop. Tree tenue was modofoed in the 2010s and farmers have in 2024 the right to cut and sell timber trees which changes drastically farmers strategies on timber.

Rubber drove *"a shift from a tribal political economic formation to a peasant formation"*, as defined by Dove (1995) in Kalimantan. In other words, technically, it implied a shift from gatherers to real rubber planters after a stage of extensive rubber cropping through a fallow enrichment process. Politically, this situation led to a "contest" between the State and farmers in 1997, which in 2024 is still reflected in policies concerning rubber, wood, timber, oil palm, and in policies implemented in transmigration areas where tree crops were forbidden (such as food crop-based transmigration schemes in West Kalimantan until 1991). These policies did not take into account the fact that local traditional systems have proven their sustainability and their ability to adapt to economic development. In this respect, they are in fact "modern", at least in the opinion of the authors of this work. Policies have been focused on monoculture (oil palm, rubber, coconut, etc.) as they are far easier to develop using the well-known "technological package" concept.

It is important to note that, historically, farmers moved to rubber not because they were forced to in any way or were under pressure to move to another or more intensive system (like Javanese farmers during the green revolution), but because it suited the local environment and was sustained by a constant market, and consequently offered farmers the opportunity to easily increase their income. Rubber has given local farmers an opportunity to improve their livelihoods. At the same time, it enabled migrants to settle in these areas in increasing numbers thereby triggering a change in population density and putting pressure on available resources. Average population density in Sumatra is in 1997 35 inhabitants/km^2 and land is becoming scarce in some provinces (North and South Sumatra, Lampung).

According to Dove (1993), *"the comparative ecology and economy of rubber and upland swidden rice result in minimal competition in the use of land and labour, and even in mutual enhancement, between the two systems.* The notion of "composite system" was developed by Dove (1993).

The consequences of this low-level farm management are (i) slow and heterogeneous rubber growth and long immature period or late reaching tappable size (8 to 12 years after rubber planting), and (ii) rapid forest regrowth.

Rubber and fertility

Rubber increases nutrient content in the upper soil layer due to leaf littering (4 to 7 tons/year/ha; Sethuraj, 1996) and low nutrient export through latex (Between 20 and 30 kg N.P.K.Mg/year/ha; Tillekeratne, 1996; Compagnon, 1986). Of course, rubber wood extraction involves high nutrient exports that will need replacing through high fertilisation rates at replanting. Soil moisture is very high under rubber, probably also accelerating decomposition and improving nutrient turnover. Mature rubber is a nutritionally self-sustaining ecosystem, unlike for instance, oil palm. Nutrient cycling is likely to approach that of forest ecosystems (Shorrocks, 1995, cited in Tillekeratne, 1996).

Biodiversity

With rapid deforestation underway in Sumatra since the 1970s, rubber agroforests are becoming the most important forest-like vegetation cover in substantially large areas in the lowlands (Joshi et al., 2000). While jungle rubber cannot replace natural forest in terms of conservation value, the question whether such a production system could contribute to the conservation of forest species in a generally impoverished landscape is very relevant. Jungle rubber is itself a major reservoir of forest species and provides connectivity between forest remnants for animals that need larger ranges than the remaining forest provides. This leads to a diversified tree stand dominated by rubber, similar to a secondary forest in structure (Gouyon et al., 1993).

Michon and de Foresta (1996) concluded that overall, vegetation diversity is reduced to approximately 50% in agroforest and to 0.5% in plantations (Figure 1.1); but these estimates are based on plot-level assessments. Similar findings were reported for plants, birds, mammals, canopy insects and soil fauna by Gillison and Liswanti (2000), who, in their investigation, covered a wider range of land-use types, from forest to

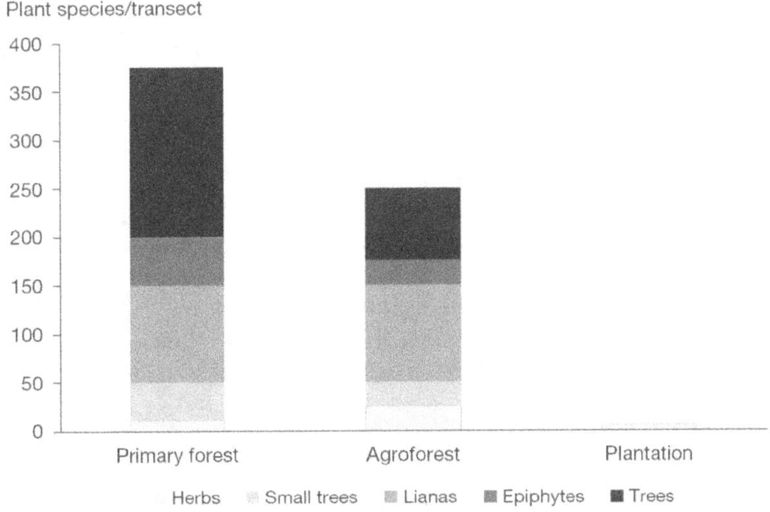

Figure 1.1. Comparison of plot-level plant species richness in higher plants between natural forest, rubber agroforest and rubber plantation (de Foresta and Michon, 1995)

Imperata grassland. Studying terrestrial pteriodphytes, Beukema and van Noordwijk (2004), also found that average plot-level species richness did not differ significantly in forest, jungle rubber and rubber plantations, however at the landscape level the species-area curve for jungle rubber had a significantly higher slope parameter, indicating higher beta diversity.

Of all the plants that are abundant in traditional rubber gardens, be they spontaneous or managed, about one third are used (Table 1.1), including timber species and non-timber forest products (Table 1.2). "Timber" uses are divided into fuelwood (mainly low-quality timber) as well as house construction and furniture making. In areas where there is no more natural forest within reach of the villages, however, traditional rubber gardens have become the main source of timber for the local population (de Foresta, 1992a). In these areas, timber from rubber gardens is already being sold, pointing to a prospective source of income that could be increased by the planting of valuable timber species.

Non-timber uses include food (i.e. fruit and vegetables: edible shoots and pods). Planted fruit tree species include durian, petai/stinkbean, jengkol, rambutan, mango, jackfruit and mangosteen (see Table 1.2 for Latin names). Petai and jengkol, both members of the family Mimosoidae, do not yield sweet, juicy fruits, but pods whose seeds are eaten raw or cooked as a vegetable. Legumes and many fruits fetch high prices in urban markets and could probably be sold if transportation could be provided. Some fruit tree species, like langsat and carambola, are only planted in the village area because they are said not to grow well in shady forest conditions. In Sumatra, as opposed to Kalimantan, some mango species (macang, kwini, mangga golek, mempelam) were also mainly found within the village area.

Other NTFPs are medicinal plants and handicraft materials, especially rattan, pandanus and tree bark, but also timber used to craft special items (e.g., machete sheaths). Latex and resin from rubber agroforestry systems are also sold (e.g., Hevea-latex, the latex of some Sapotaceae (nyatoh/*Palaquium* spp.) and Apocynaceae (especially *Dyera costulata)*. Apart from these, few products are harvested to sell for cash. Worth mentioning, however, is tengkawang, or illipe nut, harvested from Dipterocarpaceae and cultivated in West Kalimantan by the local Dayak population. Forest gardens, including tengkawang, are named *tembawang.* They usually contain fruit trees and sometimes rubber (Penot et Werner, 1997). Other uses of plants growing in rubber gardens are for ceremonial purposes, as ornamentals, thatching materials for field huts, fruits used as fish feed, or latex used to trap birds.

The data presented above prove the strong relationship between rubber garden bio-diversity and the presence of useful species. About two-thirds of all species present in rubber agroforestry systems have one or more uses. In the quest to increase the yield of rubber gardens, it is therefore important to search for systems that provide optimal growing conditions for improved rubber varieties, but still allowing a major part of the biodiversity of traditional gardens to persist: one of the objectives of SRAP activities Cirad/ICRAF (1994/2007).

Modern rubber agroforestry systems have to be able to integrate local wisdom about useful plants because in times of shrinking forest reserves, these systems might soon be the only ones still harbouring these species over large areas. Preserving biodiversity, therefore, also means guaranteeing the access of local people to these plant resources for their daily needs (Werner, 1993).

Table 1.1. Number of useful plants found in traditional rubber gardens in Jambi, West Sumatra and West Kalimantan

Province	W-Sum	Jambi	W-Sum	Jambi	Jambi	Jambi	Jambi	W-Sum	Jambi	Jambi	W-Kal	W-Kal
Plot N°.	LM 9	DB 2	LM 7	P 8	P 9	DB 16	P 5	LM10	P 6	P 16	E 1	S 1*
Plot age	65	25	20	20	60	50	20	65	60	60	60?	70?
Cleared/not cleared	(yes)	yes	yes	no	no	(yes)	no	(yes)	no	no	no	no
TIMBER												
Construction, furniture	3	6	3	5	6	6	6	5	8	9	17	35
Fuelwood	3	14	14	6	3	20	6	11	6	6	n.a.	n.a.
NON-TIMBER												
Fruits	7	2	3	8	5	2	6	7	8	11	20	25
Vegetables		1	2	2	3	1	1	1	2	2	1	4
Medicinal	4	2	6	2	3	5	4	12	3	4	2	3
Handicraft		1			2	1	2	1	3	2	5	4
Latex and resin	1	1	1	2	2		2	1	2	2	5	-
Sold for cash						1					1	1
Other	7	7	5	10	8	9	9	10	11	12		1
Total**	23	30	30	33	32	37	35	44	40	45	37	49
Total Biodiversity	40	40	48	48	50	61	55	73	61	73	69	126

*Plot size 2,500 m² as opposed to 1,000 m² for other plots. *Tembawang*, no abundant rubber. **Less than the sum of uses, because some species have more than one use.

Table 1.2. Useful spontaneous vegetation found in rubber gardens not cleared by farmers in West Sumatra and Jambi

Fruit tree species		Medicinal plants	
Durian	*Durio zibethinus*	Sicerek	*Clausena* c.f. *excavate*
Nangka	*Artocarpus heterophyllus*	Sidingin	*Kalanchoe pinnata*
Rambutan	*Nephelium lappaceum*	Jirak	*Eurya acuminata*
Macang	*Mangifera foetida*	Sitawa	*Costus speciosa*
Mango	*Mangifera indica*	Bidaro	*Eurycoma longifolia*
Langsat and Duku	*Lansium domesticum*	Daun kasai	*Pometia pinnata*
Jambu	*Eugenia aquea*	Sikarau	*Cyrtandra* sp.
Petai	*Parkia speciosa*	Kunyit	*Curcuma domestica*
Mangosteen	*Garcinia mangostana*	Kunyit balai	*Zingiber purpurteum*
Jengkol	*Pithecellobium jiringa*	Sikumpai	indet.
Kabau	*Pithecellobium bubalinum*		

Timber species		Plants with other uses	
Sungkai	*Peronema canescens*	Rimbang	*Solanum torvum*
Meranti	various genera and families, but esp. *Shorea* spp.	Daun kayu sibuk	indet.
Kulim	*Scorodocarpus borneensis*	Damar	*Dipterocarpaceae*
Petaling	*Ochanostachys amentacea*	Kopi	*Coffea robusta*
Kumpabok	Indet.	Jambu monyet	indet.
Maraneh	*Elaeocarpus palembanicus*	Sitarak	*Macaranga* c.f. *nicopina*
Tamalun	Indet.	Dalo	*Macaranga javanica*
Kawang	Indet.		
Madang	Various genera and families but esp. Lauraceae		
Surian	*Toona sureni*		

Jungle rubber is a balanced, diversified system derived from swidden cultivation, in which man-made forests with a high concentration of rubber trees replace fallows with a structure and biodiversity similar to that of secondary forest. Jungle rubber has accommodated increasing population densities, while preserving a forest-like environment. Yet farmers' income from jungle rubber has been declining since the 1990s due to the exhaustion of forest reserves, reduced land availability, low rubber prices, low yield and competition with oil palm. Short-term, small-scale credit schemes could help farmers adopt high-yielding rubber varieties which already emerged as a real necessity in 1997 with the introduction of oil palm in rubber areas. New options are thus needed to improve farmers' incomes with minimal call on government funding. In fact, the Indonesian government already stopped providing any help in 2001 at the end of the SRDP[18]/TCSDP rubber development projects. This led to the development of research on RAS based on clonal rubber which could be achieved by trying to preserve some of the advantages of jungle rubber including low maintenance and income diversity.

▶▶ The development of clonal rubber-based agroforestry plantations: a new challenge

From improved or enriched fallow to a complex agroforestry system: the importance of rubber planting material

In 2002, as since the beginning of the 1900s, most farmers still relied on unselected rubber seedlings for their jungle rubber system, whereas estates and project farmers had all started using clones. Rubber clones[19] have proven to be the best planting material in terms of yield and secondary characteristics[20].

In the 1930s, researchers tried to compare "estate" monoculture and smallholder jungle rubber. Some even tried to integrate limited weeding through the "bikemorse system" in Malaysia (cited by Sivanadyan et al., 1992) or through "jungle weeding" in Indonesia (mentioned by Dijkman, in 1951, referring to a researcher with a private company in the 1930s). Experiments of this type were considered as failures in both cases, which resulted in rubber monoculture being considered as the only relevant technology for both estates and smallholders. This was the prevailing view until today in most private and public research centres.

The importance of the adoption of clones

Yields of clonal rubber obtained by estates in Indonesia or by the best farmers in the SRDP rubber scheme (In South-Sumatra, Prabumulih, DGE) ranged between 1,400 and 1,800 kg/ha in 1997. In 2024, non-project farmers who use clones obtain yields ranging from 800 to 1,500 kg/ha/year depending on tapping quality and leaf diseases impact on their plantations.

Other improved rubber planting material available in the 1990s were clonal seedlings (seeds from plots planted with 1 clone), which are not often used due to poor performance, and polyclonal seedlings (seeds from an isolated garden planted with several selected clones). In Indonesia, there is only one estate (London Sumatra in North Sumatra) able to produce real polyclonal seedlings (BLIG)[21]. Polyclonal seedlings (from Bah Lias Isolated Garden in North Sumatra) , which were popular with

---

18. SRDP: Smallholder Rubber Development Project, a World Bank scheme that lasted from 1980 to 1990 and was replaced by TCSDP: Tree Crop Smallholder Development Project (same scheme) from 1990 to 1998.
19. Rubber clones have been selected in national research stations (Bogor, Medan in Indonesia, Prang Besar and RRIM in Malaysia and RRIC in Sri Lanka are the best known). The minimum required for the multiplication of clonal planting material through grafting is a budwood garden, a rootstock nursery and grafting skills.
20. The first "generation" of clones was released in the 1930s, the second in the 1950s, the third in the 1960s and 1970s. The fourth generation of clones, which was released in the 1990s, is still under investigation or undergoing preliminary small scale testing by estates. Currently, estates and rubber development projects are still using the best third generation clones, such as RRIM 600, PB 260, RRIC 100. However the most widely planted GT 1 clones in Indonesia date from the first generation (released in 1922 in Bogor).
21. BLIG: Bah Lias Isolated Garden, London Sumatra, North Sumatra.

the estates in the 1950s and 1960s, have generally been abandoned and replaced by clones, which are more homogeneous, better adapted to high levels of production and which have good secondary characteristics (resistance to diseases), in particular the clones of the 3rd generation that have been available since the 1970s. Clonal rubber is therefore the first most important innovation to be adopted by farmers (also the case for improved varieties used in other systems). In other words, the IGPM (Improved Genetic Planting Material) revolution has not yet stopped giving rubber farmers a reliable reservoir of productivity.

A key question: Is there a specific best clone for RAS?

Historically, there has been no choice of a specific clone best adapted to agroforestry. Most farmers developed their RAS with locally available clones, mainly GT 1 and PB 260 in the 1980s and 1990s. This is still the case in 2024 as one of the the main constraint is still clonal planting material availability and clone adaptability.

In 2024, varietal improvement is focused on two major objectives, improving yields on the one hand, and on the other hand, sustainable production, i.e., resistance to disease and reduction in the length of the immature period. It takes about 25 years to recommend new clones. Adaptation to new climate conditions and in particular drought during the dry season through the global climatic change might be also a new priority. The main problem for smallholders when replanting, aside from choosing the right clone for their local conditions, is universal access to good quality clonal planting material, and for nursery owners, access to good quality clonal budwood gardens. In the meantime, farmers need more information about new clones or those more adapted to RAS, climate change, etc.

Rubber breeding has been devoted to promoting latex and latex/timber clones for monoculture through selection based on latex yield, tree growth to reduce the immature period, and disease resistance. There are several obstacles to the development of efficient rubber production systems: (i) the low level of adoption of innovation by farmers; (ii) the low quality of planting material available to farmers compared with that available to estate plantations; (iii) the long process of breeding and recommendation of clones (25-30 years); (iv) the spread of diseases (old and new) and climate change.

For these reasons, breeding programmes need to undertake multi-disciplinary research and to use participatory approaches to improve the design of solutions for smallholders and the impact of innovations as well as to speed up long conventional breeding programmes.

Rubber-based agroforestry systems have higher land productivity and biodiversity than monoculture but little is known about their adaptative capacity in response to climate change. Competition with associated crops may increase in the future. Consequently, breeding for RAS should integrate the most advanced technologies to predict the potential of rubber clones in such a system.

Use of physiological traits for early selection of rubber clones for agronomic traits

High-yielding clones are identified after five years of production in small-scale clone trials (SSCT). If the 5-year immature period is included, SSCT last 10 years.

Application of the latex diagnosis-based clonal typology is an interesting tool to predict both the production potential of clones and the appropriate frequency of ethephon stimulation[22] for each clone (Gohet et al., 2003).

Predicting susceptibility to tapping panel dryness (TPD) is another challenge. A small percentage of TPD-affected trees are identified every year, but after 10 or 15 years of production, this can reach up to 30% (Okoma, 2011). Putranto et al. (2015) showed that early TPD occurrence can be induced in TPD-susceptible clones after six months of tapping under a severe harvesting system (Herlinawati et al., 2022; Putranto et al., 2015). These two methodologies are being tested in 2023 in the framework of the Rubis Project[23] with the aim of reducing the duration of SSCT from 10 to 6 years and obtaining reliable results more rapidly.

Modelling tree architecture

Knowing the architecture of rubber trees is essential to predict timber production, the capacity to provide shade for associated crops, and resistance to wind damage. The Water, Nutrient, Light Capture in Agroforestry Systems (WaNuLCAS[24]) model was used to evaluate and understand the impact of crop management on inter-cropping scenarios and competition between rubber and associated annual crops. Improvement of such modelling systems is necessary using architectural traits such as height, girth, branching and canopy typologies. The susceptibility of rubber clones to wind damage was determined using ground-based mobile LiDAR[25] (Yun, 2019). These technologies have to be implemented in breeding programmes to select rubber clones that are resistant to the extreme natural disturbances linked to climate change, and also to better estimate the wood production potential.

The need to study eco-physiological traits

The frequency and intensity of El Niño and La Niña make water stress an important selection criterion for future rubber clones. The competition for water between rubber and associated crops in RAS will be further exacerbated in the future.

Many studies have been conducted to characterise drought tolerant rubber clones. Clone RRIM 600 is one of the best clones planted in drought areas in North East India and Thailand, for instance. The development of simple, robust and rapid methods of selection is necessary. In this perspective, some authors reported that the drought factor index (DFI) could be used for early selection of drought tolerant clones and contribute to adaptation to climate change (Cahyo et al., 2024). Rubber clones with high DFI need to be proven to be drought tolerant through large-scale clone trials and through characterisation using the LI-6800 Portable Photosynthesis System, which simultaneously measures photosynthetic gas exchange and chlorophyll a fluorescence.

---

22. Stimulation with etephon chemical product is necessary for rubber when reduced frequency of tapping is adopted (with D3, D4 and more) according to clone typology.

23. Rubber agroforestry breeding initiative for smallholders (https://www.rubis-project.org).

24. https://worldagroforestry.org/output/wanulcas-model-water-nutrient-and-light-capture-agroforestry-systems

25. LiDAR: light detection and ranging or laser imaging detection and ranging, is a remote senting tool using laser.

Breeding for traits related to smallholder practices

Setting up efficient rubber-based agroforestry systems can be a way to cope with socio-economic issues. Low rubber prices combined with low land productivity drastically affects farmers' incomes. Most farmers use high tapping frequency, which leads to high bark consumption and reduces the rubber production cycle. Bark damage is also a serious problem resulting from combined use of seedlings with variable bark thickness and limited tapping skills. These factors are also behind the high rate of occurrence of TPD in smallholdings.

In 2024, developing high-yielding rubber clones with specific characteristics for smallholders is possible by combining several traits including a short immature period, a good yield, resistance to TPD, to leaf diseases and bark damage. Subsequently, these clones need to be disseminated with specific training for farmers.

Planting clones that are specifically adapted to low-intensity tapping is also a major challenge given the need to improve labour productivity and to give farmers time to diversify their activity and income, particularly with the implementation of RAS. The low-intensity tapping systems require ethephon stimulation. Designing and implementing training dedicated to this system will be necessary to ensure the development of efficient RAS.

The Indonesian case

When rubber is planted using standard single-row spacing, other crops can only be planted for the first 2/3 years because closure of the rubber canopy can reduce light intensity by 55% and crop yields by 60%. At this conventional planting density, RAS are planted with seedlings (still the case in Jambi, Indonesia, in 2017 for instance by migrant farmers from North Sumatra) and with locally available rubber clones (PB 206 in Indonesia). Most of remaining jungle rubber in Indonesia is planted using seedlings. Recommended clones in RAS are PB 260 in Indonesia and RRIM in Thailand. At normal density, clone RRIM 600 has a weak canopy (< 60% shade). This clone is planted in more than 95% of all the plantations in Thailand and hence also in RAS.

Interestingly, double-row spacing enables a light penetration area reaching 3-4 m from the row of rubber trees, which is still more than 80% when the rubbers trees are 8-9 years old. Analysis showed that double-row spacing with upland rice, corn, and soybean is feasible with 1.98 of a marginal benefit cost ratio. In other words, the double-row system was technically suitable for long term intercropping, because when the rubber trees reached 8 to 9 years of age, light penetration was > 80% at a distance of about 4 m from the rows of rubber tree.

Sahuri et al. (2021) reported that some rubber clones are better adapted to RAS (Oktavia and Agustina, 2021). The International Rubis Workshop 2021[26] addressed the question of the need for specific clones for RAS. New superior rubber clones have been produced by the Indonesian Rubber Research Institute (IRRI), namely IRR 112, IRR 118, IRR 220 and IRR 230 with a potential latex yield of about 2.5-3 tons/ha/year (recorded on station). These clones were evaluated in large-scale clone trials and are currently recommended in Indonesia. IRRI is also experimenting these clones for RAS

---

26. https://www.e3s-conferences.org/articles/e3sconf/abs/2021/81/contents/contents.html

at the Sembawa Research Centre and in on-farm trials. Again, according to Sahuri et al., some clones are well-adapted to the double-row system with wide spacing. Clone RRIM 600 is also better suited for RAS with a low percentage shade (60%) than other clones. Interestingly, some Brazilian clones with small leaves may have some advantages such as providing less shade.

Conclusion

Adapting rubber clones and cropping systems to obtain the efficient agroforestry systems needed to overcome socio-economic issues and face climate change is the main challenge for this decade. The aim of research in ecophysiology is to better understand the response of rubber clones to climate change. In addition to water stress and wind damage, obtaining rubber clones adapted to high concentrations of CO_2 and high temperatures, as well as new diseases like circular leaf disease are new challenges for research.

Early selection methods based on latex physiology and ecophysiology have helped speed up breeding programmes. Molecular breeding is already being attempted in some breeding programmes thanks to next generation sequencing (NGS). Molecular-assisted selection should also reduce the time of selection by skipping the SSCT (small scale clone trial) steps and testing the selected clones directly in LSCT (large scale clone trial). NGS should also facilitate the certification of budwood gardens and commercial nurseries. The certification of planting material should help improve planting material.

Finally, genetic improvement requires access to genetic resources. Erosion of genetic resources is already underway due to deforestation of the Amazon basin, to the difficulties involved in establishing new collections and maintaining germplasm collections. Conserving *Hevea* diversity is a further challenge for both researchers and authorities, but jungle rubber may also represent a source of diversity to be characterised.

Availability of planting material to farmers through development schemes: the limits of government action

In the 1970s, the Indonesian government began to support the smallholder rubber sector, as had the Malaysian and Thai governments, as early as the 1950s in the case of Malaysia[27]. This type of policy was inspired by the green revolution for rice and was funded using income from oil after 1973. Table I.1 summarises historical relations between farmers and the government since the 19th century. The technical model promoted by government development projects for smallholders drew directly on the estate model: rubber monoculture with high labour and input requirements and no intercropping during the rubber immature stage (cover crops were promoted, but only 5% of farmers used them). The objective was to develop a simple rubber system that could be used over a vast area without requiring major adaptations to local conditions (adaptation was generally limited to the choice of clone and the fertilisation rate). This model proved to be efficient but costly. So far, 16% of Indonesian farmers have been directly or indirectly affected by projects, and only some of the projects resulted

---

27. In 1990, around 80% of smallholders in Malaysia (65% in Thailand) had been reached by various rubber schemes and adopted the estate model with clonal rubber.

in full production plantations. Several "partial approach" (ARP and GCC)[28] and "full approach" (NSSDP and WSSDP)[29] projects were implemented between 1973 and 1980.

The "partial approach" consisted in providing farmers with certain components of the cropping system, i.e., planting material, fertilisers, and a small credit with limited extension. The "full approach" was based on a complete technological package provided to farmers, generally under a full credit scheme. In 1979/80, the government decided to launch two types of projects: the NES/PIR projects that targeted transmigration areas with the settlement of migrants in virgin areas, and SRDP/TCSDP projects for existing local farms[30]. As a general rule in "full approach" projects, farmers were provided with a whole credit package, which was supposed to be refunded within 15 years, and included the following components: clonal rubber plants, fertiliser, pesticides, cash to help farmers with terracing, a land certificate and a monthly wage for the first 5 years (in NES/PIR only for transmigrants). Table 1.3 lists the distribution of rubber planting among the various projects.

Table 1.3. Planting of clonal rubber through projects between 1970 and 2000

	TCSDP	TCSSP	SRDP	NES	PRPTE	GCC/ ARP	NSSDP WSSDP	TOTAL
Period	1990-2000	1990-99	1980–90	1978–90	1980–90	1974–80	1973–79	
Surface area	75,000	78,000	110,000	168,571	15,697	112,600	20,019	501,887
% of total	15%	15.5%	21.9%	33.5%	3.1%	22.4%	4%	
Class A and B	80% estimated	80% estimated	89%	60%	39%	< 25%	80% estimated	309,330 estimated

Source: Gouyon (1995) and Penot (2001). Class A and B: plantations are good quality. productiveplantations.

Historical analysis of innovation processes in rubber farming

In this section, we analyse the production of innovation and the process of its adoption in the three following steps:

– *innovations in the jungle rubber system by non-project farmers*: smallholders produced their own innovations mainly though the development of agroforestry practices, resulting in what can be defined as "indigenous knowledge". Between 1900 and the 1980s, the farmers shifted from slash and burn agriculture to enriched fallows, then to a type of complex agroforestry system called jungle rubber;

– *innovations introduced into the "estate-like" rubber monoculture system by former project farmers*. After having adopted rubber monoculture (as an external technical innovation) in the 1980s and 1990s (Table 1.4), in the case of most farmers as a result of development schemes, smallholders used innovations to adapt the system to their own needs and strategies including the reintroduction of agroforestry practices by some of them (20-40% of farmers depending on the project; Chambon, 2001);

---

28. ARP: Assisted Replanting Project; GCC: Group Coagulating Centre.

29. NSSDP: North Sumatra Smallholder Development project; WSSDP: West Sumatra Smallholder Development project.

30. Former projects, as well as SRDP-like schemes funded directly by the Indonesian government are grouped together under PRPTE.

— recombination of knowledge: in the 1990s, the RAS developed by farmers involved in research project combining endogenous innovations with exogenous innovations provided by SRAP/Cirad/ICRAF[31].

Table 1.4. Past relations between the Indonesian government and smallholder commodity producers

Date	Action	Result
1870	Government passes the Agrarian Act claiming all fallow land belongs to the State and granting it to European estates	Swidden cultivators decide to plant more perennial crops in their fallow fields.
1910-1943	Government restricts the collection of forest latex by smallholders to protect European concessions.	Smallholders decide to cultivate rubber instead.
1910-1930	Smallholders out-plant estates and increase their market share.	Government decides to protect the estates.
1935-1994	Government imposes punitive export taxes on smallholders to force them to reduce their production	Smallholders increase the quantity and quality of their production to maintain a stable income
1951-1983	Smallholders increase their market share from 65% to 84% by extending their cultivated area	Government focuses all capital and technical assistance on the estates to minimise their loss of market share and to Increase yields.
1980-1990	Government promotes nucleus estate schemes to provide markets for cloves, oranges, and coffee. Smallholder cultivation brought under estate control	Smallholders resist the loss of autonomy implicit in these schemes.
Present	No more Government support provided	Smallholders abandon each commodity in turn as prices drop

* The Agrarian Act of 1870 classified as state dominion any land not kept under constant cultivation. Source: Dove (1995).
**This gave swidden cultivators in disputed areas a strong incentive to plant perennial crops in their swidden fallows (Potter, 1988).

In the shift from jungle rubber to improved RAS, farmers looked beyond the limits of jungle rubber and integrated external components either through the SRAP project or endogenously with systems called "RAS *sendiri*" or "endogenous RAS". After experimenting with RAS, up to 60% of SRAP farmers developed their own systems between 1997 and 2002. Such systems proved to be economically competitive with alternative crops (rubber or oil palm monoculture). In the three above-mentioned stages, the innovation process resembles an "innovation elaboration process" rather than simply an adoption process consisting of step-by-step integration of different technical components or agricultural practices, or the re-appropriation or adaptation of technologies. The traditional endogenous/exogenous division of innovations does not apply here as

---

31. SRAP: Smallholder Rubber Agroforestry Project: a research programme based on farm experimentation using a participatory approach with: Cirad (*Centre de coopération Internationale en Recherche Agronomique pour le Developpement,* France), and ICRAF (International Centre for Research in Agroforestry).

innovation processes include technologies, the transfer of techniques, management and the development of specific "know-how". Re-combining knowledge, techniques, and "learning by doing" is the basis for the development of know-how.

The beginning of the agroforestry system and innovation in the jungle rubber system

Farmers initially adapted the "estate model", which then became a complex agroforestry system. Farmers introduced five major technical innovations in the jungle rubber system.

The first innovation concerns the planting material and its use in an agroforestry system. Clonal rubber stumps are currently relatively expensive and are often not available in many rubber-producing areas. Initially, access to planting material was through seeds collected from nearby estate plantations. After the 1930s, estates started the massive use of clonal planting material. Farmers collected these "clonal seedlings" (generally the product of GT 1, which is the most widely planted clone). The innovation lies in the fact that rubber was not used by smallholders in monoculture (a "copy effect"), but as jungle rubber in an agroforestry system for which it proved to be highly suitable.

The increase in productivity in jungle rubber using clonal seedlings is low although yields can reach 700/800 kg/ha in the case of pure GT 1 seedlings (Gouyon, 1995; Dijkman, 1951)[32]. The real proportion of clonal seedlings (like GT 1 seeds) in the "unselected rubber" population after several generations of jungle rubber is not known (the lifespan of the jungle rubber system is between 30 and 40 years). For the first replanting cycle, farmers may use seeds from jungle rubber that are already mixed with clonal seedlings but these seeds do not conserve the parents' characteristics. No jungle rubber system includes clones, as unless they are planted in rows, clones cannot survive competition with secondary forest for light.

The second innovation concerns planting techniques. In the 1970s, farmers began to plant rubber trees in rows in their jungle rubber systems to facilitate tapping and improve the return on labour.

The third innovation concerns weeding. In the 1980s, farmers tended to weed once a year using selective cutting to conserve useful timber and fruit trees along some other species like rattan. Even with such limited weeding (compared to weeding 6-12 times a year in the estate model), the rubber trees can be tapped in the 7th or 8th year after planting, instead of after 10 years in Sumatra (or 10-15 years in Kalimantan) in traditional jungle rubber.

The fourth innovation is intercropping. Many farmers traditionally intercropped for several reasons: (i) the fact there was a market for some products, for instance chilli and pineapple in Palembang in South Sumatra, (ii) the need to grow food crops where land is scarce, which is the case in the transmigration areas, or (iii) some farmers required continuous very intensive upland food cropping, which is the case of Minangkabau farmers in the East Pasaman district in West Sumatra. Before 1993, such practices were very rare in project areas due to a ban by project management authorities. However, research

---

32. Yields from original unselected seedlings were around 350 kg/ha/year in the 1920s. Yields from jungle rubber are now around 500 kg/ha/year (including an unknown proportion of clonal seedlings).

programmes in several countries (IRRDB[33] annual meeting, Colombo 1996) showed that in fact, intercropping favours rubber growth and has no negative impact on rubber.

The last innovation concerns the control of *Imperata cylindrica* particularly in transmigration areas and in West Kalimantan where this noxious weed is rampant. *Imperata* control is very time and labour consuming. Due to competition from *Imperata*, production can be delayed up to the 8th or the 9th year. As soon as 1997, and still in 2024, farmers very often use Roundup®, a glyphosate-based herbicide, at a rate of 2 to 5 litres/ha to kill *Imperata cylindrica* and enable rice to grow. The cost is largely compensated by savings in the cost of labour (between 50 and 70 man-days[34]) required for rice crops in the 4th to 5th months of cropping.

Farmers are gradually adopting some of the components of the estate model, at least those which seem to be advantageous for jungle rubber such as a reduction in the length of the immature stage (thanks to weeding), an improvement in the return on labour (planting in rows reduces the amount of labour needed for tapping, the use of herbicide reduces the labour needed for weeding of *Imperata*). So far, these innovations have been integrated in the jungle rubber system with no external help (herbicide is an "external" technical innovation but its use is a labour-saving strategy on the part of the farmer). The "production" of these innovations enabled the transition from the one-by-one production/adoption of selected technologies or practices to the building of more complex real agroforestry systems that are more sustainable, in other words, moving from fallow enrichment to a real cropping system.

When questioned about the main reasons they chose agroforestry systems instead of monocropping, smallholders gave the following answers:
– they did not have enough cash to purchase the complete estate rubber package or enough labour for that system;
– for the savings in time and money for weed control. Farmers said they weed only once a year and this had proved to be sufficient in the jungle rubber system;
– the returns to labour per farm plot are far higher during the immature rubber stage;
– land was, and in many areas, still is available, making a reasonably extensive rubber cropping system possible;
– smallholders observed that agroforestry systems offered efficient erosion control, as well as being a sustainable source of biodiversity through timber and fruit species.

These practices cost little and require only a very limited amount of additional labour except for intercropping, which still is an important step towards intensification. Intercropping is used by farmers who are progressively abandoning shifting cultivation. In fact, intercropping may not require cash or inputs, only labour. However, without any inputs, particularly fertilisers, yields may remain very low and intercropping can thus be considered as relatively risky due to the required investment in labour.

The reasons it will be impossible to maintain this system (except in remote and pioneer zones) at the end of the 1990s are the following:
– other perennial crop alternatives emerged in the 1980s and 1990s such as oil palm, cinnamon (in Jambi and West Sumatra) and, more recently, pulp trees and pepper;

---

33. IRRDB: International Rubber Research and Developemt Board
34. Labour cost is generally around 3,500 Rp/day (2 US $ in 1997) so the weeding cost for 50/70 man-days is 175 000/200 000 Rp

– other off-farm opportunities are becoming available with industrialisation, expanding city markets and the expansion of trade;
– jungle rubber productivity is limited and farmers all know that rubber clones can double or triple yields (one very positive outcome of rubber development schemes), which means that farmers who use jungle rubber will eventually have to use clonal rubber, irrespective of the system they choose.

The first step in the transition was from improved fallow through enrichment with rubber to jungle rubber. The second step will be from jungle rubber to improved rubber-based agroforestry systems with a high rate of productivity and reasonable initial investment costs. In other words, the jungle rubber system has reached its limits and needs to be upgraded as soon as the end of 1990s. The only areas where it still can be considered as a possible alternative are remote or pioneer areas inhabited by poor farmers who have no capital at all.

The future of the jungle rubber system can be secured by planting clonal rubber to boost rubber production while conserving agroforestry practices that not only provide diversified sources of income and are better suited to farmers' limited resources but also benefit the environment and biodiversity. These aspects are discussed in the third stage of innovations based on RAS, and are the subject of the research implemented by ICRAF/Cirad from 1994 to 2007.

Process of innovation of the rubber monoculture system by former project farmers

Some farmers realise that productive complex agroforestry systems are made possible by re-introducing certain agroforestry practices in monoculture plots.

How farmers re-introduce agroforestry practices in monoculture

Rubber development projects are widely described in the literature (Gouyon, Barlow, etc.). For project farmers, one major innovation was the planting and/or the selection of trees that resulted from natural regrowth in what were originally monoculture plots. Personal observations of such trends in North Sumatra, South Sumatra and West Kalimantan Province (Sanggau area) in 1993-1998 were evidence for such practices. B. Chambon[35] investigated the frequency of this practice in the West Kalimantan Province (Table 1.2) and her results showed it was not an isolated phenomenon but a real trend. Although in 1997 the trend is still limited to 18% in NES projects (transmigrants) due to the influence of extension, the proportion rises to 45-50% in SRDP and in the "partial approach" projects concerning local farmers. In the latter case, 24% of the plots are in fact replanted with a sufficient number of associated trees to be able to describe them as complex agroforestry systems. Table 1.5 shows that up to 65% of farmers use clones in 1997 when establishing new plantations in agroforestry systems and Table 1.6 describes the type of replanting.

In this case, the innovation is clearly diversification through planting or through selection of fruit and timber species in the rubber inter-rows, resulting a tree-tree

---

35. B Chambon, Cirad/University of Montpellier, France, did her PhD field research in 1997-2000 under the supervision of the author.

Table 1.5. Agroforestry practices used in clonal project plots in West Kalimantan (1997)

Practices/type of projects	NES/PIR	SRDP/TCSDP	Partial approach
Re-introduction of agroforestry practices	18%	44.5%	51%
Type of trees:			
Fruit trees	72%	85.7%	54.5%
Fruit trees + cash crop trees	28%	4.7%	18.2%
Cash crop trees	0%	9.5%	27.3%
Number of associated trees per ha			
2 to 10: no RAS	62.5%	56%	36.7%
11 to 100: simple agroforestry system	25%	34%	40%
> 100: complex agroforestry system	12.5%	10%	23.3%
Age of rubber trees when associated trees were introduced:			
< 3 years	0	45.5%	57.5%
4 to 7 years	20%	27.3%	42.5%
> 7 years	80%	27.2%	0

Source: survey by B. Chambon, 2001, SRAP.

Table 1.6. Replanting by project farmers in West Kalimantan Province (1997)

Type of plantation		Percentage	Average area planted	Type of cropping system
No replanting		42%		
Jungle rubber		8.5%	1.3 ha	Traditional system.
Replanting with seedlings		27.5%	1.5 ha	47% with associated trees 53% monoculture
Replanting with clones (22%)	New plantation (project)	7.5%	1.5 ha	45% monoculture
	Purchase of a plantation	6%	2.25 ha	78% monoculture
	Setting up of a new plantation	8.5%	1.5 ha	69% monoculture

Source: survey by B. Chambon, October 1998 to April 1999 and April to June 2000.

association, which was strictly prohibited by both rubber researchers and extension services[36]. Farmers have always been told by extension services that clonal rubber should be cropped in monoculture. In projects, the farmers were generally obliged to maintain clean inter-rows (at least during the rubber immature stage).

The village of Sanjan in the Sanggau area in West Kalimantan Province – Agroforestry innovation in SRDP: the cradle of RAS

In Sanjan, in 1995, 13 years after introduction of monoculture, 15 out of the original 50 farmers (30%) had re-introduced associated trees in what were originally

---

36. In Thailand at that time, RRIT (Rubber Research Institute of Thailand) had been experimenting with associating fruit and timber trees with rubber for a decade (Sompong, 1996). ORRAF (Rubber Extension Service for Rehabilitation) and RRIT have been promoting such systems since 1991.

monoculture clonal rubber plots (Figure 1.2). The ratio of associated trees was 94-291 trees/ha (average 167) to 500 rubber trees/ha, mainly using the following species (ranked in decreasing order of importance: pekawai and durian (*Durio* spp.), belian (*Euxyderoxylon zwageri*), rambutan (*Nephelium lappaceum*), cacao, assam (*Tamarindus indica*), cempedak (*Artocarpus integer*), petai (*Parkia speciosa*) and nyatoh (*Palaquium* spp.). Pekawai, durian and rambutan were present in all the plots, underlining the farmers' preference for fruit trees (Figure 1.3). Sixty-four percent of the trees were planted, the rest resulting from natural regrowth and selection. In the study area, income diversification and reintroduction of financially profitable plant diversity in former monoculture plots are only two of the strategies applied by Dayak farmers.

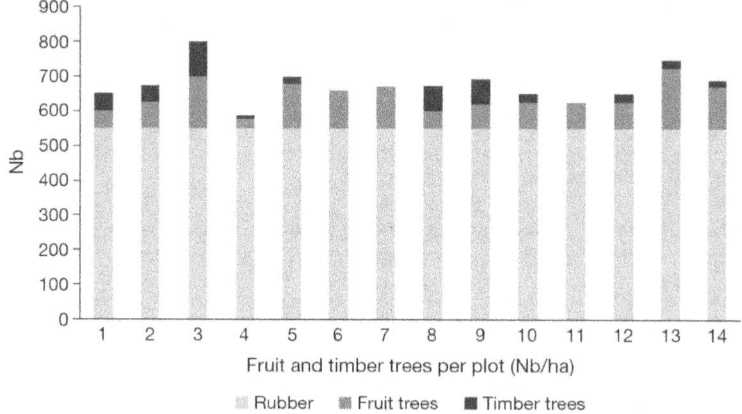

Figure 1.2. Re-introduction of associated trees in former rubber monoculture plots: the case of Sanjan village in West Kalimantan

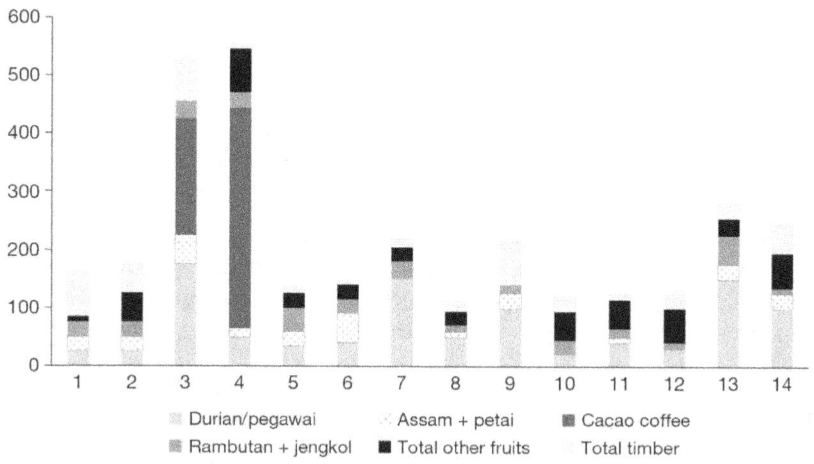

Figure 1.3. Type of associated tree distribution in former TCSDP rubber monoculture

In Sanjan, 35% of project farmers have re-introduced "associated trees". The fruit trees species include meranti (*Shorea* spp.), teak (*Tectonia grandis*), nyatoh (*Ganua* spp.) for timber, durian (*Durio zibethinus*), pegawai (*Durio* spp.), rambutan (*Nephelium lappaceum*), duku (*Lansium domesticum*), petai (*Parkia speciosa*), jengkol (*Archiden-dron pauciflorum*), jackfruit (*Artocarpus heterophyllus*), cempedak (a wild jackfruit, *Artocarpus integer*) (Shueller et al., 2003). In Sanjan, Dayak farmers have also already integrated traditional agroforestry practices in jungle rubber and in their *tembawang* system (a fruit/timber based complex agroforestry system).

This innovation is remarkable for two reasons. Firstly, because farmers always believed it was possible to grow perennial intercrops (trees) with rubber, as is the case in jungle rubber, and consequently decided to proceed. Their problem was knowing what percentage of associated trees can be combined with rubber without causing a serious reduction in latex yield. From the 1920s to the 1950s, unselected rubber seedlings produced the same yield in estates as in jungle rubber, which led researchers to hypothesise that the same was probably true for clonal rubber based on the same density of associated trees. Experiments on RAS were based on this hypothesis, which has been partially confirmed by observations made in the village of Sanjan, where no decrease in yield has been observed (Penot, 2001).

Agroforestry practices increase the sustainability and flexibility of cropping systems

In addition to income diversification and biodiversity, another major advantage of combining rubber with associated trees is that it is possible to change crops when rubber reaches the end of its lifespan (35 years). The plot can then be converted into a fruit and timber agroforest (*tembawang*) with the progressive disappearance of rubber trees. At age 35, clonal rubber wood can also be sold and will give the farmer enough capital to fund replanting.

Rubber trees can also be grown for timber, but in this case no tapping is possible. This is not yet the case in the smallholder sector. In other words, for a short return on investment, the farmer has to choose between latex or wood production. No clones can provide the two products at the same rate, but the economic lifespan of rubber can be considered from two different viewpoints: production of latex or wood, and growing rubber for timber is not economically viable. Apparently, the best economic option in monoculture is to grow clonal rubber at a rate of 550 trees/ha (this is the usual planting density used in Indonesia) first for latex and then to extract the timber as a residual product at the end of a 15-year cycle (Gan Lian Liong et al., 1994). With RAS, farmers have the choice of cutting and extracting all their timber in the 15th year (as mentioned above) or in the 35th year after planting (the end of the rubber lifespan) or of leaving the plot as it is and shifting from a rubber-based to a fruit and timber-based agroforest with a total lifespan of 45 to 50 years. Agroforestry gives farmers a range of options that can be adapted to the market and to their own needs. In other words, agroforestry practices also offer flexibility to change systems.

Farmers' constraints and the slowdown factor in the process of innovation

From an institutional point of view, there are no major constraints to associating fruit and timber trees with rubber as long as project officials no longer have authority over farmers' plots. On the other hand, problems of competition between rubber and associated trees may arise after 10/15 years in the case of fruit trees (such as rambutan) and after 15/20 years in the case of timber trees (meranti or even durian trees) if the planting density of the associated tree is too high, and the density varies with the species. Even in 2014, no scientific data are available on this type of competition, as no trials have been conducted using long-term associations of this type. Some experimental plots do exist (RRIC/Rubber Research Insitute of Cambodia and SOGB/Société des caoutchouc de Grand Bereby in Côte d'Ivoire with timber for instance[37]) but no data are published. The planting density of associated trees observed in Sanjan suggests that farmers are aware of this risk and keep the planting density of the associated trees fairly low, i.e., between 100 and 200 trees/ha vs. an average of 550 rubber trees/ha, while limiting the number of tall trees that have a canopy above that of rubber trees, such as durian. Other important constraints are land and tree tenure. In 1997, planting rubber is, under the traditional "*adat* law", a factor that ensures land acquisition similar to that of ownership. This is still the case in 2024. As far as tree tenure is concerned, it appears that farmers were not officially permitted to cut and sell their timber trees in 1997, but this has changed in 2020. A tax is also collected on rubber wood.

To conclude, in the past, rubber farmers came up with a series of innovations that allowed them to incorporate rubber in their extensive agroforestry practices (i.e. in the jungle rubber system) and later, in the "estate" monoculture model, by associating rubber with annual or perennial crops. But by the end of the 1980s, a point had been reached where further innovation was limited and any additional increase in productivity could only be obtained by including rubber clones and applying other external technologies that required a different management strategy. After passing through two intermediary stages, the first between shifting cultivation and improved fallow, and the second between improved fallow and a complex agroforestry system (jungle rubber), they faced the challenge of how to significantly improve the productivity of their system. Levang recall in 1996 that "*Complex agroforestry systems can no longer compete with other agricultural systems which may be more risky but are more profitable in the short term*". Agroforestry systems based on improved clonal rubber can meet this challenge with reduced risk and an increase in environmental benefits. Farmers have shown their ability to develop remarkable innovations, endogenously or through participatory experimentation in the case of the SRAP projects. In 2002, jungle rubber covered more than 2.5 million ha in Indonesia. The challenge now is to help farmers continue to acquire suitable innovations and to encourage them to adopt RAS.

Indonesia is still going through a stage of "late agricultural transformation". Historically, political instability up to the 1960s and subsequently the priority given to a policy for self-sufficiency in rice production (achieved in 1984) did not allow farmers to acquire improved technologies for rubber on a large scale. In 1997, jungle rubber is still the most widely used system in Indonesia, while sustained economic growth and new crop opportunities, in particular oil palm, invite farmers to increase the productivity of their rubber systems and also diversify. The introduction of external technical innovations (improved availability of good-quality planting material), taking indigenous knowledge (agroforestry practices) into account, providing micro credits and relevant technical information will play key roles in the future of the rubber sector.

---

37. Current experiments (in Malaysia, Thailand and SRAP in Indonesia) have been underway for less than 15 years on average.

A further important challenge in 1997 is ensuring that all the different types of farmers have access to improved technologies suited to their own particular strategies as well as access to available local resources; in other words, to promote equity as well as sustainability whether through agroforestry or monoculture. In a country that has been able to develop millions of hectares of different types of sustainable complex agroforests, agroforestry still has a great potential as long as environmental concerns are taken into account, and, if necessary, considered as a priority.

The organisation of rubber farmers and the availability of a wide range of rubber cropping systems, from semi-intensive rubber-based agroforests (RAS 1 type as defined later) to intensive monoculture systems, are the main preconditions in terms of policy and technology development that will give environmentally friendly systems a chance to survive and to maintain balanced regional development with other crops. Rubber agroforestry systems may only be options amongst others, but these systems do not entail the risk of crop failure, or uncertainties (in terms of the rubber market and output) that affect other crops, as there is a steady and reliable demand for natural rubber.

Definition of modern RAS

Because of their physiognomic and ecological resemblance to forests, their sustainability, and the well-known environmental attributes of forests, agroforests enjoy a good reputation. The agroforestry literature abounds with favourable judgements such as *"a unique combination of high levels of productivity, stability, sustainability and equitability"* (Soemarwoto and Conway, 1991); *"eminently sustainable agroforestry systems"* (Torquebiau, 1992); *"traditional systems of exceptional merits"* (Nair, 1993); *"[...] agroforestry successfully simulates the forest environment in the form of home gardens and 'analog forests'"* (McAdam, 2000); *"[...] structure of natural forest habitats... imitated"* (Scherr and McNeely, 2012); and *"epitome of sustainability"* (Kumar and Nair, 2004).

Most authors who recognise the quality of agroforests, including recent studies, refer to their ecological attributes, in particular biodiversity conservation and their long-term benefits for soil fertility and water management (Penot, 2001; Gajaseni and Gajaseni, 1999; Kaya et al., 2013), even in somewhat harsh environments (e.g. the Soqotra Island in Yemen; Ceccolini, 2002). Socio-economic variables are taken into account in some studies (e.g. Penot, 2003; Wezel and Bender, 2003) to analyse how agroforests function, but most authors do not describe socio-economic attributes in the same way as they do ecological variables. Some studies that use bio-economic modelling (e.g. with the Beam model) are only performed at cropping system level (e.g. Purnamasari et al., 2002). Labour requirements and return to labour, investments and returns to investment in the medium and long term, product benefits, income generation, are sometimes described, but are seldom presented as arguments for adoption or even taken into account in the innovation process behind the adoption of agroforests. In other words, global advantages as well as positive externalities of agroforests are widely recognised as a whole but are not properly valued. The direct benefits of agroforests are recognised at farm level but not entirely valued either, on the contrary, they are widely under-assessed and sometimes not even taken into account at the community level. To put it in another way: going beyond individual farmers, the impact and use of resources as well as the generation of income and product benefits needs to be considered at the level of what French agronomists call the "territory", i.e., as "anthropic land" and for some components — including biodiversity — at global level.

The only two economic variables which appear to be convincing arguments are: (i) diversification linked with risk spreading, with diverse sources of income and with labour spreading (e.g. Torquebiau, 1992; Penot and Chambon, 2003; Wezel and Bender, 2003), and (ii) income generation as a whole. The large number of products and uses of agroforests make it difficult to go beyond mere description and economic quantification. Similarly, the links between diversification, risk buffering and the long-term economic and ecological sustainability have not been sufficiently taken into account. The role of risk and uncertainty has been studied in the context of agroforestry adoption (Mercer, 2004) but not as an innovation process in its own right.

Deforestation, development of oil palm and other opportunities in Indonesia

Rubber plantations are considered by different authors as a major source of deforestation in tropical areas (Global Witness, 2013; Assembe-Mvondo et al., 2015; Hauser et al., 2015; Roberts, 2016; Vongkhamheng et al., 2016; Fern, 2018; Fritts, 2019; Higonnet et al., 2019). According to Costenbader et al. (2015), rubber expansion is one of the six most important drivers of deforestation in the Greater Mekong Subregion. According to Cowie et al. (2018), forest losses due to the establishment of rubber plantations since 1980 is estimated at 4.5 million ha. However, in 2024, most new rubber trees are no longer planted on forest land, mainly because no such land is available in accessible areas. Forest losses increased after 2000 with the increase in the price of rubber. The total extent of rubber plantations increased by 5 million ha, with major absolute changes in areas in Thailand, where there has been an increase of around 1.6 million ha since 2000, and in China, Vietnam, and Côte d'Ivoire, but not in Indonesia, where the increase has only been 0.3 million ha since 2000 (IRSG, 2018). Is the increase in rubber plantations observed since 2000 only due to deforestation? A field study is needed to answer this question. The example of Côte d'Ivoire where rubber has replaced cocoa is worth thinking about.

In Africa, industrial companies (estates) are responsible for most deforestation undertaken for rubber production (Penot et al., 2020; Fritts, 2019), whereas in Asia, most forest degradation or deforestation, very limited in 2024, takes place to create village rubber plantations, at least in the case of land on which rubber was actually planted (the same is not true for oil palm). The policy of large land concession to oil palm and rubber plantation is over in South-East Asia. It is important to realise that most recent growth of the sector that relies on the increase in area is due to the establishment or extension of village plantations, mostly at the expense of old jungle rubber, but also to a small extent, due to deforestation in remote areas. A major challenge to implementing a "zero deforestation policy" will be including smallholders in the process.

Chapter 2

Rationale for RAS and impact of agroforestry systems

Éric Penot, Bénédicte Chambon, Gede Wibawa, Karine Trouillard,
Elok Mulyoutami, Ilahang, Diah Wulandari, Laxman Joshi,
Stephanie Diaz Novellon, Isabelle Michel, Aude Simien,
Laetitia Stroesser, Vichot Jongrungrot, Somiot Thungwa, Didier Snoeck,
Uraiwan Tongkaemkaew, Marion Theriez, Phillipe Courbet

As the result of a joint analysis performed by ICRAF, Cirad and GAPKINDO, the three institutes pooled resources in a development-oriented research project (SRAP[38]) which began in April 1994. The goal of the project was to improve rubber agroforestry productivity by optimising labour and by reducing the use of inputs and costs, while conserving the benefits of agroforestry practices and not shifting too far from current practices in order to increase the farmers' rate of adoption of technical innovations. Even though agroforestry systems are very similar to what the farmers were practicing at the time, in our opinion it is important to base innovations and technologies on analyses of the constraints to and opportunities offered by existing farming systems, to be sure farmers' strategies and trends are taken into account, and to incorporate them in an operational classification of farming systems. In this case, irrespective of the innovation concerned, the viewpoint of a farming system is relevant, as it helps ensure that both apparent and hidden farming constraints are incorporated in strategies that result in the adoption of innovations. In 1994, the main innovation was implementing Rubber Agroforestry Systems (RAS) as clonal rubber-based agroforestry alternatives to both the jungle rubber system (low productivity but low cost) and the estate system (high productivity but high cost).

▸▸ The need for improved rubber agroforestry systems (RAS)

The objective of this new approach in 1994 was to demonstrate the advantage of conducting trials in real farming conditions using a participatory approach, to show that rubber agroforestry systems (RAS) are an improvement over traditional jungle rubber practices or standard rubber-based monoculture development schemes

---

38. SRAP: Smallholder Rubber Agroforestry Project, a research project based on farm experimentation using a participatory approach with Cirad, ICRAF (International Centre for Research in Agroforestry) and GAPKINDO (the rubber association of Indonesia).

based on estate technology. The main challenge for research was to test improved clonal planting materials to identify the optimum level of inputs and labour under which the planting materials grow and produce best in these agroforestry systems, and which were most appropriate – and affordable – for smallholders (Penot, 1996a). In other words, it meant trying to optimise the natural trend of endogenous farmers' experimentation with RAS *sendiri* (the farmers' own RAS experiment or "of his own").

An on-farm experimentation network was set up with 120 farmers in three selected provinces: Jambi and West Sumatra in Sumatra and West Kalimantan in Borneo (Table 1.1). All the innovations tested were first discussed with the farmers to improve their adoptability and to match RAS technologies with farmers' resources and requirements. Experimentation was based the maximum possible reduction in inputs and labour while conserving agroforestry practices and their advantages, i.e., income diversification, obtaining an income during the rubber immature stage through intercropping, conservation of a certain level of biodiversity and the use of an environmentally friendly approach. SRAP was based on a participatory approach to on-farm experimentation with three main kinds of RAS. The suitability of each system was tested in local agro-ecological conditions to identify associated labour requirements and costs, and the optimum level of intensification.

Three different RAS types were tested: (i) RAS 1, which involved planting clonal rubber with forest regrowth in the interline (the most extensive system), (ii) RAS 2 in which clonal rubber was associated with fruit and timber trees and intercropping during the immature period (the most intensive system), and (iii) RAS 3, which was the same as RAS 2 but with the addition of fast growing shade trees and of a cover crop (mainly *Flemingia congesta*) to get rid of alang-alang (*Imperata cylindrica*) in invaded plots (Penot, 2001). The main aim was to determine whether the different combinations of trees and crops associated with clonal rubber had a long-term impact on income diversification and on the adoption of agroforestry practices.

In SRDP[39] plots in the village of Sanjan (Penot 1997) where, before 1994, local farmers were already implementing what ultimately became the RAS 2 type of agroforestry, 25% of the SRDP (Smallholder Rubber Development Project) farmers in the village successfully implemented agroforestry associated with fruit production and very limited timber production (Shueller, 1997) and according to Chambon (2001), 46% of SRDP farmers did develop agroforestry in one form or another. The SRDP RAS plots in Sanjan showed that agroforestry practices were possible with no significant decrease in rubber production (the main economic output). The idea, through SRAP, was to test several combinations of trees to provide a wide range of technical solutions adapted to a variety of local situations.

The main problems were the following: (i) making sure that agroforestry really had no negative impact on rubber production and in which conditions, but also no effect on rubber growth during the immature period, to enable the rubber trees to be tapped as soon as possible after planting (generally between 5 to 7 years), and (ii) to identify the best combinations of trees and other plants to achieve the desired results in terms of competition with *Imperata cylindrica*, among others.

---

39. SRDP: Smallholder Rubber Development Project, developped by the World Bank.

Each trial was replicated in 2 or 3 villages with 7 to 10 replications/farms in each trial using the same planting density, association of trees and practices on the same type of soil and in the same climatic conditions (Tables 2.1 and 2.2). Each trial comprised 6 to 8 sub-plots in which a different treatment was applied (type of clone, type of fast-growing associated trees, type of intercrop, type of cover crop, etc.). All the trials were managed by the farmers using the same agronomic practices, which were defined before planting (Boutin et al., 2001). In all, 60 trial plots/farmers were involved in West Kalimantan, in 2 main zones: (i) on Dayak smallholdings (mainly following the jungle rubber system) in local traditional zones, and (ii) on Malayu

Table 2.1. Characteristics of benchmark sites in West Kalimantan in 1997

Factors	Forest margins with poor soils: traditional jungle rubber	Forest margins with poor soils: jungle rubber + SRDP	West Kalimantan transmigration areas.
Villages	Kopar, Engkayu	Embaong, Sanjan	a) Pariban Baru (Sintang) b) Trimulia c) Sukamulia
Type of farm population	Dayak (Christians)	Dayak (Christians)	a) Dayak (Christians) b) Javanese transmigrant (Muslims)
Population density	Low with plenty of land	Medium: land was becoming scarce	High with limited land (2.5 ha/household)
Ecological environment	Secondary forest, jungle rubber and *tembawang**, poor soils	Secondary forest, jungle rubber and *tembawang*, poor soils	Degraded *Imperata* land, poor soils risk of fire
Farmers' behaviour and strategies	Extensive systems, slash and burn (S&B) for local upland rice only grown for wine rice. Accept a low level of intensification.	Extensive and intensive systems (rubber monoculture), S&B for local upland rice. Accept a medium level of intensification.	Intensive on *sawah/ irrigated rice*; extensive on rubber on uplands.
Main constraints	Low productivity of jungle rubber	Low productivity of jungle rubber. Wrong choice of rubber clone in SRDP: leaf disease, limited production.	Very degraded land with *Imperata* on a very limited cropping area (2 ha). Risk of fire. Remoteness.
Opportunities	Land is plentiful. Oil palm and wood pulp. Existing old complex agroforestry practices.	Presence of SRDP/ TCSDP project: rubber monoculture in the 1980s. Oil palm and pulp. Existing old complex agroforestry practices.	*Sawah* off-farm activities.
Type of on-farm trial	RAS I and 2	RAS I and 2	RAS 2 and RAS 3

Tembawang are indigenous fruit and timber-based complex agroforestry systems where the main tree species is usually the illipe nut tree.

farms in transmigration[40] areas (where some *Imperata cylindrica* was present) through dedicated programmes or the relocation of people from Java.

The first type (RAS 1)[41] resembles the jungle rubber system, but unselected rubber seedlings are replaced by clones selected for their potential adaptation capacity. These clones must be able to compete with the natural secondary forest growth. Different planting densities (550 and 750 trees/ha) and weeding protocols were tested to identify the minimum management needed for the system to succeed. This is always a key factor for farmers whose main concern is to maintain or increase labour productivity. Biodiversity is presumed to be very similar to that of jungle rubber, which is quite high and resembles that of secondary forest at the same age. This system is probably the closest to the concept of fallow enrichment and suits a vast number of farmers because of its simplicity.

The second type, RAS 2, is a complex agroforestry system in which rubber trees (550/ha) and perennial timber and fruit trees (92 to 270/ha) are planted after slash and burn. It is very intensive, with annual crops intercropped in the first 3 or 4 years, with emphasis on improved upland rice, using different rates of fertilisation as well as dry season cropping with groundnuts, for instance. Several different combinations of crops were tested including food crops and cash-crops such as cinnamon. Several planting densities of selected species were tested according to a pre-established tree typology, in particular with the following species: rambutan, durian, petai and tengka-wang. Biodiversity is limited to the planted species (between 5 and 10) and those that regenerate naturally and are consequently preferred by farmers.

The third type, RAS 3, is also a complex agroforestry system with rubber and other trees planted in the same way as in RAS 2, except that this system is used on degraded lands invaded by *Imperata cylindrica,* or in areas where *Imperata* is a major threat. The main constraints are labour or cash to pay for herbicides to control *Imperata.* In RAS 3, annual crops, generally rice, are only grown in the first year, with non-vine cover crops planted immediately after the rice harvest (*Mucuna* spp., *Flemingia congesta, Crotalaria* spp., *Setaria* and *Chromolaena odorata*), multi-pur-pose trees (wingbean, *Gliricidia sepium*), or fast-growing trees for use as pulpwood (*Paraserianthes falcataria, Acacia mangium* and *Gmelina arborea*) can be planted (several combinations were tested). The objective was to eliminate the need for weeding by providing a favourable environment for the rubber and associated trees to grow, while preventing the growth of *Imperata* with limited labour. The aim of associating non-vine cover crops and MPT[42]'s for shade was controlling *Imperata.* Biodiversity was expected to be similar to that of RAS 2.

These RAS types were tested from 1994 to 2007 and surveyed again in 2019 and 2021. The clones tested were PB 260, BPM 1, RRIC 100, and RRIM 600, compared to rubber trees grown from seedlings.

---

40. Transmigration was a Indonesian government programme to resettle people from Java in less populated areas of Indonesia (known as the "periphery"), mainly Kalimantan, Sumatra, Sulawesi, Maluku and West Papua (Irian Jaya).
41. The description of RAS types has been published in Didier Babin (ed), 2004. *Beyond tropical deforest-ation. From tropical deforestation to forest cover dynamics and forest development,* UNESCO/Cirad, 488 p.
42. MPT: multi-purpose tree.

Table 2.2. Specific constraints to the adoption of RAS in 1996

Topic	West Kalimantan	Jambi	West Sumatra
Previous and/or current projects Access to information	SRDP/TCSDP	ASB (Alternatives to Slah and Burn Project)	Pro-RLK (Project for land Rehabilitation for rubber)
Indigenous knowledge and agroforestry practices	+++	+++	+/-
Clone availability	+	+/-	-
Availability of Bah Lias Isolated Garden (BLIG) seedlings	-	-	+++
Fertiliser use	+	-	-
Availability of high yielding varieties (HYV) of upland rice	-	---	--
Seed quality	-	-	-
Availability of cover crop seeds	-	-	-
Pests and diseases	-	-- Monkeys, pigs	- Pigs
Weeds	*Imperata*	*Mikaenia*	*Imperata*
Rubber diseases	*Colletotrichum*		possibly *Colletotrichum*
Land constraints	Very low fertility, land scarcity in transmigration areas	Steep slopes in pioneer zones	Very low fertility and steep slopes, altitude: 550 m - close to upper limit for rubber
Upland rice production	Average potential with selected local rice varieties	May succeed in peneplains	Excellent weed control, requires soil and water conservation techniques
Potential for the adoption of RAS			
RAS 1	+++	+++	0
RAS 2.2/rice	++	+	+++
RAS 2.5/cinnamon	0	+++	++
RAS 3	+++	0	+

▸▸ Main results of RAS

RAS in Indonesia

The performance of clones in RAS 1 environments was encouraging 6 years after planting. Compared to plants originating from seedlings, clones perform better in terms of growth from establishment on. Among the clones tested, BPM 1 grew best up to 40 months, followed by other clones, while trees grown from seedlings grew the most slowly. After 40 months, due to white root disease, the growth of the two clones BPM 1 and RRIM 600 was reduced whereas the growth of the other two clones, RRIC 100 and PB 260, was very good and the trees were ready to be tapped at 5 years of age. However, trees grown from seedlings can also be tapped about 5.5 years after planting. In this trial, the frequency of weeding in the rubber rows was 3-4 times per year.

Farmers know that rubber growth will be affected by competition with other vegetation. In West Kalimantan, farmers did not fully respect the trial protocol, rather, to adapt to local conditions, they slashed the vegetation in intra-rows once a year from the second year on and kept only a few tree species, mostly those of monetary value. This resulted in slightly slower rubber growth than in Jambi. No significant difference in rubber growth was observed due to the level of weeding. The effects of perennial intercrops on rubber growth varied from year to year, but, except for the treatment involving durian, no significant difference due to intercrops was observed at 54 months. The difference in rubber performance was more due to the site and/or to the practices used by the farmers who took part in the trial than to the different intercrops used.

Due to shading by rubber trees, fruit trees cannot produce fruit of the same quality as the fruit of trees planted in open areas. The RAS 2 trials in West Kalimantan were not as intensive as we expected. Annual intercrops (mainly upland rice) were only planted in the first two years. It is also clear that if the rubber tree spacing is 6 m × 3 m, planting perennial plants under rubber is not optimal in terms of fruit production. Under RAS 3, creeping legumes were clearly the top performers in controlling *Imperata*. *Pueraria* was slightly but statistically significantly better than *Mucuna* for rubber growth. Both *Pueraria* and *Mucuna* grew well and managed to prevent regrowth of *Imperata*. However, the creeping legumes had to be regularly removed from the rubber rows as they entangled the rubber trees. Among the erect legumes, *Flemingia* was good for rubber; *Crotalaria* was disappointing. Rubber trees with no cover crops but with *Imperata* or *Chromolaena* had not yet reached tapping size by the end of the trial. This finding is consistent with the results of earlier work done in Sembawa research station where it took more than 10 years for rubber trees to reach tapping size in the absence of proper control of *Imperata* (Mulyoutani et al., 2006).

Farmers very often do not follow all the protocols designed by and proposed to them by researchers. This kind of problem was encountered both in Jambi and in West Kalimantan. Again, establishing a close relationship with farmers and trying to understand why they do not follow a protocol is one of the objectives of participatory on-farm trials. In addition, intensive discussion is important so as to choose the technical options that best match the farmers' needs. Our results showed that the trade-off between inputs (fertilisers, labour, chemicals) and growth or plant diversity interests most farmers. Due to the many constraints that farmers face, especially lack of cash for most Indonesian farmers, they have to choose between spending money and allocating family labour. Maximum rubber growth is not always the objective farmers have in mind when choosing between different forms of RAS. The main challenge for researchers is consequently offering farmers technologies that account for their real constraints and opportunities.

RAS 1

Effects of different levels of weeding on rubber growth under RAS 1

In West Kalimantan, farmers did not fully respect the weeding protocol (Table 2.3). The level of weeding used by farmers was slightly below that specified in the trial protocol. It was thus logical that there was no significant difference in rubber growth between plots classified as "medium", "intensive" or "intensive with legumes cover crops (LCC)" plots because the weeding frequency was the same in all of them from the 1st to the 5th year. Nevertheless, rubber growth in this group of treatments was better than that in plots with "low weeding frequency" (Figure 2.1).

Table 2.3. Frequency of weeding within the row of rubber trees specified in the protocol and the frequency actually implemented by farmers in RAS 1 trials in West Kalimantan

Treatment	Expected frequency per year	Actually, implemented by farmers				
		1st year	2nd year	3rd year	4th year	5th year
Low	4 then 2	2	1	0	0	1
Medium	6 then 4	2	2	1	1	1
Intensive	8 then 6	2	2	2	1	1
Intensive+LCC	8 then 6	2	2	2	1	1

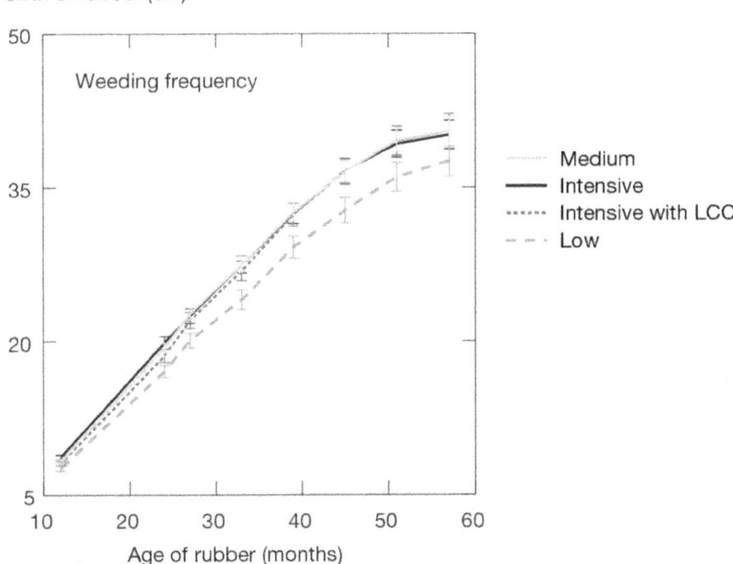

Figure 2.1. Effects of weeding frequency within rubber rows on the girth of rubber trees in West Kalimantan (RAS 1)

LCC: legumes cover crops.

Variation in growth was based more on the location of the plot (farmers) than on weeding frequency (Figure 2.2). The slowest relatively good rubber growth was observed in plots in Loheng and Sidon. In Loheng, particularly after the third year, the rubber rows were not cleared and were consequently infested by *Melasthoma, Chromolaena, and Mikania,* all noxious weeds for rubber. Between the rows, the vegetation was dominated by the same weeds and by a variety of trees that reached more than two metres in height. Many plants died in the second year due to white root disease and continued to die in the third year. The height of the different types of vegetation in the inter-rows can reduce rubber girth, as shown in Figure 2.2. The other four farmers in Sidon controlled weeds (which did not include noxious species and were dominated by grasses) in the rubber rows up to year 3 and the height of the vegetation in the inter-row was less than two metres.

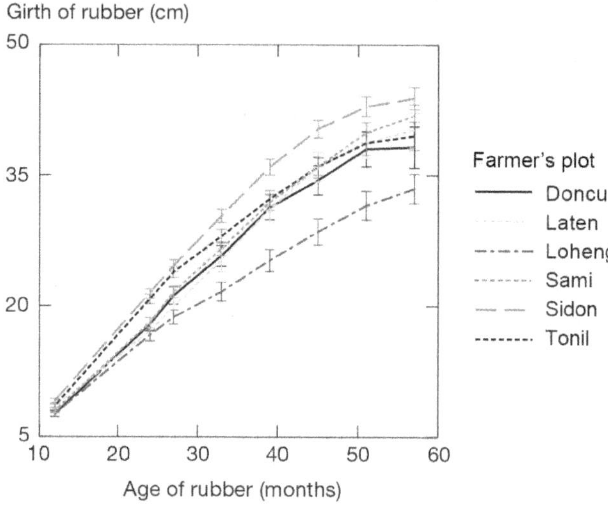

Figure 2.2. Variation in rubber growth in plots belonging to different farmers in West Kalimantan (RAS 1)

Performance of different rubber clones in RAS 1 environments

There is widespread belief among farmers in Sumatra and Kalimantan that, compared to rubber trees originating from seedlings, rubber clones cannot perform well in agroforest environments. A series of trials were carried out in Jambi and West Kalimantan starting in 1996 to test the performance of different rubber clones in agroforest environments (RAS 1 series). Clones PB 260, BPM 1, RRIC 100, and RRIM 600 were compared with rubber trees originating from seedlings. According to the RAS 1 principle, the land was previously jungle rubber or secondary forest, prepared using slash and burn. In the 1st year, a variety of food crops were planted as intercrops. In the rows of rubber trees, weeding was only carried out up to a distance of 1 m on each side of the rubber row, 3-6 times in the first year (considered as low and medium levels of weeding) and 3-4 times in the second year; and only once in the third year. Vegetation growing between the rubber rows was expected to be left in place by the farmer to conserve a certain level of biodiversity.

Results of the trials in Jambi suggested that weeding frequency has a positive influence on rubber growth starting in the early stage of establishment. The trials clearly showed that by limiting weeding to the rubber rows (at a frequency of every two months in the first two years; every six months in the 3rd year, and only one weeding in the 4th year), and letting the vegetation between the rubber rows (in this case *Micania, Melasthoma, Chromolaena*) grow to a height of 1.5 m, rubber reached tappable size 5 years after planting. However, when the frequency of weeding was reduced to 3 times a year or once every 4 months, then rubber only reached tappable size between 5 and 7 years after planting.

As mentioned previously, farmers know that rubber tree growth is reduced by competition with other vegetation. Like in RAS 1, the farmers did not fully apply the trial protocol in West Kalimantan. The weeding frequencies they used are listed in Table 2.2. The trial protocol clearly stated that farmers should let vegetation grow in the intra-rows and respect certain weeding frequencies on the rubber line. Most slashed the vegetation in the intra-rows once a year, starting in the second year. Only a few tree species were kept in the plots, especially plants that had monetary value at the time. This resulted in a slightly slower rubber growth than in Jambi.

The performance of clones in RAS 1 environments compared to the performance of trees originating from seedlings was encouraging, clones performed better in terms of growth immediately after establishment (Figure 2.3). Up to 40 months, clone BPM 1 showed the best growth rate. The rubber rows in the plots in this trial were weeded between 3 and 4 times a year, again confirming that rubber seedlings grow more slowly than clones. Using this plot as a demonstration plot for farmers was very effective as the performance of the rubber clones was significantly better than that of seedling rubber. In Jambi, except in the plot affected by white root disease, there was no significant difference in rubber growth linked to farmers' performance.

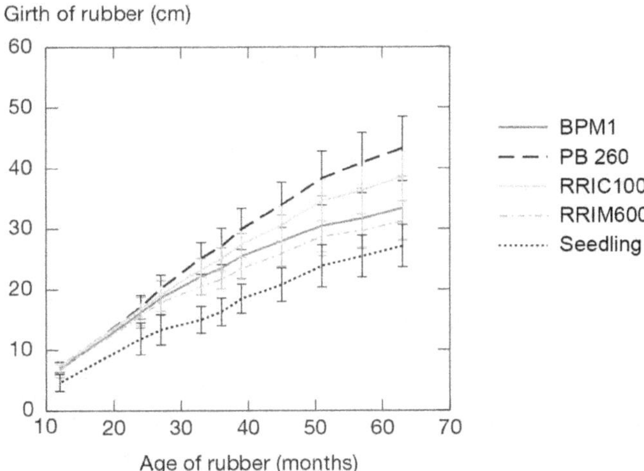

Figure 2.3. Growth performance of different clones in the RAS 1 environment in West Kalimantan

Fertilisation requirements in RAS

In Indonesia, most farmers use no fertilisers in their rubber plantation, or if they do, they apply less than half the recommended doses. Trials to study the effects of fertilisation on rubber growth in monoculture rubber plantations are very well documented in all rubber producing countries worldwide, but many questions concerning the application of fertilisers in RAS remain unanswered. A study was thus undertaken to compare the effects of additional doses of urea, SP36, KCl (see Table 2.1) with the effects of the application of basal fertiliser (200 g urea, 160 g SP36, 100 g KCl per tree in the first year; 100 g urea, 80 g SP36, and 50 g KCl in the second year), on rubber growth. The fertiliser was applied four times a year. The doses tested are listed in Table 2.4.

Table 2.4. Doses of fertilisers (g/tree/year) applied in different treatments based on RAS 1 and RAS 3 in West Kalimantan

Type of RAS	Treatment	First year (g/tree/year)			Second year (g/tree/year)		
		Urea	SP36	KCl	Urea	SP36	KCl
RAS 1	Year 1	300	160	100	100	80	50
	Year 2 to 5	200	160	100	100	80	50
RAS 3	Year 1	300	160	100	100	80	50
	Year 2 to 5	200	160	100	100	80	50

In RAS 1, rubber responded positively to additional urea in the first months after establishment. The additional urea, i.e., increased from 50 g/tree/application to 75 g/tree/application, was needed to increase rubber growth by about 7% in the 30 months after planting. Even the statistical test showed no significant difference in girth resulting from the treatments, but the growth of rubber with additional urea was consistently better than without (see Figure 2.6). These results indicate that additional urea (nitrogen) is needed as additional fertiliser (rather than P and K) to increase rubber growth. Indeed, this result has been put into practice by farmers who have to choose between fertilisers. They choose urea before other fertilisers. In this way, farmers who practice annual intercropping provide additional benefits to their rubber especially when they cultivate horticulture species that require intensive fertilisation (including organic fertilisers). Combining perennials and intensive horticulture species as intercrops creates a positive relationship between rubber and intercrops (Wibawa et al., 2006).

RAS 2

The growth of rubber under different treatments that associate perennial intercrops in RAS 2 conditions showed that variation within a farm was higher than variation within treatments (see Figure 2.7), especially after the second year. The effects of growing perennial intercrops on rubber growth varied from year to year, and except with durian, no significant difference due to intercrops, was observed at 54 months. Differences in rubber performance were more due to the site or to the practices used by the farmers who took part in the trial than to different intercrops (Figure 2.4). Due to shading, fruit trees do not produce as much fruit as fruit trees planted in full sun. The RAS 2 trials held in West Kalimantan were not as intensive as expected. The annual intercrops (mainly upland rice) were only planted in the first two years. With normal spacing of the rubber trees (6 m × 3 m), fruit trees will produce less due to more intensive shading. If farmers want to plant trees, double-row spacing is a better option, in which case rubber will reach tappable size 6 -7 years after planting.

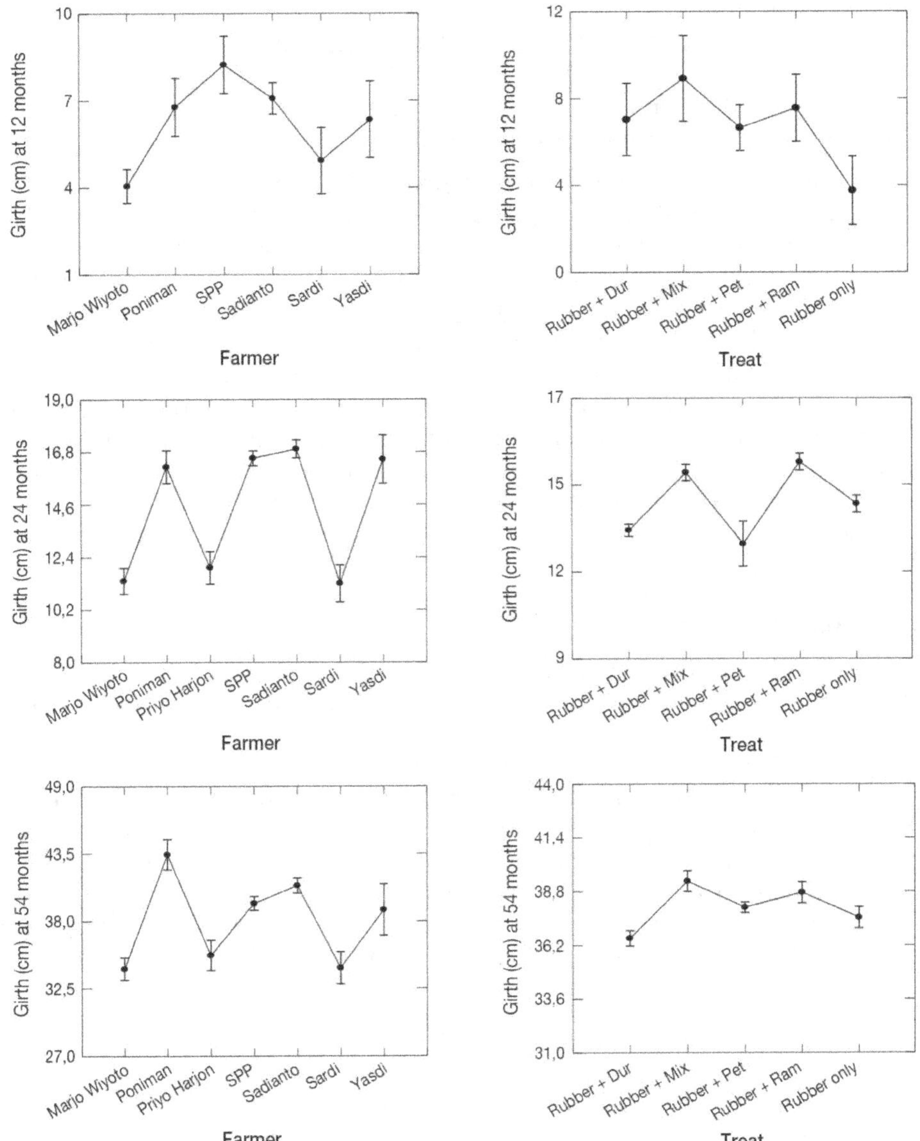

Figure 2.4. Variation in the girth of rubber trees at different ages, sites and treatments in RAS 2 in West Kalimantan

RAS 3 type to kill *Imperata cylindrica*

One idea behind RAS 3 (Mulyoutami et al., 2006; Penot, 2001) was to enable the establishment of rubber agroforests on land previously infested by *Imperata* by using legume cover crops, (*Pueraria javanica, Mucuna utilis*), shrubs (*Flemingia congesta, Crotalaria anagyroides*) and fast-growing trees (FGT – *Paraserianthes falcataria, Gmelina arborea, Acacia mangium*) that are capable of shading out *Imperata* regrowth, particularly in the first few years of rubber establishment. Planting cover crops is usually recommended when establishing rubber monoculture. FGT density was kept under 100 stems per ha under the assumption that a higher density seriously affects rubber tree growth. Natural seedling mortality meant that in some plots, only a few individual FGTs remained three years later (Penot, 1997).

The first RAS 3 trials were planted in 1996 in three farmers' fields in Kopar village in West Kalimantan (Boutin et al., 2000). High yielding clonal (PB 260) rubber plants raised in polybags were planted in the field after clearing using slash and burn. *Mucuna, Pueraria, Flemingia* and *Crotalaria* were planted in four rows between the rows of rubber at varying densities depending on the crop. Naturally occurring *Imperata* and *Chromolaena* were also left in place (i.e. not weeded out) for the purpose of comparison. In the village of Trimulya, located in a Javanese transmigration zone, FGT were planted between rows of rubber trees themselves planted at their usual density. All the plots, i.e., both those with cover crops and those with FGT, were weeded (manually or using herbicides) at 3-month intervals, but only in the rubber rows; limited fertilisers (rock phosphate and urea) were applied only in the first two years. Regular measurements of the girth of rubber trees, and the presence and dominance of ground vegetation formed the basis of our analysis.

Cover crops

The combined results of more than 6 years of monitoring the 3 experimental sites in Kopar village indicated that legume cover crops have different potential for the control of *Imperata* and hence for influencing the growth of young rubber trees (Figure 2.5). Creeping legumes were clearly the top performers, with *Pueraria* topping the list, followed by *Mucuna*. Among erect legumes, *Flemingia* was the best, while *Crotalaria* proved disappointing. Plots containing rubber trees with no cover crops that were invaded by *Imperata* or *Chromolaena* had not yet reached tapping size at the end of the 6-year monitoring period. This finding is consistent with the results of earlier work done in Sembawa Research Station where, without proper control of *Imperata*, it took more than 10 years for rubber trees to reach tapping size (Joshi et al., 2001).

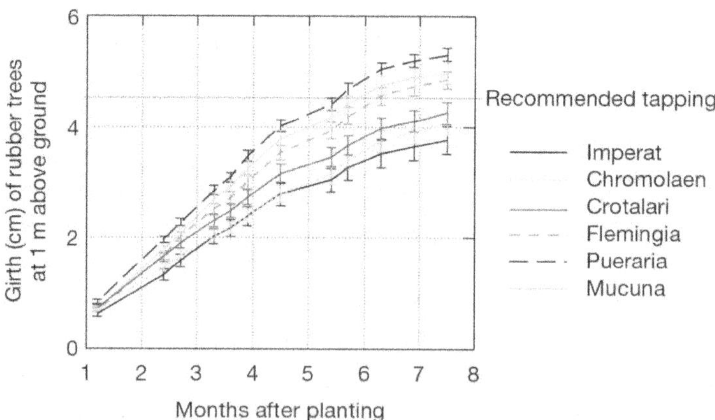

Figure 2.5. Rubber tree growth in the RAS 3 trial plot with cover crops against the weed Imperata

Although *Pueraria* and *Mucuna* grew well and succeeded in preventing regrowth of *Imperata*, they required regular interventions to remove the climbing vines from the trunks of the young rubber trees. Another major problem with *Mucuna* is the need to plant its seeds repeatedly as its life cycle is shorter than six months, which consequently requires more labour than the other species. On the other hand, the seeds produced by the previous *Mucuna* crop can be sown to maintain the cover. *Pueraria* seeds cannot be produced locally and are not easy to obtain on the local market. Likewise, the supply of *Flemingia* seeds was problematic in 1994/1997 (and still is in 2023).

Fast growing trees

The FGT trials in Trimulya village showed that all the FGT were only partially successful in controlling *Imperata* regrowth, i.e., *Imperata* manged to regrow in more than half the plots. There was no significant difference between the FGT species tested, *Acacia*, *Paraserianthes* and *Gmelina*, either in controlling *Imperata* or in their influence on rubber growth. In the early years, the effect of *Acacia* on rubber trees was slightly less positive than that of other species. However, the rubber trees in *Acacia* plots soon caught up when the *Acacia* trees were cut down after three years. The rubber trees in the FGT mixed plots took nearly six years to reach tapping size, i.e., a girth of 45 cm measured 1 m above the ground.

Comparison of rubber data from cover crop trials and FGT trials yielded quite interesting results. While rubber growth in FGT mixed plots was better than in *Imperata* or *Chromolaena* infested plots, growth was far behind that of rubber grown in plots with legume crops. Rubber trees needed more than a year longer to reach tapping size than rubber grown with cover crops (*Pueraria* and *Mucuna* plots; Figure 2.6).

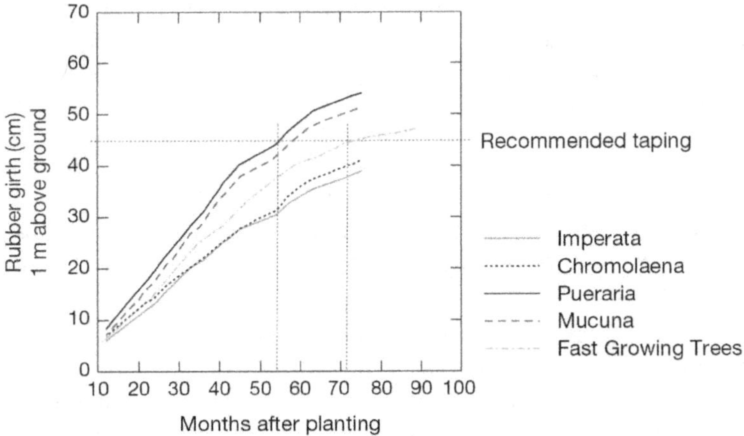

Figure 2.6. Comparison of rubber tree growth with cover crops and fast-growing trees against the weed *Imperata*

While FGT mortality was high, the surviving trees, particularly *Acacia*, grew rapidly and started affecting rubber tree growth in year 2 and year 3. The farmers who obviously preferred rubber were concerned, and, after three years, all remaining FGT had been removed from their fields. The harvested wood was only useful as firewood.

White root disease of rubber is becoming a serious problem in West Kalimantan and is known to be more severe in areas where the previous system was jungle rubber. However, in the RAS 3 trial, there was no evidence for a higher incidence of rubber tree mortality due to white root disease in plots converted from jungle rubber (2-6%) than in plots that were previously *Imperata* grassland (1-7%). The survival rate of planted rubber trees was more than 90%.

The FGT trials in Trimulya village showed that all the FGTs were relatively successful in controlling *Imperata* regrowth, even though *Imperata* was still encountered in nearly half the plots. This is not surprising as in their early stages, trees only have small crowns and are consequently unable to efficiently shade out *Imperata*. There was no significant difference between the FGT species tested – *Acacia*, *Paraserianthes*, and *Gmelina* – either in the control of *Imperata* or in their effect on rubber growth. The negative effect of *Acacia* on rubber trees was apparent from the early years, but, as mentioned above, the rubber trees caught up rapidly after *Acacia* trees were cut down after three years. Analysing the results of on-farm participatory trials is difficult due to uncontrolled factors that could interfere with the main factors described above. An inventory of any factors that could influence rubber tree growth needs to be undertaken very carefully. Participatory trials depend on a close relationship and continuous communication with farmers. Planning, implementing and any modification to the trials need to be preceded by detailed discussion with the farmers. Trust between researchers and farmers has to be built from the very beginning of a project if the objective of the on-farm trial is to be achieved. Once trust was established, the SRAP programme and associated activities were carried out more efficiently.

RAS 3 is a rubber agroforestry system whose "technologies" were tested and promoted only in West Kalimantan (Penot, 1997). The primary aim of the RAS 3 series was to establish productive rubber agroforests in degraded *Imperata* grasslands using legume cover crops or FGTs to shade out *Imperata*, combined with limited labour and limited use of chemical inputs. Legumes improve soil fertility thus benefiting the rubber trees in addition to controlling *Imperata* and *Chromolaena*. The RAS 3 trials confirmed that the cover crops alternative is the best. The proof that rubber trees can be tapped less than five years after planting, require less intensive weeding (generally only half that needed in standard monoculture plantations), and less fertiliser is certainly encouraging for small-holder farmers. The results obtained in these trials are comparable with those obtained in intensive monoculture plantations. However, lack of seeds of these useful legumes and the need to replant *Mucuna* are serious drawbacks that remain to be addressed.

On the other hand, FGTs were planted at the same time as rubber and were expected to control *Imperata* and *Chromolaena* in the early years of establishment. The sale of timber from these trees to the pulp industry in seven or eight years was predicted to provide extra income for farmers. Neither of these expectations was completely fulfilled. While all the FGTs tested proved partially successful in controlling *Imperata*, they also had a negative impact on the rubber trees. The farmers were reluctant to accept any negative impact of these FGT on rubber trees and consequently after 3 to 4 years, they cut down their FGTs, especially *Acacia mangium*, due to very high competition for light, even though *Imperata* was effectively controlled by the shade provided by *A mangium*. An interesting point is that more Javanese and Dayak migrants, who have fewer land resources, consider FGT as a viable source of income than local Dayak people.

One major problem that emerged after establishment of the cover crops was the poor quality of the seeds, a widespread problem in Indonesia in the case of varieties that are not indigenous. Both cover crops and FGT have roles to play, albeit slightly different, in improving the chances of successful and rapid establishment of clonal rubber in a low input system, and modifications to enable the combined use of cover crops and FGT may be a better solution than choosing one or the other. Based on the results obtained so far, it appears possible to control *Imperata* by planting a cover crop (*Mucuna* or *Pueraria* when the seed problem has been solved) within the first two or three years. FGTs can then be planted when rubber is already established. The effect of FGTs on rubber is significantly reduced when the FGTs are planted too late. The selection of FGT and other associated trees (such as fruit and timber species) will require careful thought as the choice depends on the local context and on demand for their product. Both smallholder farmers and rubber agroforests will then be able to profit from optimal use of previously degraded *Imperata* grassland.

Biodiversity observed in RAS

The two main advantages of jungle rubber (and subsequently of clonal RAS) were: (i) biodiversity conservation, as biodiversity is close to that of primary forest or old secondary forest in the case of old mature jungle rubber (de Foresta and Michon, 1992, 1995; Werner, 1997), (ii) environmental benefits in terms of soil conservation (Sethuraj, 1996) and water management due to its forest-like characteristics. The biomass of a 33-year old rubber plantation is very similar to that of a humid tropical evergreen forest.

Previous results on jungle rubber biodiversity that were available to the author (Werner, 1997; de Foresta, 1997) as well as a guidebook on plant uses (Levang and de Forestia, 1991) provided very useful preliminary information for this chapter.

The data presented in this section were collected between August and October 2001 in 4 villages in the West Kalimantan province and included 23 rubber agroforest plots. It has been originally published in 2004[43]. The nature of the previous vegetation, neighbouring vegetation and soil characteristics were recorded in addition to standard data (rubber growth, etc.) collected from the plots used for on-farm trials.

The "transect" method was used to assess existing biodiversity, with a sampling size per transect of 1 m × 0.2 m, and 15 replications of each treatment. The transect method was chosen to cover as wide a range of situations as possible, but the results obtained using this method do not allow direct comparison of biodiversity between RAS and jungle rubber systems because the plots — and hence the transects — are too small. Further research is therefore required but this has never been completed.

Correspondence analysis

Correspondence analysis was chosen because it makes it possible to focus on the different effects that influence plant biodiversity in rubber inter-rows under the different systems. The five first axes were taken into account in each analysis. Data were

---

43. Diaz-Novellon S, Penot E, Arnaud M, 2004. Characterisation of Biodiversity in Improved Rubber Agroforests in West-Kalimantan, Indonesia: Real and Potential Uses for Spontaneous Plants. *In*: Gerold, G., Fremerey, M., Guhardja, E. (eds), *Land Use, Nature Conservation and the Stability of Rainforest Margins in Southeast Asia*. Springer, Berlin, 426–444.

collected from RAS as well as in selected fallow plots with different densities of existing vegetation. Our analysis included 23 RAS plots, 1, 2, 3, 4 and 5-year-old fallow plots, 6-year-old jungle rubber plots (i.e. the same age as RAS plots), some *tembawang*[44] plots and some secondary forest plots located near the study villages. Complete results are reported in Diaz-Novellon's MSc.Thesis (2001).

Assessment of biodiversity under rubber agroforestry systems: a comparison of jungle rubber and RAS 1 and 3

Species diversity, i.e., the number of species, appeared to be higher in jungle rubber than in improved RAS. However, in RAS 2 and 3 in which fruit and timber trees were planted in the inter-rows, the biodiversity of a number of species per transect was comparable with that of jungle rubber. A similar result was observed with RAS *sendiri*. It thus appears that different methods of cultivation can have a direct influence on the spontaneous diversity of plants in the inter-rows, and in practice, experimental RAS, "RAS *sendiri*" and jungle rubber are managed differently, resulting in a significant "farmer effect".

According to trial protocols discussed with project farmers each year, the inter-rows are weeded more frequently in RAS. In practice, in RAS, weeding is limited to selective cutting of trees and shrubs that grow taller than the young rubber trees, whereas in jungle rubber, no weeding is done in the first few years. Weeding appears to be the main factor that influences plant diversity. When the cutting of spontaneous vegetation of the inter-row is spread out over time, the result is more species. On the other hand, the type of rubber trees (clonal or seedlings) does not influence the type, diversity or the quantity of vegetation. Inter-row biodiversity is therefore more influenced by farming practices and in particular by the frequency of selective cutting or by the number of weeding operations. Species distribution and biodiversity of RAS plots is shown in Table 2.5.

Table 2.5. Distribution of each type of plant across all plots

	Trees	Herbaceous	Ground lianas	Climbing lianas	Bamboo rattan	Shrubs
Number of plants	1,138	2,480	368	128	54	231
Number of species	55	24	7	6	1	3

One important question concerning the comparison between RAS and jungle rubber is whether jungle rubber has a higher specific plant density, i.e., number of plants per unit area, than improved rubber agroforests. Our results show that jungle rubber does indeed have more individual plants in inter-rows than RAS, although the density of plants is very similar to that of "RAS *sendiri*". The different agroforestry practices (and in particular the frequency of selective cutting) explain this difference. Compared to the biodiversity found in secondary forest or *tembawang*, the number of species appears to be similar to the number found in RAS even though the ground-level density of species is considerably lower (see following tables). In other words, the difference is mainly quantitative.

---

44. *Tembawang* is the name of fruit and timber agroforestry system developed by Dayak people.

What are the most significant factors that explain the variation in biodiversity? Discriminant analyses showed that previous farming practices play a significant role. An area that had been cultivated for at least 3 years hosted higher species biodiversity than an equivalent fallow area. One possible explanation is that areas cultivated as "open systems" have a bigger seed bank and can collect seeds from surrounding forests or agroforests. Environmental factors probably also influence biodiversity. The presence of jungle rubber in the immediate vicinity results in greater biodiversity. One to 5 years of fallow around plots probably also increases biodiversity. As far as agricultural practices are concerned, the number of selective cuttings per year appears to be the main factor that influences plant biodiversity in the inter-row (see Diaz-Novellon's MSc thesis, 2001 for details).

Smallholders' perception of plant biodiversity

It was clear that local populations know the plant species in their fields and their specific uses perfectly well. More than 300 species needed indexing during field surveys and interviews with farmers. The most common uses of spontaneous biodiversity (in forest and agroforests) ranked in decreasing order of importance are health (medicinal plants), food (fruit, vegetables), construction (wood and timber), firewood and others (Table 2.6).

Table 2.6. Existing and potential uses of biodiversity by the Dayak population (1997)

Uses	Number of species
Timber for construction, housing	83
Firewood	40
Timber for sale or furniture making	2
Fruits	112
Vegetables	68
Medicinal plants	179
Animal feed	24
Pulp (for paper making)	1
Cosmetics	1
Colouring properties	2
Use as paper	9
Weed control	14
Insecticide	6
Handicrafts	66
Latex	4
Oil	7
Fertilisation	14
Spices	55
Others	8

However most spontaneous vegetation is not yet used by the local population and is thus available for "potential uses". Medicinal plants have considerable potential (Table 2.7), they are not widely used because some farmers prefer "modern" drugs, which are thought to be far more effective against malaria, diarrhoea and other illnesses as long as their incomes do enable such expenses.

Table 2.7. Uses of medicinal plants

Health disorders treated with local plants	Number of species identified
Coughs	12
Fever	23
Itching	15
Tiredness	11
Malaria	2
Dysentery	1
Sore throat	13
Toothache	1
Stomach ache Nausea	44
Burns	9
Headaches	11
Others	11

In the case of timber and wood, the most valuable species (Table 2.8) are becoming scarce in local forests in the study area in 1997 and the situation is worse in 2024 leading to a real new demand on such products.

Table 2.8. Timber species that are becoming scarce in remaining forests (1997)

Local names	Latin names	Village
Belian	*Eusideroxylon zwageri*	All villages
Tapang	*Koompassia excelsa*	Embaong, Kopar
Tekam		All villages
Benkirai	*Shorea* sp.	Embaong
Meranti	*Shorea* spp.	Engkayu, Trimulya
Terenak		All villages
Jeluntung	*Dyera costulata*	Trimulya
Kayu Raya	sorea spp.	Kopar
Majau	*Shorea palembanica*	Embaong
Omang	*Hopea dryobalanoides*	Sanjan, Engkayu
Medang	*Litsea elliptica*	Kopar, Engkayu
Tunam	*Shorea lamellata*	Kopar
Nyatuh	*Palaquium* spp.	Engkayu
Owan		Engkayu

Local names	Latin names	Village
Ubah	*Glochidion* sp.	Sanjan, Engkayu
Taba	*Aquilaria malaccensis*	Kopar
Keladan	*Dryobalanops beccarii*	Engkayu
Tengkawang	*Shorea macrophylla*	Kopar

Most farmers are interested in 2024 in including particular timber species (Table 2.9) in their agroforests for both housing (construction) and sale.

Table 2.9. Timber species preferred by farmers (1997)

Local names	Latin names	Local names	Latin names
Belian	*Eusideroxylon zwageri*	Mentibu	
Keladan	*Dryobalanops beccarii*	Medang	*Litsea elliptica*
Tekam		Nyatuh	*Palaquium* spp.
Ketuat		Oman	*Hopea dryobalanoides*
Meranti	*Shorea* spp.	Owan	
Terindak	*Shorea senimis*	Jonger	*Ploiarium alternifolium*
Tengkawang	*Shorea macrophylla*	Taba	*Aquilaria malaccensis*
Mengkirai	*Trema orientalis*	Tantang	*Buchania sessifolia*

Prices of timber species vary considerably, indicating that this market was already well developed in 1997 (Table 2.10). However production has seriously decreased with the loss of the forest, the deman is still high in 2024 for quality timber.

Table 2.10. Prices for local timber species in 1997 (just given as an example)

Timber species	Latin name	Price in rupiah in 1997
Belian	*Eusideroxylon zwageri*	50,000 Rp/board
Raya		3,500 Rp/board
Jonger	*Ploiarium alternifolium*	4,000 Rp/board
Owan		8,000 Rp/board
Medang	*Litsea elliptica*	8,000 Rp/board
Paku		5,000 Rp/board
Tapang		20,000 Rp/board
Tengkawang	*Shorea macrophylla*	4,000 to 10,000 Rp/board
Tantang	*Buchania sessifolia*	200,000 Rp/m2

Note: Exchange rate in 1997: US$1 = 10,500 Rp. Prices are given to show the difference in price for different types of timber.

Some local species have always been maintained or preserved by replanting or by favouring regeneration from natural regrowth in the different types of agroforests (Table 2.11) and have a range of different uses.

Table 2.11. Spontaneous timber species maintained in local agroforests and their uses (1997)

Local names	Latin names	Uses
Leban	*Vitex pinnata*	Timber, wood, spice, medicinal
Medang	*Litsea elliptica*	Timber, latex
Ramboutan	*Nephelium lappaceum*	Fruits, timber
Jengkol	*Pithecellobium jiringa*	Fruits, vegetable, timber, medicinal
Durian	*Durio zibethinus*	Fruits, timber
Pingam	*Artocarpus* sp.	Fruits, timber, vegetable
Cempedak	*Artocarpus integra*	Fruits, medicinal, vegetable
Lengsat	*Lansium domesticum*	Fruits, medicinal, handicrafts
Pekawai	*Durio* c.f. *dulcis*	Fruits
Mentawa	*Artocarpus* c.f. *anisophyllus*	Fruits
Nyatuh	*Palaquium* spp.	Timber, latex
Owan		Timber, handicrafts
Bungkang	*Polyalthia rumpfii*	Timber, spice
Belian	*Eusideroxylon zwageri*	Timber
Ubah	*Glochidion* sp.	Timber
Kemenyan	*Styrax benzoin*	Timber, latex, animal feed
Tantang	*Buchania sessifolia*	Timber
Bidara	*Nephelium maingayi*	Fruits

Some of these species have been re-introduced in agroforests (Table 2.12), in particular in *tembawang*, or are protected when they emerge in natural regrowth in jungle rubber and RAS.

Table 2.12. Local species reintroduced in agroforest (1997)

Local names	Latin names	Uses
Jengkol	*Pithecellobium jiringa*	Fruits, vegetables, timber, medicinal
Mangga	*Mangifera indica*	Fruits
Ramboutan	*Nephelium lappaceum*	Fruits, timber
Manggis	*Garcinia mangostana*	Fruits
Durian	*Durio zibethinus*	Fruits, timber
Cempedak	*Artocarpus integra*	Fruits, medicinal, vegetables
Coklat		Cocoa
Kopi		Coffee
Petai	*Parkia speciosa*	Fruits, vegetables
Lengsat	*Lansium domesticum*	Fruits, medicinal, handicraft
Kedupai	*Mischocarpus pentapetalus*	Fruits
Sibau	*Xerospermum norotanum*	Fruits

Local names	Latin names	Uses
Mentawa	*Artocarpus anisophyllus*	Fruits
Pekawai	*Durio* c.f. *dulcis*	Fruits
Melinjo	*Gnetum gnemon*	Fruits, vegetables
Nangka	*Artocarpus heterophyllus*	Fruits
Tengkawang	*Shorea macrophylla*	Fruits, oil, timber
Tekam		Timber, handicraft
Ketuat		Fruits, timber
Tempuih	*Baccaurea* sp.	Fruits
Pisang	*Musa* spp.	Fruits, vegetables, medicinal

Other species farmers do not consider suitable in RAS inter-rows were also identified (Table 2.13). These species are in fact still used in that their products are still collected in true forests, but are not specifically selected in agroforests due to the fact that – at least in the farmers' opinions – they might have a negative effect on rubber growth during the immature period. For example, after 20 years of growth, the number of durian trees per ha has to be less than 20 to reduce shading when the durian canopy begins to outgrow that of rubber. Another example is *tengkawang* (Illipe nut tree) which is considered to "dry out" soils and consequently to limit rubber tree growth (but this observation has not been scientifically confirmed).

Table 2.13. List of species not specifically chosen for agroforests and their uses (1997)

Local names	Latin names	Uses
Belangai	*Eurya nitida*	Timber, medicinal, handicraft
Tucet	*Alstonia angustifolia*	Timber
Plaik	*Alstonia scholaris*	Timber, latex, medicinal, handicraft
Bamboo		Housing, handicraft, other uses
Todoh	*Phrynium capitatum*	Wrapping
Ringkan	*Ficus grossularoides*	Fruits, wrapping, timber
Resak	*Melastoma malabathricum*	Timber, fruits, vegetables, medicinal
Pakis		
Semolang	*Euodia aromatica*	Medicinal, timber
Siyet	*Sceria prupurescens*	Medicinal
Entiup	*Artocarpus sericicarpus*	Fruits, oil, handicraft
Leban	*Vitex pinnata*	Timber, spices, medicinal
Jambu america	*Bellucia axinanthera*	Fruits, wrapping, timber
Alang-Alang	*Imperata cylindrica*	Medicinal
Marade		Timber

Certain species (Table 2.14) may be selected to limit invasion of *Imperata cylindrica* in young agroforests.

Table 2.14. Species specifically used to limit *Imperata cylindrica* (*alang alang*) in young agroforests

Local names	Latin names	Type of action
Semenput		Provide shade
Beringing		
Melastoma	*Melastoma malabathricum*	Cover crop
Coklat		Cover crop
Nenas	*Ananas comosus*	Root competition
Gmelina	*Gmelina arborea*	Shading
Orok-Orok	*Crotolaria mucronate*	Competition with alang[2]
Gamal	*Gliricidia sepium*	Shade (limited)
Akacia	*Acacia mangium*	Shade
Albizia	*Albizia* sp.	Shade

Table 2.15 summarises the different species in agroforests, *tembawang*, jungle rubber, RAS *sendiri*, RAS 1 and RAS 3, as well as in home gardens (*pekarangan*) belonging to the local population and consumed and sold on local markets. It gives an idea of the wide variety of products that have an impact on both the household food supply and on the economy.

Table 2.15. Species and products already sold on local markets (price system of 1997)

Indonesian/local names	Latin names	Sale price	Origin
Pisang/Banana	*Musa* spp.	1,500 Rp/lot	Agroforest
Pakis piding/ferns		500 Rp/lot	Agroforest
Kangkong	*Ipomea aquatica*	500 Rp/lot	
Cangkok manis		500 Rp/lot	Agroforest
Daun kacang/bean leaves		500 Rp/lot	Home garden
Daun ubi/cassava leaves	*Gnetum gnemon*	500 Rp/lot	Agroforest
Bunga pisang/banana flower	*Musa* spp.	1,000 Rp/fleur	Agroforest
Jengkol	*Archidendron pauciflorum*	1,000 Rp/kg	Agroforest
Maram	*Eleiodoxa conferta*	2,000 Rp/kg	Agroforest
Kacang panjang/bean	*Vigna unguiculata*	2,000 Rp/kg	Home garden
Timun/cumcumber	*Cucumis sativus*	2,000 Rp/kg	Home garden
Bunga jagung/maize flower	*Zea* sp.	500 Rp/flower	Pontianak
Bayam	*Amaranthus hybridus*	500 Rp/lot	Home garden
Petai	*Parkia speciosa*	2,500 Rp/kg	Agroforest
Labu air/pumpkin	*Lagenaria siceraria*	2,500 Rp/kg	Home garden
Jahe/gingember	*Zingiber officinale*	2,500 Rp/kg	Home garden
Kelapa/coco nuts	*Cocos nucifera*	1,000 Rp/fruit	Home garden
Peringgi		4,500 Rp/kg	Home garden
Kecambah		1,000 Rp/portion	
Ubi/cassava	*Manihot esculenta*	2,500 Rp/kg	Agroforest

Indonesian/local names	Latin names	Sale price	Origin
Kedondong	*Spondias pinnata*	500 Rp/lot	
Pekawai	*Durio c.f. dulcis*	10,000 Rp/lot	Agroforest
Terong	*Solanum melongens*	5,000 Rp/kg	Home garden
Cabe/pepper	*Capsicum annuum*	20,000 Rp/kg	Pontianak
Buncis	*Phaseolus vulgaris*	3,500 Rp/kg	
Gambas	*Luffa acutangula*	2,000 Rp/kg	
Jeruk/lemon	*Citrus* sp.	3,000 Rp/kg	Home garden
Nangka/Jacqj fruit	*Artocarpus heterophyllus*	2,500 Rp/kg	Agroforest
Kencur	*Kaempferia galanga*	10,000 Rp/kg	
Kunyit	*Curcuma longa*	5,000 Rp/kg	Agroforest
Serai	*Cymbopogon nardus*	500 Rp/lot	Agroforest
Keladi	*Colocasia esculenta*	1,000 Rp/lot	Agroforest
Kundur	*Benincasa hispida*	2,500 Rp/kg	
Asam	*Tamarindus indica*	500 Rp/fruit	Agroforest
Labu siam	*Sechium edule*	2,500 Rp/kg	Home garden
Pane	*Momordica charantia*	5,000 Rp/kg	Pontianak
Wartel/carott	*Daucus carota*	9,000 Rp/kg	Pontianak
Jeruk nipis/lemon	*Citrus aurantifolia*	4,000 Rp/kg	Pontianak
Kol/cabbage	*Brassica oleraceae*	5,000 Rp/kg	Pontianak
Kentang/potato	*Solanum tuberosum*	4,500 Rp/kg	Pontianak
Tomat/tomato	*Lycopersicon esculentum*	6,000 Rp/kg	Pontianak
Bawang merah/red onion	*Allium cepa*	7,000 Rp/kg	Pontianak
Bawang putih/white onion	*Allium sativum*	7,000 Rp/kg	Pontianak
Kayu manis/cinnamon	*Cinnamomum burmanii*	2,000 Rp/lot	Agroforest
Nenas/pinepale	*Ananas comosus*	2,000 Rp/fruit	Agroforest
Sawih/cabbage	*Brassica rugosa*	5,000 Rp/kg	Pontianak
Jambu air	*Syzygium aquaeum*	1,500 Rp/kg	Home garden
Pepaya/papaya	*Carica papaya*	2,500 Rp/kg	Home garden
Kenikir	*Cosmos caudatus*	500 Rp/lot	
Lengkuas	*Alpinia galanga*	1,000 Rp/lot	Agroforest
Daun salam/leaves	*Eugenia polyantha*	500 Rp/lot	Agroforest
Daun sop/celery leaves	*Apium graveolens*	1,000 Rp/lot	Home garden
Daun pepaya/ papaya leaves	*Carica papaya*	500 Rp/lot	Home garden
Mangga	*Mangifera indica*	8,000 Rp/kg	Agroforest
Petai	*Parkia speciosa*	2,000 Rp/lot	Agroforest
Kacang tanah/peanut		3,000 Rp/kg	Home garden
Cempedak hutan	*Artocarpus integra*	500 Rp/fruit	Agroforest
Kumis kucing	*Orthosiphon aristatus*	1,000 Rp/lot	Home garden

NB: The Latin names of the species should be interpreted with caution because of the difficulty in identifying the species and correspondence between vernacular names and scientific names.

Conclusion: Market potential for associated trees in the 1990s and today

Some products were of obvious economic interest in the 1990s (see Table 2.4) and are still of interest in 2023. Smallholders tried to domesticate some of these species in their agroforest inter-rows (RAS and jungle rubber), by replanting or facilitating regeneration from natural regrowth, which has the advantage of being almost cost-free.

Timber species and fruit trees are particularly appreciated when they emerge from forest regrowth because they do not require planting and very little additional labour is needed to maintain them, but they may also be replanted to enrich the vegetation in the inter-rows.

Fruit trees have the most obvious potential market value, in particular durian which is already sold everywhere in Indonesia as well as in other countries in Southeast Asia (e.g. Thailand and Malaysia), rambutan and duku, for which demand is high on the Indonesian market. National markets did not appear to be saturated in the 1990s but in 2023, export would be the best market option for smallholders, particularly in the case of durian. The lack of larger organised marketing channels other than the traditional Sino-Indonesian one is still a serious obstacle to the expansion of fruit markets and exports.

As a result of the high demand for timber and wood products such as plywood in consumer countries (Japan, USA, and Europe), there may well be a shortage of timber in the very near future. Smallholders in West Kalimantan would be well advised to anticipate this trend and include species in their agroforest inter-rows that can be used to supply demand from the plywood industry. Some species (particularly nyatoh/ *Palaquium* spp.) have a life span similar to that of rubber (30 to 40 years). The final life cycle of RAS could then be extended with the exploitation of timber trees such as belian (*Eusideroxylon zwageri*, life span 60 years) or meranti (up to 90 years). In this way, old rubber-based agroforests could develop into *tembawang*. Finally, at the end of rubber lifespan, rattan could prove to be a useful crop, as indicated by the strong demand for furniture for export.

One major obstacle is Indonesian legislation on land and tree tenure that needs to be re-examined and adapted to the context of smallholder production, whose future potential could be high. Current regulations concerning timber exploitation practically preclude trade in timber from forests or agroforests by smallholders.

Other forest products with future potential are without doubt medicinal plants. Local sales of these products are already limited, as they have gradually declined due to the effectiveness and availability of pharmaceutical products. However, pharmaceutical firms could be interested in several forest and agroforest species in Borneo and perhaps undertake research projects that could indirectly benefit local populations. Examples of this type have been already observed in other countries in Amazonia, as well as in Côte d'Ivoire where a product to control hypertension was discovered growing under rubber.

Irrespective of the future potential of agroforest products, and even if it is high for fruit, timber, rattan and medicinal plants, most products are under-exploited in 2024, and hence represent a major challenge for the very near future. Several constraints persist in terms of both market organisation and official regulations.

⯈⯈ Farmers in West Kalimantan and RAS

Local Dayaks and Javanese immigrants are the two main ethnic groups in the area whose characteristics differ and who use different farming practices. Local Dayak populations are scattered and occupy more agricultural land. Javanese migrants are concentrated in villages and have limited land, which was often previously invaded by *Imperata* grassland after deforestation and was distributed through the government transmigration programme. Some Dayak families have also migrated to other areas within the region and the country as a whole. Like the Javanese, these Dayak migrants have limited land, but their access to local communal natural resources is not as limited as it is among the Javanese. The two groups have quite dissimilar land holdings, different access to other resources, different constraints and opportunities, which have important implications for the adoption, adaptation or rejection of RAS 3 (and other) technologies for their fields. Table 2.15 summarises the attributes of the three groups (local Dayaks, Dayaks who have migrated, and Javanese transmigrants) directly or indirectly related to rubber agroforestry.

Labour and modelling

Data on the inputs and outputs of major rubber-based systems were collected in West Kalimantan and Jambi with the aim of developing a prospective analysis tool to model fluctuating prices and yields of the different farming systems. The *Olympe* model (developed by Cirad) was used to input the data including detailed information concerning labour. RAS technologies were included in the survey and used as input data to enable comparison of these technologies and other technologies that were already available at the time. Here we only present data from Jambi. The level of maintenance refers to a combined parameter depending on the application of fertiliser and the frequency of vegetation slashing and weeding mainly during the establishment phase, i.e., in the first 6 years. In some areas where the risk of damage caused by pests (deer, boar, and monkeys) is high, considerable labour may be required to build fences; but for the purpose of our comparison reported here, labour needs were excluded as labour is independent of technology.

Much of the labour required prior to planting goes into land preparation and includes cutting down trees, slashing ground vegetation, burning and building fences. The following task is planting rubber. Other regular management tasks include applying fertiliser, manual and/or chemical weeding, tapping latex and harvesting other products.

Low maintenance RAS 1 requires only infrequent manual or chemical weeding, and only between the rows of rubber. Paid outside labour is generally not used but may be needed for land preparation. The RAS 1 high maintenance category requires more weeding and slashing during the establishment stage (Figures 2.7 and 2.8); the use of chemical herbicides is limited to the first two years. Minor weed slashing is carried out during tapping. In the RAS 2 low maintenance category, the use of both external labour and of chemical fertilisers is rare. The RAS 2 high maintenance category involves intense weeding in both the rows of rubber and in the inter-rows.

Table 2.15. Farmers and their characteristic in 1997

Group to which the farmers belong	Ethnicity	Farming system	Ecological characteristics	Constraints	Farming system
Local	Dayak	Extensive and intensive rubber	Upland Poor soil	Frequent rubber damage White root rubber disease	*Ladang* 1 ha – pulut (sticky rice for wine) one crop/year *Sawah* 0.7 ha, Oil palm 3 ha *Tembawang* 0.7 ha based on illipe nut or durian
Local	Dayak	Intensive	Upland Poor soil	Frequent rubber damage White root rubber disease	*Ladang* – pulut once a year *Sawah* 0.7 ha – paddy one crop per year *Tembawang* 1.3 ha
Local Migrants	Dayak	Extensive	Low land Poor soil Imperata	Plenty of land but limited knowledge/skills Lack of capital to buy clonal stumps and fertiliser	Upland field 0.85 ha – pulut once a year *Sawah* 1 ha _ paddy one crop per year *Tembawang*
Transmigrants	Javanese	Extensive and Intensive	Lowlands Poor soil Imperata	Limited land area (2 ha per household)	*Sawah* 0.7 ha 50% are intensive Upland field 0.5 ha Herbicide used for *sawah* and *ladang*
Transmigrants	Javanese	Intensive	Lowlands Poor soil Imperata	Limited land area (2 ha per household)	*Sawah* 0.2 ha 30% are intensive

Ladang is a upland crop plot. Sawah is an irrigated rice plot.

Conclusion

Observations made in 1993 in Sanjan (in SRDP plots) and in SRAP plots (Smallholder Rubber Agroforestry Project: a research projet from Cirad/ICRAF) dedicated to RAS trials, showed that in specific conditions, clonal rubber can be associated with other trees in complex agroforestry systems, and enable good productivity of both the rubber and the associated trees. Rubber production data concerning these systems are comparable with data on intensive monoculture. RAS 1 technology requires less labour and fewer chemical inputs and allows natural regrowth between the row of rubber including regrowth of timber and fruit species and medicinal plants. RAS 2 combines rubber trees with other high value timber and fruit species. RAS 3 is suitable for rehabilitation of *Imperata* grassland using a mixture of rubber, non-rubber

External influence	Oil palm opportunity	Use of Imperata	Land use before Imperata	Preferred Imperata control	Constraint to reclaiming former Imperata land
PPKR or SRDP SRAP – RAS 1, 2, 3	Oil palm on private land	None	Rubber Ladang	Timber trees Local and clonal rubber Herbicide	Lack of capital
PPKR or SRDP SRAP – RAS 1, 2, 3	Oil palm	None	Rubber Ladang	Timber trees Only clonal rubber Herbicide	Lack of capital
PKRGK rubber project	-	None	Rubber Ladang	Herbicide Roundup Timber trees Clonal rubber	Lack of capital
RAS 2, 3	Oil palm	Thatching Mulching	Ladang Rubber Sawah	Herbicide Clonal rubber Timber trees	High use of herbicide, lack of capital
RAS 2, 3	Oil palm Private oil palm	Thatching Mulching	Ladang Rubber Sawah	Herbicide Timber trees Clonal rubber	High use of herbicide, lack of capital and labour

and cover crops. While in the periods 1993/1996, 2007/2012, the attractive price of rubber encouraged farmers to adopt intensive monoculture, diversification of rubber agroforests was a better option than monoculture for rubber smallholders because it enabled them to diversify the economic basis of rubber agroforests, with value accruing from rubber wood and other timber while fruit trees provided an incentive for maintaining diversity plus ensure the farmers receive tangible benefits.

An improvement strategy investigated in earlier rubber agroforestry research revealed the technical possibility of running rubber plantations with less intensive management. While the financial gains from latex are considered the priority, the profit to be obtained from non-rubber components of the systems should not be ignored. The production of timber from rubber trees and the cultivation of other high value timber species will almost

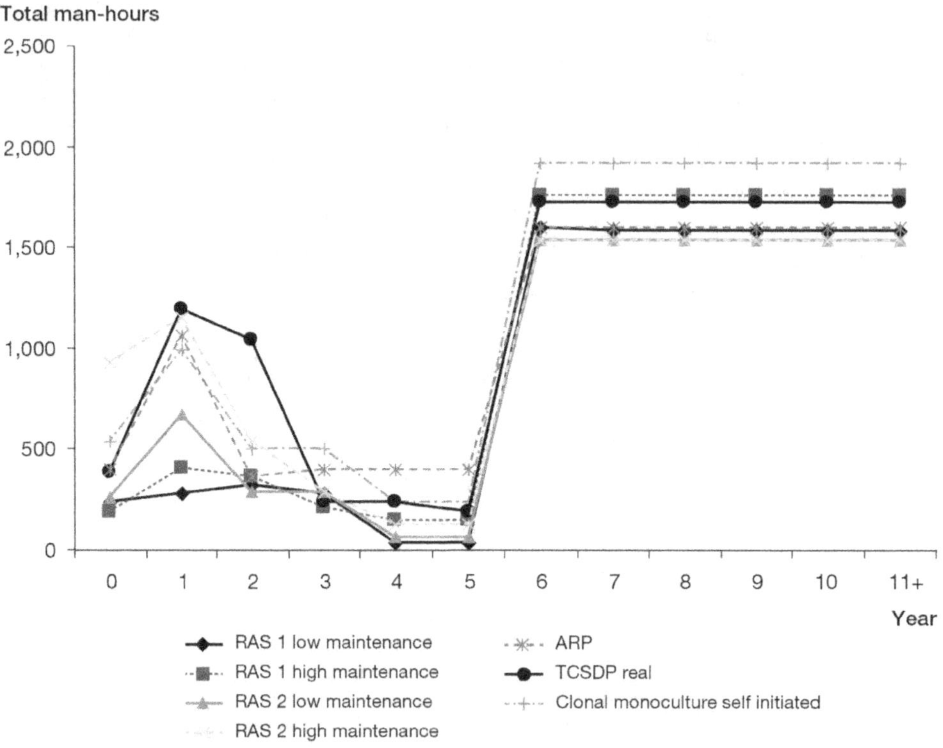

Figure 2.7. Manpower (hours) required by the different rubber systems

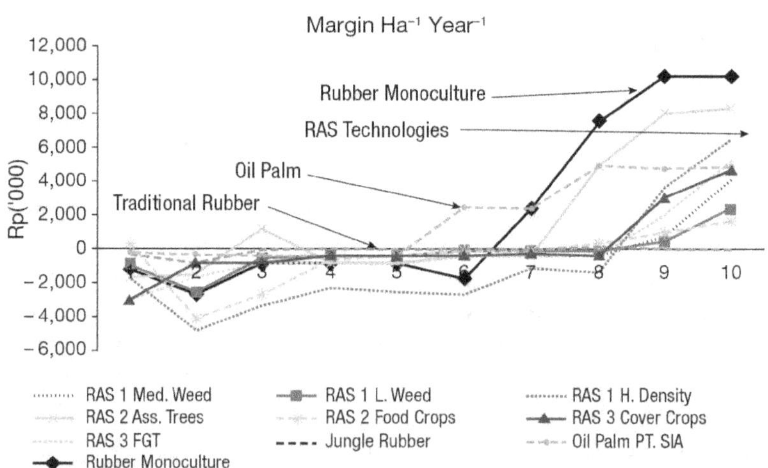

Figure 2.8. Changes in gross margin/ha under different cropping systems

certainly increase: it was true in 1997 and still is in 2024. High value fruits (both local and exotic) for local and export markets have huge potential to increase the farmers' income (as is already the case in southern Thailand). In 2023, it is clear that certain problems concerning double-row spacing have been partially solved, especially optimal spacing in certain RAS. In terms of rubber growth and the possible prolonged use of wider inter-rows for annual intercrops and tree crops, e.g. 6 m × 2 m × 14 m double-row spacing, is a very encouraging model if fast growing rubber clones such as RRIC 100, PB 260 and BPM 1 are the main tree crop. The same process of combining rubber and fruit/timber trees or other permanent crops occurred in Thailand in the 1990s.

In 2002, results obtained at the end of the immature stage showed RAS to be very well suited to local constraints and easily adopted by farmers. Rubber tree growth was excellent and most trees were tappable before 6 years of age. In 2024, there is still a considerable demand from surrounding farmers who want to join the project or to develop similar systems on their own (RAS *sendiri*); however there is also strong pressure from local private oil palm estates for local people to plant oil palm. Impact analysis conducted in 2000 (Trouillard, 2001) showed that 60% of SRAP farmers had replanted in the preceding 5 years and that 60% of the farmers concerned replanted using RAS *sendiri* systems. RAS *sendiri* can be considered as a "type of RAS" entirely re-appropriated by farmers, some of which were originally developed by them. Non-SRAP farmers in the area also began to adopt RAS *sendiri* after witnessing the efficiency of RAS (demonstration effect).

Since then, most farmers have planted oil palm, which in 2022, accounted for around 2/3 of tree-crop plantations. In 2023, farmers are all aware that clonal rubber requires more weeding and more inputs than unselected seedlings under jungle rubber, even in an improved agroforestry system such as RAS. They sometimes underrate the minimum requirements tested at different levels in RAS trials. One constraint is farmers' unwillingness to incorporate the minimum amount of inputs and labour in their current practices, which lies between what is currently used in jungle rubber (very low) and that used in the estate model (very high). Research is now underway to discover what level of capital for investment and labour would be acceptable to farmers during the immature stage.

A major challenge in development terms is also to decide which is best: a "complete approach" (as used in current development projects) or a "partial approach" based on the supply of only key components of RAS. Surveys by Chambon (1997-2000, published in Chambon, 2001) showed that a "partial approach" can work well if farmers' awareness has already been raised by previous development projects.

▸▸ Diversification of perennial crops to offset market uncertainties in West Kalimantan

This section has been partially published in 2001[45] as a result of a local study.

In less than one century, Dayak farmers in Indonesia first shifted from traditional hunting and gathering of forest products to slash and burn agriculture with progressive

---

45. This section was co-written with Karine Trouillard, and originally published in 2001 in a working document : Penot E, Trouillard K, De l'intégration à la substitution : histoire sur période longue des stratégies des producteurs hévéicoles en Indonésie : le cas de Ouest Kalimantan.

incorporation of rubber in agroforestry systems called "jungle rubber", then to rubber monoculture in the 1980s based on the use of clonal planting material, and finally to oil palm in the 1990s. The farming systems used by Javanese transmigrants in official transmigration programmes underwent different changes due to the weed *Imperata cylindica,* and to land scarcity. Local farmers progressively integrated export crops and now have access to international markets. The 1997-1999 economic crisis in Indonesia increased the need for development and technical change. A significant degree of coherence was maintained between technical systems and social systems. The example of the Sintang and Sanggau areas in West Kalimantan Province (Borneo) enabled characterisation of the different farming systems, and the identification of a situational framework and of pathways for future change. Here we discuss the different strategies from the perspective of a regional approach to development. Two major challenges characterise the rubber sector: the transformation of existing jungle rubber (2.5 million ha, 85% of smallholders' plantations) into clonal plantations (either in agroforestry or as monoculture) and partial substitution by — or complementary cultivation of — oil palm.

A study conducted in 2000 identified two situations: (i) the planting of new plantations in a parallel process of land acquisition, and (ii) the replanting of old jungle rubber (renewal of productive capital and beginning of intensification). For smallholders, these structural changes implied both technical change and innovation. Here, technical change refers to the adoption of clonal planting material, either in monoculture or in an agroforestry system. At the same time, official and/or spontaneous transmigration as well as the expansion of oil palm estates tended to increase pressure on remaining available land. Dayak communities thus felt the need to secure their land by expanding their plantations. In a context of uncertainty, the use of clones helped reinforce land ownership. It also led to an effort to rehabilitate degraded land. At the end of the 1990s, smallholders profited from a variety of on-farm and off-farm alternatives to diversify their sources of income, e.g., rubber and oil palm monoculture, agroforestry systems, running a nursery, off-farm jobs.

The two ethnic groups have followed distinct courses of action in terms of land use and agricultural practices. The traditional Dayak production system is based on extensive slash and burn rain-fed rice cultivation (*ladang*), with, in the past, the progressive incorporation of jungle rubber in 1997 and in still in 2024 of clonal rubber systems. This system gradually became more intensive (line plantation, maintenance before tapping, etc.). Jungle rubber became economically obsolete. Partially inspired by Javanese transmigrants, the Dayaks also adopted flooded rice. Old fallow, jungle rubber and local *tembawang* (timber and fruit agroforests) are also a valuable reserve of forest products. Originally, Dayak villages did not have to face the problem of limited land[46]. From the beginning of the 1980s, rubber projects gave some villages access to clones and monoculture techniques. At the end of the 1990s, the creation of oil palm estates had the same effect, offering new opportunities based on oil palm, which, at the time, was a new crop for local farmers. After 1997, farmers in villages belonging to the SRAP network also started nurseries and new improved clonal agroforestry plantations. The Javanese who settled as a result of the transmigration programme only had access to a very small area of cultivated land (2.5 ha). They originally focused on intensive

---

46. The population density is still relatively low with an average of 20 to 30 inhabitants/km².

irrigated rice (*sawah*), which allowed them to be self-sufficient – as long as local plots allowed planting of irrigated rice. Initially, the cultivation of fruit, timber and forest species was forbidden as the Javanese were officially supposed to specialise in food crops. Today, in 2023, they are establishing perennial plantations (rambutan, rubber, oil palm, pepper) in addition to food crops[47] (rice, peanuts) on the remaining uplands (dry land or *ladang*). The majority of Javanese planters also own a few cows, which is a good way to accumulate capital. However, most Javanese are obliged to take on off-farm work for 3 or 4 months a year to meet their family's needs (e.g. purchase complementary food, pay off loans). Javanese farmers are very open to agricultural intensification and, whenever possible, will rapidly incorporate perennial crops and seize any other opportunities for income diversification. Their main constraints are lack of land, limited labour, and high pressure from *Imperata cylindrica* in their deforested plots. Rice cultivation remains a strategic and sometimes social crop in both farming systems. It uses up family labour but does not guarantee complete self-sufficiency. The extent of production of clonal planting materiel (nurseries), which represents a relatively new opportunity, varies from village to village, depending on the social, economic and technical status of the farmers.

Thus, different strategic groups with different innovation objectives may co-exist in the same village (Trouillard, 2001). Concerning improved planting material, at the time, we distinguished five behavioural types: (i) rubber smallholders developing nurseries as their main activity, (ii) high status smallholders who invest in monoculture, (iii) smallholders-purchasers who buy clonal planting material, (iv) autonomous smallholders who produce their own clonal planting materiel, and (v) private nurseries (without a plantation). Some villages specialised in one or other of these categories, and were then generally referred to as "nursery villages" (i.e. villages that produce planting material) or "purchasing villages".

The study described here was implemented within the framework of SRAP[48], based on the concept of participatory action research. The project depended on a series of technical and organisational innovations (rubber-based cropping systems, the production of planting material, the organisation of farmers[49] around activities, etc.) that concern pre-established groups of producers. These groups were characterised within a situational framework according to different constraints. Each situation corresponded to a village that was considered representative of a homogeneous situation. A situational framework was established comprising 6 types of villages.

We observed diverse behaviours in the face of similar innovation processes in relatively homogenous zones, and sometimes even within the same village. Farmers may have similar medium- and long-term objectives but different short-term objectives that justify different choices among available opportunities. This led us to use a "constructivist" approach (Chauveau, 1999). In our situational framework, we disregarded geographical and social entities that had previously been defined as operational, such as

---

47. Up to 80% of transmigrants abandoned their land when they were obliged to only grow food crops, mainly due to lack of control of *Imperata cylindrica*. Those who stayed on all adopted perennial crops.
48. SRAP: Smallhollder Rubber Research Project, implemented by Cirad, ICRAF, GAPKINDO (the Indonesian rubber association) and local NARs (IRRI-Sembawa, Indonesian Rubber Research Institute).
49. Prior to 1998, in Indonesia there were no independant farmers' representatives or organisations, i.e., that were not controlled by the government. Farmers' organisations are still lacking in 2023.

villages, in order to consider smallholders as a "strategic unit". In this way, we were able to emphasise the process from individual decision-making to collective decision-making. However, at the village level, a collective decision may have a significant impact on the farmers' decision-making process with respect to a given problem. Within this framework, we were able to identify behaviours and actions based on similar logic, as well as decisive collective choices or differentiated strategies and were consequently able to identify different groups from those that had been apparent in our first sample of villages.

From a methodological point of view, the qualitative analysis of farmers' strategies led us to use the analytical approach of Yung and Stravinsky (1994), which consists in classifying behaviours according to a "defensive-offensive" gradient of strategies. "Offensive strategies" are defined as behaviours whose objective is economic growth, the accumulation of wealth, and the desire to transform and improve the household's welfare. Defensive strategies are defined as actions aimed at minimising risks, and securing the family's current welfare (for instance, food security as an objective). We then tried to distinguish the strategic groups, the relations that exist between the groups (through a study of the networks and family links) and the innovation processes implemented by the groups.

Identification of the strategic groups

Based on these criteria, behaviour analysis led to the identification of 7 strategic groups according to K. Trouillard (2001):
– *Smallholders who were becoming increasingly specialised in clonal rubber.* These smallholders were gradually replacing their ageing jungle rubber with clonal rubber (38% of those interviewed). Of these smallholders, 35% continued to practice *ladang* but the majority preferred to buy rice rather than to grow it, 70% still tapped their old jungle rubber. *Ladang* was maintained as long as land was available to avoid losing their "right of avail" (usage);
– *Clonal rubber smallholders who specialised in the production of planting material (in a nursery).* This group was composed of Dayak farmers who originally belonged to the first group. They created nurseries. They replanted clonal rubber under monoculture (50%), in agroforestry systems with fruit trees (25%) or with fodder intercrops (25%). These farmers were formerly leaders, heads of *kelompok* (farmers' groups) and often played the role of knowledge transmitters;
– *Traditional planters in transition.* This group consisted of young Dayaks who worked productive jungle rubber units, and who replanted using clonal rubber as far as their limited means allowed. *Ladang* was still a strategic activity in this group, but had a more social than economic function in maintaining the right of use of land. This strategic group was in transition towards group 1;
– *Young smallholders with off-farm activities.* This group of young Dayaks had access to limited labour resources and to limited areas of productive jungle rubber. They favoured off-farm activity. Some recognized the opportunity offered by nurseries. They lacked the necessary capital to invest in clonal rubber plantations;
– *Traditional "fence-sitters".* These Dayak farmers continued to rely on jungle rubber and *ladang* and did not replant with clonal rubber. They represented the most conservative group with respect to food security. They did not succeed in using grafting as a means of producing planting material. Lack of capital and technical skills as well as the

absence of appropriate information discouraged them from investing in clonal rubber. If they had access to full credit and a fully identified technological package, they generally changed to oil palm. Most had off-farm employment, mainly as workers on private estates or in local gold mines in order to increase their annual income in the short term;

– *Opportunist owners of private nurseries and people with multiple activities.* These were mainly employees on private estates. At that time, production of planting material was a marginal activity but in a few years, it would replace off-farm jobs. Using their own limited means, they planted clonal rubber in agroforestry systems, as these systems require less labour and less capital investment. They also planted oil palm and viewed any new crop opportunity very favourably. Those who pursued multiple activities were mainly Javanese transmigrants who found themselves in a very precarious position, or people who traded as a way to diversify their income;

– *Javanese transmigrant opportunists.* This group comprised old Javanese farmers who had *sawah* but did not plant clonal rubber because of land scarcity. They pursued commercial activities, particularly cattle rearing and sale. These producers continued traditional practices and favoured short-term accumulation of wealth with the security of an immediate and regular income obtained by working on the estates.

Farmers' strategies and pathways

The strategies we identified are grouped in Table 2.16.

Table 2.16. Type of farmer strategies in 1997

Type of farmers	Strategies	Type of strategies
Rubber smallholders	Planting and replanting	Offensive
	New planting by young farmers	Offensive
	"Fence-sitters", no plans for the medium term	Defensive
People pursuing multi-activities	Diversification	Offensive
	Development of trade or production (entrepreneurs)	Offensive
Workers on estates	Diversification	Offensive
	Fence-sitters who gave priority to the short term	Defensive

The strategy of replanting with clones was slowed down (or blocked), at least initially, when farmers had access to alternatives, e.g. off-farm activities or oil palm. Consequently, there were links between short-term strategies (off-farm) and long-term strategies (new plantations or replanting). Current replanting was funded by salaries (earned off-farm) or income from new oil palm plantations. At the time, we hypothesised that in the medium term, incomes generated by oil palm plantations would fund replanting with clonal rubber.

Smallholders developed diversification strategies while maintaining traditional practices including agroforestry. The persistence of traditional practices is proof of the attachment people have to traditions and social standards, and consequently to cohesion and social structure, at least at the community (village) level. Indeed, the whole process of social organisation is concerned with maintaining these practices, in particular the deployment of labour. Farmers with an off-farm job and/or who were involved in multiple activities changed their social behaviour in the sense that working off the farm implies making concessions with respect to social standards and in particular abandoning the use of *gotong-royong* (collective labour) due to lack of availability. This social rupture, together with the economic cost of such labour, may also explain the progressive abandonment of *ladang*.

Strategic groups as the expression of different pathways/courses of action

Farmers' courses of action over time are particularly influenced by their access to projects (clonal rubber, production of planting material/nurseries, oil palm, even *Acacia mangium* on a smaller scale[50]), which were seen as new crop opportunities and as part of the global innovation process. Farmers integrated these alternatives to varying degrees depending on how appropriate the innovations were for them. We thus observed different courses of action in similar contexts. Three main pathways based on changes in practices emerged from our analysis.

– The first pathway emphasised the shift from jungle rubber to clonal rubber (monoculture or agroforestry). This pathway maintained traditional production systems based on jungle rubber (for Dayaks) and *ladang/sawah (for* transmigrants*)*, with the progressive incorporation of clonal rubber through access to government projects or by their own means (10-20% of farmers accomplished it in this way in the five years preceding the survey)[51]. This pathway was directed towards rubber specialisation and improving productivity and enabled some diversification of activities. Some Javanese farmers chose this pathway which involved changing from traditional off-farm and *ladang/sawah* to plantations (either oil palm or rubber depending on the opportunities available). This strategy aimed to secure income and intensify production. Jungle rubber was expected to progressively decrease and eventually to disappear.

– The second pathway was characterised by a move to off-farm activity and the adoption of oil palm. It involved progressive substitution of similar traditional Dayak or Javanese systems of temporary off-farm activity, which were short-term strategies, by rapid adoption of oil palm, which is a long-term strategy, for the generation of income. In this case, jungle rubber was progressively abandoned. But later on, income from oil palm could be partially invested in new clonal rubber plantations.

– The third pathway was "mixed and opportunist". It combined complementary traditional systems (jungle rubber and *ladang*) with off-farm activities and other crop opportunities, i.e., oil palm (through development projects) and clonal rubber (generally using their own money, particularly for the production of planting material). Emphasis was on intensification and crop diversification to secure an income in the medium and long term.

Conclusion

Agrarian dynamics are characterised by internal conflicts in rural society and in communities as well as by conflicts with other stakeholders (the State, private estates, etc.), by dependence on markets (export crops) and on different projects. Proposed

---

50. Forestry plantations with *Acacia mangium* are proposed to some farmers by HTI semi governmental estates (Hutan Tanaman Industri)
51. Results of SRAP surveys in Kalimantan (K Toruillard, 2001) and in Sumatra (Komardiwan/Penot, 2001).

development models are often irreversible but the strategies developed by the farmers in the face of such constraints are extremely varied, and include activities outside the agricultural sector. Innovation through intensification, diversification, and off-farm activities are some of the pathways available to farmers in a global context characterised by market uncertainty and economic crises. In 2024, these pathways represent the baseline of agrarian dynamics. They have been constantly changing since the end of the 1970s, which saw the introduction of development project policies for perennial crops. In the case of rubber, it took 30 to 40 years for clones to technically prevail over jungle rubber (but clonal plantations still only represent 15% of the total area planted to rubber), whereas less than 10 years were sufficient for oil palm to become "the new crop" thanks to the increase in private estates. Both a strong innovation process and market pressure drive agricultural dynamics.

These pathways are the result of changes and advances in farming systems that led us to design prospective scenarios for the future of the West Kalimantan province. The first "only oil palm" scenario would result in the complete abandonment of jungle rubber and *ladang* and their replacement by oil palm. This is a scenario of substitution. The second "diversification" scenario is more balanced with endogenous development of clonal rubber plantations (monoculture or agroforestry) in addition to oil palm, with temporary off-farm employment in the estate sector to guarantee income during the transition stage. This is a scenario of adjustment and complementarity.

Scenario 2 appears to be the most realistic. Indeed, the development of oil palm through private estates will probably continue for the next 10 years until the mid-2030s assuming land availability and the Indonesian economic context. Land and labour are still plentiful in Indonesia compared to in its neighbours, for instance Malaysia, and this leaves scope for smallholder development as well as for export crops. After that, the continuing development of oil palm and rubber plantations on farmers' own initiatives, in the absence of state or other projects, or alternatively, the establishment of more estates, will require an increase in farmers' organisations like *Kelompok Tani* as well as access to micro-credits. Clonal rubber systems have great potential as they are more accessible to local farmers than oil palm and also ensure ecological sustainability thanks to their agroforestry component. The availability of planting material as well as its satisfactory quality are pre-requisites for this type of endogenous development. Scenario 1 might apply in highly saturated zones with severe land scarcity such as transmigration areas or in areas entirely under the control of private estates.

These scenarios need to be discussed in detail with local stakeholders in order to account for their concerns and their vision of the future.

▸▸ RAS case studies in southern Thailand

Thailand is currently the world's number one rubber producer with 4.77 million tons in 2017, corresponding to 37.1% of global production. In the last 5 years, rubber production has continued to increase at an average rate of 4.3% per year. It is the only country where rubber has been almost exclusively cultivated by family farmers. This is largely due to the fact that the country has never been colonised, the Thai State has strongly supported family farms, and has never had a policy which encouraged private investment and large-scale industrial plantations (Fox and Castella, 2013; Chambon et al., 2018). Support for family farms in the south, the cradle of rubber cultivation in the 1950s and 1960s, was also provided for political reasons, mainly to counter the communist rebellion (as was the case in Malaysia) and represented an important source of income for local farmers (Besson, 2002). Despite the trend in some neighbouring countries (land concessions to create rubber plantations awarded to foreign investors) during the rubber boom triggered by increasing rubber prices in the late 2000s, industrial plantations in Thailand still only account for 7% of total rubber production (FAOSTAT, 2019).

Thai rubber plantations are characterised by two cropping systems: (i) monoculture, which is now the most widely-used system (85% of rubber), and (ii) agroforestry with various level of intensification (different types of RAS), based on rubber associated with different crops (fruits, vegetables, tubers, shrubs), estimated to account for 15% of the rubber growing area in southern Thailand according to surveys conducted by Chambon (FTA project – Forest, Trees and Agroforests; 2021, published in Penot et al., 2023). Almost all smallholders use rubber clones mostly the RRIM clone either in monoculture or RAS. The average yield of these rubber trees in 2000 was 1,360 kg/ha/year, in 2016, it was around 1,500 kg/ha/year, and in 2020, close to 1,700 kg/ha/year (RAOT, 2021 annual report).

Since 2016, both the institutional and ecological environments have been highly favourable for the development of agroforestry practices based not only on food inter-crops during the immature period but also for the association of fruit/timber and rubber trees in complex agroforestry systems. The vast majority of farmers use the RRIM 600 clone. The original single clone policy is somewhat risky in case of a major disease outbreak, but the policy of using clonal rubber on a large scale through the rubber replanting programme has been successful. Some partial clone diversification at small scale has occurred with RRIT 251 and BMP 24 as well as RRIM 2000/3000 series, introduced illegally from Malaysia (B. Chambon, personal communication).

The main tree species that have been tested by local farmers alongside rubber are the following:
– Timber trees: neem or thiem (*Azadirachta excelsa*), thang (*Litsea grandis*), a timber tree that regenerates naturally in rubber plots, teak (*Tectonia grandis*), mahogany (*Switenia macrophylla*), phayom or white meranti (*Shorea talura*), tumsao (*Fragacs fragans*), *Acacia mangium*, rattan (*Calamus caesius* seems to be the most promising),
– Fruit trees: salak (*Sallaca* spp.), durian (*Durio zibethinus*), longkong (*Lansium domesticum*), sator (*Indonesian petai*), *Parkia speciosa* or Nita tree, jack fruit (*Artocarpus heterophyllus*), cempedak (*Artocarpus integer*), and mangoustan (*Garcinia dulcis*).
– Other species: coffee (*Robusta canephora*), pineapple and banana.

Many studies have been conducted by Thai researchers at PSU (Prince of Songkla University), TSU (Thaksin University) and KKU (Khon Kaen University) since 1990. Here we summarise their main results. Tree diversification was found to be an important step forward by small-scale farmers to remain economically viable (Somboonsuke, 2001b). Tree diversification can also provide timber and environmental services (Joshi et al., 2006); help reduce the risk of the hevea being blown over during storms; and reduce the amount and severity of surface runoff, thereby reducing soil erosion (Kheowvongsri, 1990; Jongrungrot and Kheowvongsri, 2021). Plant diversification favours carbon fixation, and has also been shown to reduce daytime temperatures in summer in rubber-based intercropped plantations compared with mono-cropped hevea (Hughes et al., 2012).

In 1996 in Phangnga province (South Thailand), a rubber-based agroforestry system with old jungle rubber (more than 40 years old) was reported that had also been enriched with bamboo, rattan species, and multi-purpose trees (timber plus trees whose leaves are consumed) such as miang and manboo without Latin names identified (Kheowvongsri, personal communication). Old jungle rubber was still present in Phattalung and Songkhla provinces in the 2010s (authors' personal observations).

Overview of RAS in Thailand in the 2010s

First, surveys conducted by B Chambon in all rubber producing areas of Thailand in 2016/2018 provide an overview, after which key studies conducted between 2005 and 2020 are reviewed.

The data presented here were collected for projects with specific objectives between 2016 and 2018. The surveys conducted for the different projects were of rubber-based households in the South, Centre-east and Northeast, i.e., the three main rubber-producing areas of Thailand (see Table 2.8). At the time of the survey, all the respondents selected had at least one mature rubber plantation. For some projects focussed on harvesting and post-harvest practices, farmers were also selected based on the type of products they were selling (coagulum or field latex). Although each survey had specific objectives, questions were always included to characterise the rubber cropping systems and particularly agroforestry practices.

The total sample included 771 farms, but the distribution of the sample: 270 farms (35%) in the South, 348 farms (45%) in Northeast and 153 farms (20%) in Centre-east was not representative of the geographical distribution of the rubber farms in Thailand. Consequently, the results are presented here for each region and not for the sample as a whole. This also allows us to highlight possible regional specificities in agroforestry practices.

Diversification is a very common strategy in Thai rubber-based households (Chambon et al., 2021); diversification at the farm level takes the form of non-rubber crops or livestock raising and at the household level, in the form of off-farm activities by members of the household. Here we focus on another level of possible diversification i.e., at the level of rubber plot through agroforestry practices. Most farmers in all three regions had only one or two rubber plots (Table 2.17). However, some farms especially in the Northeast had up to six rubber plots.

Table 2.17. Number of rubber plots per farm in the different regions (% of farms)

Number of plots	Centre-East	Northeast	South
1	61.4	47.7	58.5
2	28.1	31.9	26.7
3	6.5	15.5	13.0
4	2.6	3.2	1.1
5	0.7	0.9	0.7
6	0.7	0.9	0.0

Agroforestry practices during the immature period of the plantations

Whatever the region, intercropping during the immature period of the plantations was common but not systematic. At least 31% of the farmers in the Centre-east and up to 58% in the two other regions had never intercropped during the immature period of their rubber plantations, whether they were immature or mature at the time of the survey (Table 2.18). Intercropping in immature rubber plantations was reported to have long been a common practice among Thai smallholders, and was already mentioned in the early 1970s (Garot, 1970). Intercropping in young rubber plantations was encouraged by the Rubber Authority of Thailand for farmers who received a replanting subsidy (from the Office of Rubber Replanting Aid Fund, ORRAF). The economic interest of the intercrops, i.e., reducing management cost compared with a monocropping system and providing a substantial source of income during the unproductive period of rubber, is well known (Laosuwan, 1988; Polthanee, 2018; Hougni et al., 2018) particularly for poor farmers (Min et al., 2017).

Consequently, we expected intercropping to be adopted by all the farmers; however, in our sample, this was not the case. Indeed, constraints to the adoption of intercrops during the immature period were also well known and could explain why all Thai rubber farmers did not systematically practice intercropping: the condition of the soil, topography, location of the plot, the availability of family labour (which is considered by Langenberger et al. (2017) as being the most important factor in the adoption of intercropping) and marketing opportunities for cash crops (Masae and Cramb, 1995; Somboonsuke et al., 2011). The farmers' perceptions (and the fear that intercropping may be detrimental to the rubber trees) has also been found to limit the adoption of intercropping in some areas (Hougni et al., 2018).

Some authors also mentioned that the level of adoption of intercropping during the immature period of the plantations varies over time depending on the socio-economic situation (Hougni et al., 2018; Jin et al., 2021). In our case for instance, farmers who owned both immature and mature rubber plantations in the very early 2010s may not have needed to intercrop in the immature plots since the income produced by the mature rubber plantations at that time when the price of rubber was very high meant additional income from intercrops was not essential. Supporting the hypothesis concerning intercropping in mature rubber plantations, Romyen et al. (2018) mentioned that the adoption of agroforestry practices was motivated by the need for alternative income, which was provided by the inter-crops. This is probably also true for the immature period and could explain the different practices we observed.

Centre-east was the region where intercropping was the most common. The percentage of farmers who planted intercrops in all their rubber plots was much higher than in the two other regions. One of the reasons for this difference could be that in the Centre-east, it was quite common that the owner of a rubber plantation let someone else cultivate the intercrop and in return, this person was responsible for maintaining the rubber trees. This eased the family labour availability constraint. But in these circumstances, the only advantage for the owner of the plantation is avoiding (or reducing) the cost of labour for the maintenance of the rubber plantation as owners do not usually receive any of product or income from the intercrops. Renting out the rubber land in the immature period was also popular in Sri Lanka where this increased the adoption of intercropping practices (Herath and Takeya, 2003) and supports the hypothesis that it may also have facilitated the adoption of intercropping in the Centre-east region of Thailand.

It should be noted that, when a farmer had several rubber plots, the same intercropping practices were not always used in all the plots. Similar observations were also made in China (Min et al., 2017a). In Thailand, the reasons could be linked to the specific characteristics of each plot (type of soil, distance from the homestead to the village, topography/water situation) or to the availability of labour on the farm which could differ for plots established at different periods (different stages in the household life cycle).

Table 2.18. Intercropping practices during the rubber immature period (with number of farms in brackets)

	Center-East	Northeast	South
Yes in all plots	56.2 [86]	27.9 [97]	32.6 [88]
In some plots	12.4 [19]	14.4 [50]	9.6 [26]
No	31.4 [48]	57.8 [201]	57.8 [156]

In about three quarters of the cases, the household planted the same intercrops in all their rubber plots during the immature period out of habit (e.g. a popular crop in the area where the household lived). The remaining farmers who had more than one rubber plot planted different intercrops: this could be linked with the characteristics of the plots, the preference of the person growing the intercrops or market concerns (Tables 2.18 and 2.19).

Table 2.19. Use of the same intercrop(s) in all the plots during the immature period of the plantation (with number of farms in brackets)

	Center-East	Northeast	South
Yes	79.4 [27]	71.7 [43]	76.9 [30]
No	20.6 [7]	28.3 [17]	23.1 [9]

In most cases in the Centre-east and Northeast, only one type of crop was associated with rubber, but in the South, almost half the farmers planted mixed crops in the inter-rows of some or all their rubber plantations in the immature stage (Table 2.20).

Table 2.20. Only one crop associated with rubber during the rubber immature period (with number of farms in brackets)

	Center-East	Northeast	South
Yes in all plots	85.7 [90]	89.1 [131]	57.9 [66]
In some plots	3.8 [4]	2.7 [4]	4.4 [5]
No	10.5 [11]	8.2 [12]	37.7 [43]

The most frequently associated crops during the rubber immature period were (Table 2.21):
– short term crops (rice, corn, peanut, watermelon, melon, vegetables such as cucumber, pumpkin, calabash, long bean, aubergine, rosella, chili), particularly in the Northeast (peanut, rice, corn) and the South (the other short-term crops). Short term crops were less common in the Centre-east;
– tubers (almost only cassava) in the Northeast and to a lesser extent in the Centre-East. In the Northeast, cassava was intercropped in both upland and lowland plantations, whereas rice was only intercropped in lowland plantations (Hougni, 2018). Polthanee et al. (2016) reported that cassava intercropped with immature rubber produced a five times higher net income than banana. Indeed, the market for bananas was not much of an incentive for farmers in the Northeast, where the crop was more considered as an indicator of soil fertility and/or soil moisture content (Hougni, 2018). A market and the farmers' habits were probably important drivers of the choice of cassava as an intercrop in Northeast. Tubers were rare in the South;
– multi annual crops, mainly in the Centre-east (pineapple) and in the South (mainly banana and to a much lesser extent pineapple). Pineapple provides a very high income (380,000 THB/ha/year), much higher than banana (21,500 THB/ha/year; Somboonsuke et al., 2011). We did not find much intercropping with papaya even though it is a potentially interesting intercrop in Southern Thailand (Choengthong et al., 2014). Multi-annual crops were rare in the Northeast for reasons that need to be explored;
– fruit trees and parkia species mainly in the South; in some cases, these long-term crops were planted at the same time as the rubber trees whereas in others, perennial crops were planted before rubber and were continued during the immature period of the rubber trees (e.g. in the case of conversion of another plantation to a rubber-based agroforestry system). The conversion of fruit tree plantations into rubber plantations was also observed in the Centre-east (author's personal observation).

Table 2.21. Percentage of farms in each region where intercrops were observed during the immature period (number of farms in brackets)

	Centre-East	Northeast	South
Short-term crops	10.5 [11]	38.8 [57]	45.6 [52]
Tubers	34.3 [36]	64.6 [95]	2.6 [3]
Multi-annual crops	57.1 [60]	4.1 [6]	39.5 [45]
Fruit trees and parkia	7.6 [8]	0.7 [1]	20.2 [23]
Timber tree	2.9 [3]	0.0	1.7 [2]
Pak miang	0.0	0.0	6.1 [7]
Other*	2.9 [3]	6.8 [10]	7.0 [8]

* rattan, palm, betel nut, bamboo, lemongrass, galangal, curcuma, napier grass, ruzi grass, jasmine, tobacco, coffee, eucalyptus.

Although they were not present in our sample, some crops including marigold that procure a very high income but only in niche markets have been reported in Northeast Thailand (Hougni, 2018) and could represent opportunities for farmers who have the necessary connections.

Farmers in all the three regions rarely planted cover crops (Table 2.22). This is not specific to Thai rubber farmers, but applies to rubber smallholdings in general (Langenberger et al., 2017). Leguminous cover crops have long been recommended notably to prevent soil erosion (Baulkwill, 1989 cited by Langenberger, 2017) Research was conducted many years ago to identify potential cover crops adapted to Thailand (Sukviboon et al., 1986). In Northeast Thailand, climbing cover crop species that have to be removed from the rubber tree trunks continuously for three years require additional labour which probably limited the adoption of these species (Hougni, 2018). In addition, research conducted in this region showed that using leguminous cover crop in rubber plantations located in dry areas does not necessarily benefit the rubber trees (Clermont-Dauphin et al., 2018), which may have further limited the use of cover crops, particularly in dry areas.

Table 2.22. Use of cover crops during the immature period of the plantations (number of farms in brackets)

	Center-East	Northeast	South
Yes in all plots	0.7 [1]	0.9 [3]	0.7 [2]
In some plots	0	1.1 [4]	0.7 [2]
No	99.3 [152]	98.0 [341]	98.5 [266]

Agroforestry practices during the mature period of the plantations

Intercropping during the mature period of the rubber plantations was rare in the Centre-east (fruit trees and timber trees) and the Northeast (bamboo); and in the South, less than 20% of the farms had intercrops in at least some rubber plots (Table 2.23). The use of agroforestry practices in mature rubber plantations was low compared with in Indonesia (Penot, pers. comm) but quite high compared with previous reports in Thailand. Indeed, according to Charernjiratragul et al. (2014) cited by Romyen et al. (2018), the percentage of farms in which rubber agroforestry practices were used in the two southern provinces they studied was only around 2%, which seems very low compared to the survey by Stroesser and Chambon.

Table 2.23. Intercrops during the mature period of the plantations (number of farms in brackets)

	Centre-east	Northeast	South
Yes	2.6 [4]	0.3 [1]	13.3 [36]
Some	0.7 [1]	0.0	4.1 [11]
No	96.7 [148]	99.7 [347]	82.6 [223]

Even if few publications (at least in English) describe the rubber cropping systems used when rubber was first planted in Thailand, agroforestry systems in the South have a long history. Before 1960 and the implementation of the rubber replanting programme, farmers used a "conventional rubber production system" also called "rubber forestry or rubber community forestry" (Somboonsuke, 2001) which corresponds to Indonesian jungle rubber. It was not the only cropping system at that time, as other authors identified other rubber monocropping systems (e.g. Besson, 2002), suggesting that rubber agroforestry practices are long-standing. With the implementation of the rubber replanting scheme by the Office of Rubber Replanting Aid Fund (ORRAF), jungle rubber has been progressively replaced by rubber monoculture, which was the technical model promoted by the scheme for a long time. Today, very little jungle rubber remains (Penot and Ollivier, 2009; Stroesser et al., 2018), and most clonal plantations are monoculture.

However, rubber agroforestry systems reappeared with the economic crisis in 1997 and started to be more widely used (Somboonsuke, 2001). Since 2008, Thai government policy has changed and a maximum of 15 intercrop tree species can be planted per rai, the equivalent of 94 trees per ha (Romyen et al., 2018) and agroforestry practices were even promoted from 2014 on (Stroesser et al., 2018). These developments encouraged farmers to establish RAS. However, implementation is still limited for two main reasons: 1) the government policy encouraged rubber monoculture through the Office of Rubber Replanting Aid Fund scheme where permanent agroforestry systems were forbidden for decades, and 2) farmers lacked incentives and often the knowledge they needed to adopt RAS (Romyen et al., 2018). Earlier, Somboonsuke and Shiratoki (2001) also identified capital, labour investment, and marketing issues and sometimes water shortage as further constraints to the adoption of RAS. Based on our personal field work and on other surveys of farmers, we could also add that, as long as land pressure is not too high, farmers seem to prefer to separate crops (which is consistent with the on-farm diversification observed we mention in the previous section). This is probably linked to the farmer's lack of knowledge about RAS, i.e., that RAS could improve their margin per ha and their farm's resilience (Stroesser et al., 2018), plus contribute to food security, especially for the poorer farmers, as well as to some extent increase the biodiversity of the rubber plantations (Warren-Thomas et al., 2020). Wider adoption of RAS by farmers will certainly take time, but tools like innovation platforms could help (Theriez, 2017). Labour is undeniably a major constraint (and will probably increase) for the implementation of RAS as most associated crops (except timber trees) require additional labour. Thus the promotion of RAS would need to be combined with other technical innovations to improve labour productivity in rubber plantations such as low intensity tapping systems. Additional research is also needed to strengthen the rubber authority's recommendations to the farmers concerning agroforestry.

In the South, the most common intercrops were *Gnetum* (local name pak liang or pak miang) present in 38% of the farms with intercrops and fruit trees (36%). *Gnetum* is a shade tolerant shrub that provides a regular income all year round; fruit trees generate an annual income but only for a few months per year. These two types of agroforestry systems were found to produce a high gross margin per hectare (Stroesser et al., 2018). Other intercrops were *Parkia* species, timber trees, different species of palms (bamboo, oil palm, betel nut, coconut) and banana (all present on between 8.5% and 13% of the farms).

Farm typology according to agroforestry practices

Based on the farmers' intercropping practices on rubber plantations, we made four groups: 1) farmers who had never planted an intercrop, 2) those who had intercrops in at least some plots but only during the immature period of the rubber trees, 3) those who had intercrops at least on some plots but only during the mature period of the plantation, and 4) those who had intercrops at least on some plots during both the immature and mature period of the plantation.

Table 2.24 shows the very significant difference in the distribution of farms based on their agroforestry practices in the different regions (statistically confirmed). The total absence of intercrops was much more common in the Northeast and the South than in the Centre-east. In the Centre-east, the proportion of farms in the group with inter-crops during the immature period was much higher than in the other regions. Farmers in the group with intercrops during both the immature and mature periods were mainly located in the South. This is also the case for farmers who only grew intercrops during the mature period of the plantations, quite a marginal practice.

Table 2.24. Distribution of groups of agroforestry practices according to the region (number of farms in brackets)

	Centre-east	Northeast	South
None	29.4 [45]	57.1 [198]	51.9 [140]
Immature period	67.3 [103]	42.7 [148]	30.7 [83]
Mature period	0.7 [1]	0.0	4.8 [13]
Immature and mature period	2.6 [4]	0.3 [1]	12.6 [34]

Analysis of agroforestry practices in Songkhla area in 2005 when rubber prices were high

In 2005, producers considered the rubber price to be "acceptable". The results of a 2005 study on 20 farms in southern Thailand (Phattalung and Songkhla areas)[52] indicated that it was advisable to diversify and to cultivate another crop in addition to rubber to be able to survive periods of crisis, which then happened in 2012 and continues today (2023). The larger the share of income from the other crop, the better it would help the farmer withstand a decline in the price of rubber. Durian in particular plays an important role in the study area as a way of diversifying farm income; based on a solid value chain and with a very good price. To grow durian at the same time as rubber on the same plot enables farmers to minimise the impact of a decrease in income when rubber prices drop. Durian and rubber are very complementary crops; the market for durian is currently very good and the long-term prospects are very promising (Figures 2.9 to 2.11).

However, the rubber/durian system has a number of drawbacks: it is intensive, requires a lot of both labour and inputs, and the farmers require a good knowledge of the necessary technical itineraries. Diversification, intercropping, and associating timber or fruit trees with rubber for the purpose of income diversification and risk management, seem to be a good alternative to the current trend towards specialisation in rubber. Some farmers cultivate fruit trees as an intercrop, or in agroforestry systems that appear to be a promising way to cope with rubber price volatility as there is a good market for fruit in Thailand, thanks to high urban demand particularly for duku

---

52. This study was conducted by Aude Simien in 2005 under the supervision of Éric Penot, Cirad and Professor Dr Buncha Somboonsuke and Dr Vichot Jongrungrot from PSU (Prince of Songkla University).

langsat, in the study area. Some trials have been carried out but few results have been obtained so far, and a complete analysis (including a long-term economic analysis) has not yet been undertaken. More research is needed on large-inter-row intercropping and double tree line systems (i.e. double spacing with large inter-rows) in southern Thailand. The study described above was implemented in a period when rubber was profitable due to relatively good prices compared to prices during the 1997-2002 slump. Farmers' behaviour and strategies are closely linked to – and may even depend on – their type of production system as well as on opportunities for diversification (fruit trees, particularly durian). The smaller farms grow either rubber in monoculture or rubber combined with some upland rice plots. Both are relatively efficient as far as intensification is concerned.

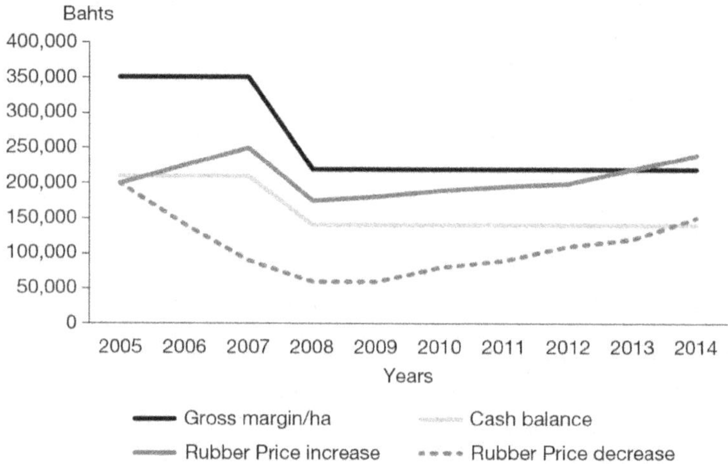

Figure 2.9. Economic results of farming systems based on rubber monoculture in 2005 and future prospects

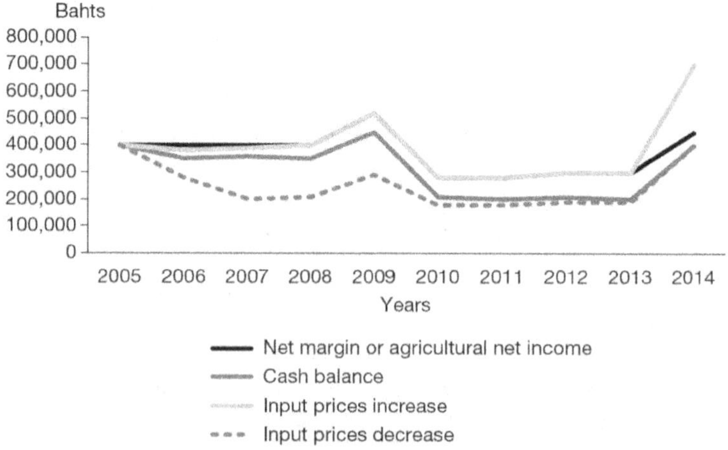

Figure 2.10. Economic results of farming systems based on rubber-durian RAS in 2005 and future prospects

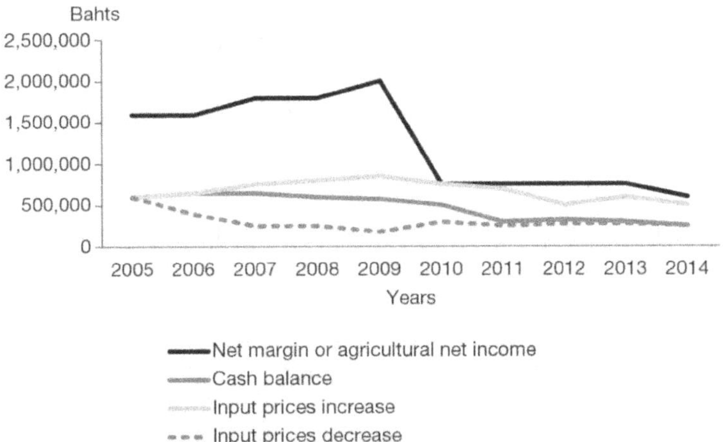

Figure 2.11. Economic results of farming systems including rubber-durian RAS plus mixed activities in 2005 and future prospects

Durian clearly plays the role of economic buffer in the eventuality of a new drop in rubber prices. In other words, for some time now, southern Thailand has been diversifying rubber systems and farming systems in order to strengthen its economy and to be more resilient in the face of possible future crises that affect commodity prices.

Economic analysis in Songkhla and Phattallung in 2012

Jongrungrot et al. (2014a) conducted a study in Songkhla and Phattalung provinces in Southern Thailand. Twelve farmers who practiced rubber-based intercropping were selected based on their social characteristics (the farmer was a member of a group or a network of farmers who practiced and promoted rubber-based intercropping) and the diversity of their agroforestry practices. The 12 farmers had a total of 19 rubber-based intercropped plots that were used to record the socio-economic characteristics of the farms concerned. Eight of the 19 plots were selected for a prospective analysis for the decade 2012-2021.

Selection was based on economic outputs (margin per ha) in 2012; potential to generate a higher income thanks to intercropping; four groups were created based on the age of the rubber trees (< 7 years old, 7-15 years old, 16-25 years old, > 25 years old); and species diversity.

Rubber was associated with different kinds of timber or fruit trees. In all, the sample contained 21 different timber species, 10 kinds of fruit trees, and 9 kinds of other plants. The most popular intercrop species was Ironwood, which was found in seven plots, followed by *Gnetum gnemon* and bamboo, each found in five plots. Next came eaglewood, white meranti, and salacca (fruit palm tree), each found in four plots. Regarding plant diversity, between 2 to 12 species were observed per plot, at densities ranging from 368 to 5,125 trees per ha. This is consistent with the results of several studies conducted by scholars in Thailand and overseas. For example, Joshi et al. (2006) found that rubber-based agroforestry systems could generate income from a variety of species including timber, increase food security, and provide environmental benefits, including biological diversity, carbon dioxide fixation, watershed protection and soil conservation.

The plots were classified in 3 groups based on simulation of the margins of the eight plots for the decade 2012-2021. The price of rubber selected for the simulation was the price in 2012, which was still a very good price compared with the top price recorded in 2011. The gross margins/ha trajectory of the eight most representative RAS for the decade are summarised below (Jongrungrot et al., 2014).
– High margins with a gradual increase and a stepwise increase: Plot 9 (trajectory 4) and plot 13 (trajectory 3).
– Medium margins with a gradual increase and fluctuating development: Plot 19 (trajectory 4), plot 7 (trajectory 4), plot 16 (trajectory 3), and plot 4 (trajectory 3).
– Low margins with a gradual increase: Plot 1 (trajectory 4) and plot 14 (trajectory 1).

Overall comparison of the estimated margins showed that all plots will have a higher margin per ha in 2021 than they had in 2012; and six out of the eight plots will have a higher margin per ha during the period 2013-2021 than they had in 2012. The reasons for these two findings are that rubber and intercropped plants will continue to produce yields with age; after 2013, the rubber trees in four out of the eight plots will be more than 21 years old and their yield will remain unchanged, while the yields of the inter-cropped trees will increase with age or will start to yield; and the old rubber in one out of the eight plots will be cut down by the farmer and sold as timber. All the RAS plot patterns are described in appendices (Table S1).

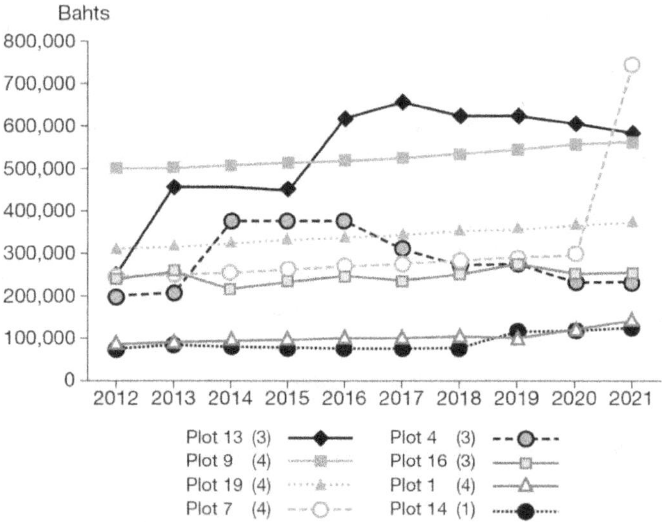

Figure 2.12. The gross margins/ha trajectory of the eight most representative RAS for the decade studied by Jongrungrot et al. (2014)

Comparison of rubber monoculture and 3 types of RAS based on timber and bamboo

Romyen et al. (2018) compared a rubber monoculture system with three rubber agroforestry systems: rubber combined with ironwood and eaglewood (S1), rubber combined with ironwood and champak (S2), rubber combined with bamboo (S3). All three systems used the same rubber density (7 × 3 m) but S1 and S2 added 18 other trees per rai[53] (112 trees per ha) and S3 added 72 bamboos per rai (448 bamboos per ha) in the rubber plantations. Tapping started slightly earlier in the intercropped systems, with the first tapping 6.6 years after planting compared with 7.2 years under rubber monoculture. Based on a rubber lifespan of 28 years in all four systems and an interest rate of 9.25%, they found that rubber combined with bamboo (S3) was the most profitable, followed by S1 and S2 (respectively, 71.5%, 70.4% and 46.3% increase of the net present value (NPV) over rubber monoculture).

Agroforestry patterns in Phattallung province in 2016

This study was conducted in Phatthalung province, Southern Thailand[54], in the framework of the ANR/Heveadapt project. The goal was to analyse how smallholder tree plantations can adapt and survive in the face of profound changes in socio-economic conditions. The study focussed on mature plantations in rubber-base agroforestry systems to understand the extent to which respectively, rubber, associated crops, trees, livestock, and off-farm activities, contributed to income stability and farm resilience. Socio-economic performances were evaluated at both the cropping system and the farming system scale, using farming system modelling with the software *Olympe*. Characterisation of the economic structure of the farms shed light on two main strategies used by farmers to maintain their income despite volatile rubber prices: income diversification through agroforestry and income diversification through off-farm activities. The best agroforestry systems, both in terms of return on land and on labour, was associating rubber trees with fruit and timber trees. Farmers also had off-farm jobs to complement their family income. Finally, prospective modelling showed that most farms were robust to rubber price volatility due to the flexibility of their agroforestry systems. Farmers with no agroforestry system were weakened by over-reliance on rubber trees.

Agronomic description of the different types of agroforestry systems

Associated plant species used in rubber plantations under agroforestry in Phatthalung province (southern Thailand) were of three types: fruit trees, timber trees and perennial vegetables. In the 64 rubber plots comprising the sample[55], fruit trees were present in 39.06%, timber trees in 31.25% and perennial vegetables 29.69% (Table 2.26). This shows that fruit trees dominated, but also revealed significant differences, i.e., fruit trees provide annual food and income, timber only provides a one-off income (when felled) while perennial vegetables provide regular food and regular income (Tongkhaenkhew et al., 2020).

A total of 44 plots belonging to 64 farmers were selected to describe the coexistence of associated plants in rubber agroforestry plots with complex (34%) and simple agroforestry systems (66%). The associated plants grown in rubber plots were divided into

---

53. 1 rai = 1,600 m².

54. This study was conducted by Laetitia Strosser, under the supervision of Benedicte Chambon, Éric Penot from Cirad and Uraiwan Tongkaemkaew from TSU (Taksin University, Phattallung).

55. This part was written by Uraiwan Tongkaemkaew from TSU (Taksin University, Phattallung), Bénédicte Chambon and Éric Penot from Cirad. Source: Laetitia Stroesser, Éric Penot, Isabelle Michel, Uraiwan Tongkaemkaew and Bénédicte Chambon, 2018. Income diversification for rubber farmers through agroforestry practices. How to overcome rubber price volatility in Phatthalung province, Thailand. Revue Internationale du Développement/Editions de la Sorbonne n°235 (2018-3), https://dialnet.unirioja.es/servlet/articulo?codigo=6537375.

seven groups based on their characteristics: four types of complex agroforest: (i) fruit trees and timber trees and perennial vegetables (26.7%), (ii) fruit trees and timber trees (33.3%), (iii) fruit trees and perennial vegetables (33.3%), (iv) timber trees and perennial vegetables (6.7%); and three types of simple agroforest: only one fruit tree species (37.9%), only one timber tree species (34.5%), only perennial vegetables (27.59%). The different associated crops were planted either using a systematic or an unsystematic system (Table 2.27).

In the agroforestry system, timber trees and fruit trees were planted in both of systematic and unsystematic systems with multiple and simple systems combined with perennial vegetables (Table I.2). After fruits and vegetables, timber trees were the third choice for local farmers as timber trees have significant advantages including low labour requirements for maintenance, but the income is generally only available at the end of the rubber life span. Fruit trees require a large labour force particularly in the harvest season and large quantities of inputs (fertilisers and pesticides) to ensure good yields.

Table 2.26. Associated species used in rubber plantations in Phatthalung province, southern Thailand

Associate species in rubber plots	No. of rubber plots*	%
Fruit trees	25	39.06
Timber trees	20	31.25
Perennial vegetables	19	29.69
Total	64	100.00

*One plot may contain more than one associated plant.

Table 2.27. Rubber agroforestry system in Phatthalung province, southern Thailand

Rubber Agroforestry systems	Systematic system		Unsystematic system		Total	
	N° rbp*	%	N° rbp*	%	N° rbp*	%
Complex agroforestry system	*6*	*40.00*	*9*	*60.00*	*15*	*34.09*
1. Rubber+fruit trees+timber trees +perennial vegetables	1	25.00	3	75.00	4	26.67
2. Rubber+fruit trees+timber trees	1	20.00	4	80.00	5	33.33
3. Rubber+fruit trees +perennial vegetables	3	60.00	2	40.00	5	33.33
4. Rubber+timber trees +perennial vegetables	1	100.00	0	0.00	1	6.67
Simple agroforestry system	*21*	*72.41*	*8*	*27.59*	*29*	*65.91*
5. Rubber+fruit trees	8	72.73	3	27.27	11	37.93
6. Rubber+timber trees	7	70.00	3	30.00	10	34.48
7. Rubber+perennial vegetables	6	75.00	2	25.00	8	27.59
Total	*27*	*61.36*	*17*	*38.64*	*44*	*100.00*

* rbp=rubber plot.

Local perennial vegetables require less labour and few inputs, indeed, they grow well in the shade and can be harvested regularly. RAS in Thailand resembled RAS 2 and RAS 3 in Indonesia (Penot, 2001; Wibawa et al., 2006). RAS technologies in Indonesia were developed between 1994 and 1998 at ICRAF, including some following a visit to Thailand in 1996 to take advantages of existing local agroforestry systems.

Species of associated plants in mature rubber plantations

Twelve families and 20 species of fruit trees were found, mostly local fruit species (Tongkaemkaew et al., 2020): 5 species of the Palmae family: coconut, salak, sala, *areca* nut palm and kelumi. The *Anacardiacea* family came in second with marian plum, plum mango and mango. *Meliaceae, Moraceae* and *Sapindaceae* families were represented by two species in each family: longkong and langsat, jack fruit and champedak, rambutan and longan, respectively. *Bombacaceae, Guttiferae, Leguminoae-Minosodeae, Minosaceae, Myrtaceae, Phyllanthaceae* and *Stilaginacea*e families were each represented by only one species: durian, mangosteen, niang, sator, wa, burmese grape and black currant tree, respectively. However, mangosteen, longkong, langsat and salack were the fruit trees most frequently associated with rubber.

The timber trees were distributed in 8 families and 15 species. The timber tree species were mainly wild varieties that are common in the southern region. Timber trees species in the family *Dipterocarceae* were takian, takina thong, payom and yang na. The *Meliace* family was represented by three species: bay wood, Siamese neem tree and mahogany. The *Malvaceae* family was represented by large and small-leaved hua. The *Labiatae, Lauracea, Leguminosae-Minosoideae, Magnoliaceae, Rubianceae and Barringtoniacea* families were each represented by one species: teak, litsea, brown salwood, champak, tuku and karuk. These timber trees species were found in all the study sites. Rubber agroforestry systems with timber trees can include several species associated simultaneously. Some of them grow naturally and are left in place by the farmers; others are planted by the farmers (Table S2 in appendices).

The species of perennial vegetables depended on the dietary habits of the people in the south (vegetables harvested once and vegetables harvested over a period of more than 3 years) belonged to six families and six species (see Table S2 for Latin names). These were phak nam (local name), phak miang (local name, *Gnetum* sp.), pineapple (used to make local dishes like *"kaengsom, pad peaw hwan"*) bamboo, pandanus palm and rattan palm. The most frequent companion crops in this group were phak maing, phak nam and pineapple, frequently consumed by southern populations. These crops also grow up very well in the shade and require little labour for maintenance (Table S2 in appendices).

The plants grown in association with rubber are popular as little labour is required for their maintenance, they are resistant, grow satisfactorily in the shade and already have good local markets. Through the development of both simple and complex rubber agroforestry systems, farmers have been able to diversify their sources of income with minimal establishment and maintenance costs. The Rubber Authority Of Thailand (RAOT) could promote such systems better and write their recommendations according to RIIT (Rubber Research Institute of Thailand). The role of RAOT is still to provide smallholders with the necessary technical and financial assistance to plant or replant rubber, to which could be added the promotion of rubber agroforestry systems.

Economic analysis

This section has been previously written in 2018 (Stroesser et al., 2018)[56] and revised.

In 2018, the following working hypotheses were formulated: (i) farmers construct RAS through progressive diversification of monoculture systems, (ii) among farms with AFS, agroforestry plots are combined with monoculture plots in different ways, (iii) AFS can effectively withstand the volatility of natural rubber prices and (iv) farmers reserve different shares of their land for AFS, because they have other opportunities to diversify their income. As the majority of studies provide qualitative descriptions of agroforestry, in this section, we provide an economic analysis of the impact of agroforestry on agricultural income.

To compare RAS cropping systems, we used economic indicators such as yield, gross margin (GM/ha) and return to labour as GM/hour of family labour. To compare the different activity systems (farm plus household), we used the following indicators: (i) net farm agricultural Income: the sum of every net margin (NM) of every product, (ii) the origin of on-farm Income: the gross margin (GM) of each product (rubber, fruits and vegetables, livestock products, etc.) divided by the sum of GM, (iii) the calculated net total income (cNTI): the sum of every NM plus off-farm income, before self-consumption, which made it possible to compare the economic efficiency of the farms, (iv) the real net total income (rNTI): the sum of every NM and off-farm income, minus self-consumption, to assess the real income including on- and off-farm incomes and (v) the cash balance: the rNTI minus all family consumption and expenses, self-consumption included (equivalent to cash flow).

By comparing farms and strategies, we provide useful up-to-date economic information for actors of the future innovation platform. Farms were classified in two operational typologies: (i) based on the AFS structure at the cropping system scale, inspired by the work of Somboonsuke (2011), Charernjiratragul (1991) and Jongrungrot (2014a), and (ii) based on the farm structure at the scale of the activity system, based on the household's incomes. The main drivers of farmers' strategic choices, which were briefly broached during surveys, complete this second typology. The main discriminant factor for farm typology was the type of AFS used as a means of diversification.

Using these typologies (cropping systems and activity systems), we modelled each type of rubber farm using *Olympe* software (developed jointly by INRA/Institut national de recherche agronomique, Cirad and IAMM/Institut agronomique mediteranéen de Montpellier; Penot, 2012). The first objective was to run simulation scenarios based on economic risk to analyse farmers' choices and feed a prospective analysis to understand current and future farmers' decisions. Modelling scenarios also enabled us to measure farm resilience.

From the 32 farmers we interviewed, we obtained an inventory of 53 agroforestry plots, of which 64% associated fewer than four different species and 36% associated four or more different species. These AFS are best classified based on the type of species associated with rubber trees: fruit trees, timber trees or vegetables. Vegetables

---

56. This section was extracted from Stroesser L, Penot E, Michel I, Tongkaemkaew U, Chambon B, 2018. Income diversification for rubber farmers through agroforestry practices. How to overcome rubber price volatility in Phatthalung province, Thailand. *Revue Internationale du Développement/Éditions de la Sorbonne*, 235(3):117–145. https://doi.org/10.3917/ried.235.0117

(*Gnetum gnemon* Linn. in this study) can be sold almost all year round, whereas fruit trees are harvested only a few months a year and timber trees are cut only once. We also identified an average technical management system for each type. The AFS based typology comprises 5 main types (all mature):

– MatAFVg: mature rubber trees only associated with vegetable species (AF for agro-forestry, V for vegetable): *Gnetum gnemon* or pak liang/pak miang. The strategy is based on diversification with pak liang.

– MatAFFr: mature rubber trees associated with fruit and sometimes vegetable species: an average of 280 trees/ha: mangosteen (*Garcinia mangostana* L.) [Mangout], stink bean (*Parkia speciosa* Hassk.) [Sator], salacca (Salacca edulis Reinw.) [Sala] and *Gnetum gnemon*. The strategy is based on fruit diversification and access to markets.

– MatAFTb: mature rubber trees only associated with timber species: an average 180 trees/ha: ironwood (*Hopea odorata* Roxb.) [Takian thong], neem tree (*Azadirachta excelsa* (Jack) Jacobs.) [Sadao tiam], tung (*Litsea grandis* L.) [Thung], mangium (*Acacia mangium*) and champaka (*Michelia champaca* Linn.) [Jumpa]. This is typically a long-term strategy where the end product (timber) is sold at the end of the life span of the rubber trees.

– MatAFMx: mature rubber trees associated with fruit, vegetables and/or timber species: with an average 310 trees/ha; mangosteen, longkong (*Lansium domesticum* Corr.) [Longkong], *Gnetum*, ironwood; neem tree, tung and champaka. This is the most diversified strategy with multiple short- and long-term products.

– MatAFLv: rubber trees associated with livestock and other plant species. (only 2 plots): rubber associated with 59 trees/ha: longkong, durian, stink bean, rambutan and 125 trees/ha: neem tree, tiam, ironwood and white meranti (*Shorea roxburghii* G. Don.) [Payom]. The diversification strategy includes livestock products.

Figure 2.13 compares return to family labour, i.e., the gross margin per hour of family labour (GM/h), for each type of AFS and for a mature rubber monoculture (family labour is not a cost). In the case of sharecropping, a worker receives 40% to 50% of the yield. The AFS types with the best return to land (AFLvA, AFLvB and AFVg) have the worst family return to labour. Taking care of the herd and harvesting *Gnetum* are very time consuming. Fruit is only harvested 2 months a year, which explains the better results obtained by AFS types AFMx and AFFr. Two categories are included in type AFMx, depending on fruit yield and the use (or not) of a tapper. In general, farmers with AFS type AFMxA have better fruit yields and hire tappers. Type AFTb provides a GM/h close to that of rubber monoculture.

Figures 2.13 and 2.14 show the wide diversity of RAS which explains why the RAS is the main discriminant factor in the following farm typology.

The list below gives the most representative features of each type linked to the discriminating criteria, and the sample distribution:

– type AR: Rubber producers who earn less than the minimum wage (6/32 farmers): with 3.9 ha of mature rubber plantations, 24% in agroforestry mainly with AFFr (0.4 ha) and AFVg (0.3 ha). Their incomes are one third lower than the minimum wage.

– type AO: Diversified producers who earn less than the minimum wage (3/32 farmers): with 1.2 ha of mature rubber plantations, 31% in agroforestry mainly with AFMx (0.2 ha) and AFFr (0.2 ha) + 1.1 ha of non-rubber crops/trees. Their incomes are half the minimum wage.

– type B: Farmers who depend on another income, earn less than the minimum wage (6/32 farmers): with 1.4 ha of mature rubber plantations, 38% in agroforestry mainly with AFTb (0.2 ha) and AFLv (0.2 ha) + financial support from family. Incomes are one third lower than the minimum wage.

– type CR: Rubber producers who earn more than the minimum wage (3/32 farmers): with 3.6 ha of mature rubber plantations, 21% in agroforestry mainly with AFMx (0.6 ha) and AFFr (0.2 ha). Their incomes are 50% higher than the minimum wage.

– type CO: Diversified producers who earn more than the minimum wage (5/32 farmers): with 1.9 ha of mature rubber plantations, 50% in agroforestry mainly with AFFr (0.7 ha) + 1.2 ha of non-rubber crops/trees. Their incomes are 50% higher than the minimum wage.

– type D: Farmers who earn more than the minimum wage thanks to their off-farm activities (1/32 farmers): with 2.1 ha of mature rubber plantations, 42% in agroforestry mainly with AFFr (1 ha) + 1.2 ha of non-rubber crops/trees + several other activities. Their income is 20% higher than the minimum wage

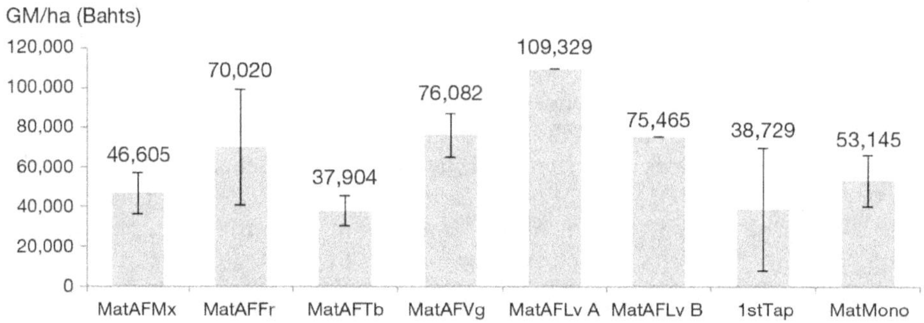

Figure 2.13. Gross margin/ha for the different types of AFS and rubber monoculture

Note all system acronyms are explained in the 5 preceding bullet points. MatAFVg = rubber + *Gnetum gnemon*/Pak liang, MatAFFr = rubber + fruits and vegetables with 280 trees/ha, MatAFTb = rubber + timber species/180 trees/ha, MatAFMx = rubber + fruits + vegetables +timber species (310 trees/ha,) MatAFLv = rubber + livestock + other plant species.

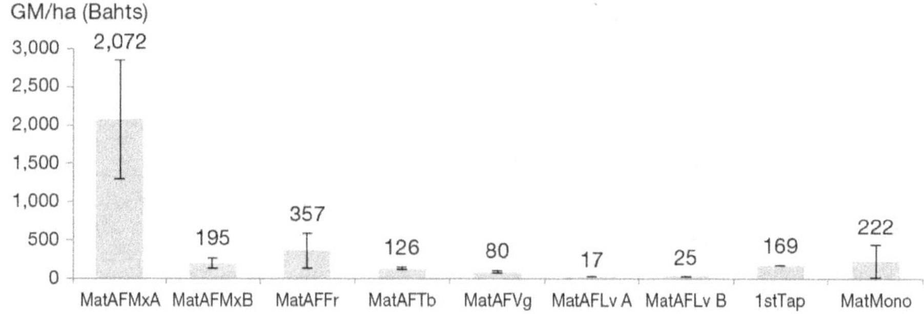

Figure 2.14. Family return to labour for each type of AFS and for a rubber monoculture (GM/h for family)

MatAFVg = rubber + *Gnetum gnemon*/Pak liang, MatAFFr = rubber + fruits and vegetables with 280 trees/ha, MatAFTb = rubber + timber species/180 trees/ha, MatAFMx = rubber + fruits + vegetables +timber species (310 trees/ha,) MatAFLv = rubber + livestock + other plant species.

– type E: Farmers who earn far more than the minimum wage thanks to their on-farm activities (4/32 farmers): with 10.5 ha of mature rubber plantations, 73% in agroforestry mainly with AFTb (5.4 ha) and AFMx (1.6 ha) + 1.7 ha of non-rubber crops/trees. Their incomes are between 2 and 28 times higher than the minimum wage.
– type F: Farmers who earn far more than the minimum wage thanks to their off-farm activities (4/32 farmers): with 2.9 ha of mature rubber plantations, 75% in agroforestry mainly with AFMx (1.6 ha) plus other off-farm activities Their income is four times higher than the minimum wage.

We observed that all three strategies (relative rubber specialisation, on-farm diversification and off-farm diversification) are applied in all the classes.

Prospective modelling of RAS

Among several scenario options, we selected variant agroforestry patterns.

First, seven variants were created for each type of farm to show the impact on economic results of a process in which the choice of agroforestry on economic results was not yet definitive (Figure 2.15). The sub-variants are based on a lower rubber price.
– Variant: Combination of AFS and monoculture plots (Comb) = farms in the current situation with land under rubber-based agroforestry that ranges from 23% (T-AR) and 65% (T-F).
– Variant: specialisation Agroforestry (AF). This refers to the previous variant taken to the extreme. Within each type, we replaced monoculture plots by their agroforestry equivalent and split the areas as a proportion of the distribution of the existing AFS.
– Variant: specialisation Monoculture (Mono). This refers to the other end of the spectrum of farmers' possible strategies. It was constructed in a similar way to the previous variant: we replaced agroforestry plots by their monoculture equivalent, while respecting the type of labour force used for tapping. Other plots were left unchanged.

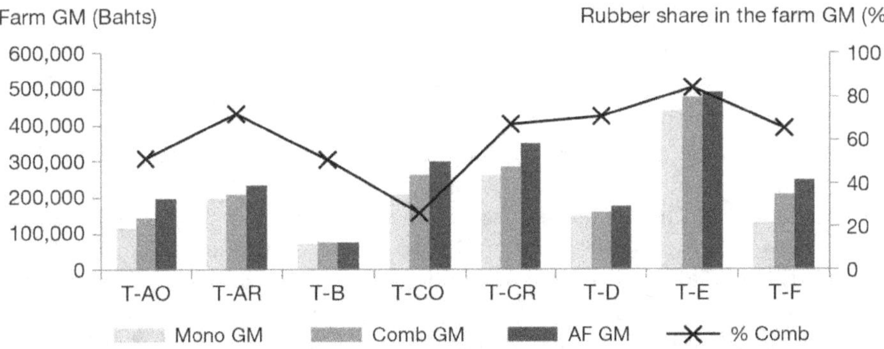

Figure 2.15. Comparison of variants Mono, Comb and agroforestry for the 8 farm types, in the context of a low rubber price (RubL)

Indicator: Farm Gross Margin = Gross Agricultural Income. Type T-AR = Rubber producers who earn below the minimum wage, Type T-AO = Diversified producers earning below the minimum wage, Type T-B = Farmers who depend on another income earning below the minimum wage, Type T-CR = Rubber producers who earn above the minimum wage, Type T-CO = diversified producers who earn above the minimum wage, Type T-D = Farmers who earn above the minimum wage + off-farm activities, Type T-E = Farmers who earn far more than the minimum wage+ diversification on-farm activities, Type T-F = Farmers who earn far more the minimum wage + off-farm activities.

– Variants in the volatility of the price of natural rubber. For each variant in the proportion of agroforestry, we created three variants based on variations in the price of rubber. The price was fixed for each 10-year simulation period.
– Sub-variant: Low rubber price (RubL). This refers to the average price cited by farmers for 2014-2015 and was used to represent the current situation: 50.0 THB/kg dry (US$1.4/kg dry). The price of rubber had dropped far lower in the past (20.5 THB/kg dry in 2001 = US$0.57/kg dry), but as some surveyed farmers had already stopped tapping, the current price was already considered low.
– Sub-variant: High rubber price (RubH). This refers to a high average price calculated based on 2010-2012 (TRA 2015): 105.4 THB/kg dry (US$2.9/kg dry).
– Sub-variant: Rubber price judged "acceptable" (RubA). This refers to the most frequently observed "average" price between 2007 and 2014. Farmers consider this to be the "normal" or "average" price: 81.0 THB/kg dry (US$2.3/kg dry).

Conclusion

The goal of the study was to understand the extent to which rubber, associated crops, trees, livestock, and off-farm activities, significantly improve household income and resilience. Diversification of on-farm activities was high: 50% of the farmers (16/32) raised small livestock, sold other farm products (tree seedlings, food, wood, etc.), collected and sold rubber as a collector. Fruits and vegetables were the second source of on-farm income for the majority of farmers. Many farmers completed their on-farm income with off-farm activities (56% of our sample). AFS were usually created by planting trees in existing monoculture rubber plantations.

Several authors mentioned that in addition to the association of different species, planting density and the timing of planting associated species are also important factors (Tongkaemkaew et al., 2020; Jongrungrot, 2014a and 2014b). Most associated plants can be planted at the same time as rubber as long as shade is provided for species that require it, for instance banana shading for timber trees. On the other hand, some plants might be planted 2/3 years after rubber to profit from the shade provided by rubber, e.g. certain timber species (*Dipterocapacees*) that are only planted three years after the rubber trees to enable cultivation of annual intercrops.

In the beginning (1980s and 1990s), farmers did not adopt AFS with a market-oriented objective, but to fulfil food (fruit) and social functions, which are very important in Southern Thailand. This social role was more important than obtaining a monetary income. However, the increasing volatility of natural rubber quickly made farmers aware of the economic advantages of these systems. The advantage of diversifying on-farm income was confirmed by the sensitivity study on the threshold rubber price required to obtain the same income without agroforestry practices: the lower the price of rubber, the greater the capacity of agroforestry (combined with fruits and/or vegetables) to maintain on-farm income while compensating for volatile rubber prices. Farms are more economically resilient in the face of fluctuating prices, but also face the price volatility of other products (mostly fruits and vegetables, such as mangosteen). Farmers are interested and motivated by RAS as members of agroforestry groups or networks that are a source of information and experience for other farmers. This sample could thus be the basis of a structured network on AFS, for instance in the framework of an innovation platform.

RAS groups and networks as a way towards an innovation platform

The study was conducted in Phatthalung province in southern Thailand in 1916[57]. The aim was to show that smallholder rubber plantations can adapt and remain sustainable despite variable climatic conditions and profound socio-economic changes. Agroforestry practices were identified as promising among the various types of cropping systems. RAS are economically more productive than rubber monocrop plantations and give smallholders more flexibility, particularly when rubber prices are low, which has been the case since 2013. However, adoption of AFS during the mature period of plantations in Thailand has been very limited. The policy to boost rubber agroforestry practices by all local stakeholders thus still requires improvement.

The objectives of the study were to identify the potential and capacity to use current AFS dynamic networks as a basis to set up a rubber agroforestry innovation platform. To this end, the research team studied: (i) farmers' collective organisations, groups or networks with full or partial RAS and (ii) the social dynamics that enable the sharing of knowledge and know-how. An individual producer's grid was created that identified original farmers or farmers with good knowledge and the ability to share. The role of local institutions involved in the promotion of RAS was also analysed.

The results enabled the design of an innovation platform and of activities suited to the socio-economic context of Phatthalung province. The main aims of the platform are to promote cooperation among innovative producers and the transmission of knowledge and know-how about RAS among them. An innovation platform is an efficient tool that Thai rubber institutions could set up to encourage the adoption of RAS by farmers.

Even if agroforestry during the first three or four years of the immature phase is rather common, i.e., is practiced on 65% of plantations spread across 10 provinces in Thailand (Delarue and Chambon, 2012), growing food crops between the rows of tree can provide an income in the period before the new plantation becomes productive. In the 1980/1990s, some farmers continued to cultivate AFS by choice despite the ORRAF ban. Located in the south, these farmers associated clonal rubber trees with on average, 2 or 3 other perennial species (fruit trees such as durian, mangosteen and longan, and timber trees such as teak and mahogany). A few rare "jungle rubber" systems still exist in Phatthalung province (first author's personal observations, 2017), as well as in Phang Nga province (Penot and Ollivier, 2009). Those farmers are usually members of associations or informal networks in order to share their knowledge and experiments and to promote their systems (Jongrungrot, 2014a). ORRAF officially lifted its ban in 1992, while maintaining interest in and funding replanting in the case of rubber monoculture. In 2001, some AFS trials were set up by ORRAF and AFS was officially promoted by the rubber act. In practice, AFS were really only promoted to

---

57. This study was conducted by Marion Theriez under the supervision of Bénédicte Chambon and Éric Penot from Cirad and Uraiwan Tongkaemkaew from TSU. It was published by Marion Thériez in 2017. Rubber production in Phattalung province, Thailand: potential of a regional innovation platform emergence to co-design innovative agroforestry systems, IRC SupAgro, Montpellier, France. The study is part of "Heveadapt," a Franco-Thai research project. Sources are Éric Penot, Marion Thériez, Isabelle Michel, Uraiwan Tongkaemkaew, Bénédicte Chambon. 2022.Agroforestry rubber networks and farmers groups in Phatthalung area in Southern Thailand and potential for an innovation platform. *Forest and society*, 6 (2), November Issue Published May 14, 2022.

deal with low rubber prices after 2015. There is an old tradition of agroforestry under specific conditions, but AFS are currently a marginal practice.

In 2015, the Rubber Authority of Thailand (RAOT) changed its policy and began to promote AFS practices in an attempt to overcome the strong negative impact of rubber price volatility on farmers' incomes. Some AFS farmers promoted a different approach to rural development, through His Majesty the King's "New Theory of Agriculture", which later became the "sufficient economy philosophy", which is socially very important for these farming communities. The downward trend of rubber prices after 2012 certainly also influenced farmers' attitudes. RAS certainly fits the scope of this new approach.

Phatthalung province in southern Thailand, the historical rubber production area, was chosen as the study site. We first present the conceptual approach and then describe the sample and surveys we conducted, after discussions with local key informants in a preliminary village information meeting. Due to dissemination of potential AFS farmers, a representative sampling method was not feasible, leading to selective sampling. The selection criteria were based on RAS representativeness and group recognition, and resulted in the selection of 54 producers who were subsequently the subject of individual interviews: 8 producers representing the Banna agroforestry community (Sri Nakarindra), 5 from the Lung Toon network (Tamod), 9 from the Lung Boonchu network (Pa Phayom) and 29 individual producers who did not belong to any group or networks and were considered as individuals independent of any structure. The idea behind the selection of groups, networks and satellite farmers was to be relatively representative of RAS in the province in order to set up a future innovation platform rapidly at the regional level. (Table 2.29).

Innovation platform

An innovation platform is an interactive tool to explore opportunities and solutions (Nyikahadzoi et al., 2012), to exchange knowledge and practices through experiments, observations, evaluation, and discussion. Such platforms enable multi-directional exchange of know-how and knowledge among stakeholders (Tittonell et al., 2012) in addition to being a social tool that stimulates collective action and discussion and increases people's ability to innovate (Tenywa et al., 2011). An innovation platform constantly evolves, along with its environment and its members.

An innovation platform has different stakeholders; in our case farmers, researchers, institutions, technicians, companies, carriers, etc. with different profiles and different objectives, but who can find a common solution to problems through discussion. Each individual defines his/her opportunities and weaknesses and his/her part in the work to be accomplished. Each individual can act on one or more points of the chain. Partners need serious collaboration to solve problems and develop innovations: they make decisions together. An innovation platform can be implemented at different scales: local, regional, or national depending on the scale of the stakeholders and their different levels of involvement (Nyikahadzoi et al., 2012). The present study aimed at creating a regional platform with a focus on RAS.

Farmers' knowledge and know-how are widely recognised in a regional innovation platform (IP). But in political terms, an institutional framework and the participation of local leadership are also essential. The challenge is to strengthen the capacity for innovation of the group by creating strong relations in the IP, and by improving everyone's understanding. Group skills evolve over time. Every stakeholder should feel concerned about and involved in the platform and its topics. At the beginning, researchers can be facilitators, defining the potential, characteristics and responsibilities of every stakeholder in order to boost farmers' participation and encourage sharing. The IP can design a stakeholder's diagram (Tenywa et al., 2011). The IP is the forum for knowledge sharing. In concrete terms, in a regional IP, sharing working days on the farm with several farmers can be implemented through sessions with a farmer-to-farmer approach, training courses, specific agroforestry events, regular meetings, etc. In the present study, we focus on what seems to be best suited to the Phatthalung area.

Main results

Table 2.29 lists the groups and networks selected in this study, as well as satellite individual farmers.

Table 2.29. Formalisation of organisations studied in Phatthalung

Name	Banna agroforestry community	Lung Boonchu network	Lung Toon network	Satellite farmers
District	Sri Nakarindra	Pa Phayom	Tamod	9 districts
Focus	Agroforestry	Diversification	Agroforestry	Diverse
Type of structure	Group	Network		Farmers who practice innovative agroforestry
Characteristics	- Established list of members - Regular meetings - Share financial expenses - Share an identity, a history and values	- Not all members know each other - No regular meetings - No delimitation		- Farmers who talk about agroforestry with neighbours, family, friends, groups but are not specialised in AF - Producers located throughout the territory who are not affiliated with any group or network

All the groups have their own network (Table 2.30) or belong to an interaction network between neighbours (only at the village level) with whom they have been able to share government support for a local project. Ultimately, they all interact with a social network, albeit often limited to the village level. Satellite producers may also have access to a DOAE (Department of Agricultural Extension) learning centre on specific practices on their farm. Finally, every producer has at least one within-village network and is not isolated. Table 2.30 describes the three local groups/networks we surveyed and considered to be representative of the different communities in the area.

The "pioneer farmers" began intensive agroforestry as early as in the 1990s. The number of network members was not originally fixed, but was tending to stabilise in 2014. The difference between networks and groups is the perception of a group of people with reciprocal interactions, in a network, new people can be easily integrated. In contrast, in a group, the framework is less flexible: there is a fixed list of members and participation in events or meetings is often a prerequisite for organisation. Finally, these groups and/or networks include several villages, but rarely extend beyond neighbouring sub-districts. Only the Lung Toon network, which is located on the border of Tamod sub-district, extends over two sub-districts. The objectives of each group/network are listed in Table 2.31.

Table 2.30. Identity card of the groups surveyed

Name	Banna agroforestry community	Lung Toon network	Lung Boonchu network
Leader	Lung Jay	Lung Toon	Lung Boonchu
Leader's age	67	69	64
Date of birth of the collective	1995	1993	2004
Subdistrict	Banna	Tamod and Kong Yai	Pa Phayom
Villages	Moo 2, Moo 5, Moo 8	Tamod: Moo 4, Moo 9 Kong Yai : Moo 2	Moo 5, Moo 6, Moo 7
Longest distance between two members	6.3 km	3.3 km	4.5 km
N° of members	8 members	> 10 members	> 10 members
Main objective	To preserve local species	Increase forest area by preserving local species and natural resources	Access agricultural knowledge in order to increase farmers' incomes

Table 2.31 Objectives and activities organised by the groups/communities

Name	Objectives	Tools and activities
Banna agroforestry community	- Develop new markets for producers - Increase producer income - Share knowledge - Use good environmental practices	- Take part in government activities - Use the group's combined production to negotiate prices - Visit farms - Lead projects to obtain funding
Lung Boonchu initial group	- Reduce dependency on inputs - Crop diversification - Find innovative species to mix with rubber trees - Develop knowledge networks	- Training organised by the government - Create local markets - Visit farms - Make organic compost - Group production to obtain good prices and new consumers - Make joint applications for funding
Lung Toon initial group	- Plant as many trees as possible - Cultivate organically - Convince as many people as possible to plant trees and use organic farming practices	- Help each other with hard tasks - Allow everyone to obtain free seedlings from the forestry department - Write a book about agroforestry - Organise crop diversification activities

These three groups are representative of groups in Southern Thailand. AFS are linked with other development activities such as poultry, fish pond, beekeeping, fruit production, planting timber trees, protecting remaining forests and diversifying crops and income. AFS differs from "integrated farming" but is very similar in terms of strategy. AFS combines cultivation of several products on the same plot, also because land is becoming scarce due to transmission of patrimony from generation to generation within a family. Among possible alternative ways to diversify sources of income, AFS appears to be a key strategy.

DOAE organises inter-village and even inter-district training in learning centres, where producers can discuss their farming practices. These activities are mostly limited to village DOAE leaders, but sometimes allow some producers with a learning centre and/or who are on a DOAE list to expand their network and even get to know members of other groups. Major interactions between networks already take place in Phatthalung. Even though we only found one formal agroforestry group, two dense networks of within-village farmers have developed in parallel. The long history of agroforestry practices is detailed in Kheowvongsri PhD thesis (1996). Originally, Lung Toon and Lung Boonchu were leaders of groups and not of networks. Lung Toon is a leader of a centre for learning on Buddhist agroforestry, and Lung Boonchu, who is the leader of the eponymous network, heads a centre for learning on self-sufficiency and subsistence farming within the framework of the King of Thailand's theory of economy. The general objectives are set out in cognitive and environmental terms. In each group, the initial goal was defined by the leader, who then tried to gather around him farmers who were thinking along the same lines. The communities aim to respond to economic and environmental concerns and to help local farmers obtain government support, but also to participate in field activities.

The leaders of the three groups are quite well known in Phatthalung province. They run small networks to organise activities, obtain funding or set up new development projects. Collectives are open to the outside through dynamic, autonomous and proactive leaders. They create rich interaction networks. Table 2.32 lists the pros and cons cited by each group of being part of an innovation platform.

Sharing knowledge on AFS is a feature shared by all these groups and is an important social dimension for the members of the groups along with an important social feeling of being "knowledge bearers" in a different way than farmers who focus on monoculture.

Six diversification categories were identified that correspond to different benefit and constraint frameworks:
- The association of forest species (20 species), not specifically for sale, and/or fruit/timber species,
- The association of local fruit species (21) and/or fruit for export,
- Combination of vegetable (11) and/or ornamental (7) species,
- Association with cash crops (coffee, palm oil or pepper),
- Livestock or fish,
- Forest wood/timber for fuel wood and valuable timber (minimum of 10 native species).

Producers use one or more diversification pathways on their plots and generally practice more than one type of AFS. Each AFS defines a framework of specific advantages and constraints for each farmer, which have to be identified and considered before planting, according to each farmer's expectations. The main reasons for choosing agroforestry mentioned in our surveys were (i) lack of land and the need to grow intercrops (21/142), (ii) the desire to experiment with new practices and develop sustainable cropping (35/142).

Table 2.32. Pros and cons for each network/group of being part of an innovation platform

Banna Agroforestry community	Lung Toon Network	Lung Boonchu Netwrok
No longer very active Sri Nakarindra Model project Nursery with native rare plants	This network is expanding The network includes learning centre facilities Tree Bank project	This was the 1st learning centre facility A success story for smallholders
High willingness to share knowledge Innovative and dynamic farmers Demonstration plots Ability to organise training in marketing, plant association, crop management techniques, and native medicinal plants	Desire to preserve natural resources Involves strong minded, convincing and innovative farmers Demonstration plots Capable of organising training in organic farming, timber production and fruit tree management	Aims to follow the king's theory of sufficiency Agroforestry systems Demonstration plots Experimental plots Capable of organising training in livestock raising, fish farming and organic farming.

Proposal for a regional innovation platform for rubber agroforestry systems as a tool to better promote AFS among monoculture smallholders

We acknowledge that the existing groups and network we have described in this study are clearly not sufficient to ensure a boom in AFS adoption. Although it is true that all the preconditions for such an AFS boom are present: (i) existing groups and farmers with AFS plots that can be used as demonstration plots for other farmers, (ii) the farmers have real knowledge and master basic AFS practices and have a real desire to share their knowledge, and (iii) there is an economic need for most rubber farmers to increase their gross margin/ha through diversification given the long period of low rubber prices, what is lacking is a regional organisation capable of transforming local opportunities into real challenges for larger communities. Political capacity does exist through the very large and active local administration (e.g. RAOT and the Ministry of Agriculture and Forests). A regional innovation platform could take up the challenge. What is needed is the political will to support it and funds to get the process underway.

The platform presented in figure 2.16 would involve many stakeholders. The main stakeholders would be farmers, donors and the government agencies (RAOT and DOAE), supported by researchers from local universities (PSU and TSU) who provide knowledge, and the local institutions and their technicians who advise farmers and the private sector (in processing and sales). The key to success is regular meetings with all the stakeholders to discuss future actions, to plan and organise actions and share the results with other stakeholders. An innovation platform is a place to share, decide and implement AFS activities, to develop value chains of products and to discuss AFS policies.

A digital centre could be created (Meta — previously called Facebook — page, website with access to documents, etc.) to pool and share reports and activities, to keep people informed about activities and training courses, with e-learning, published articles, demonstration videos about their agroforestry plots, etc. A forum could also be created in the website.

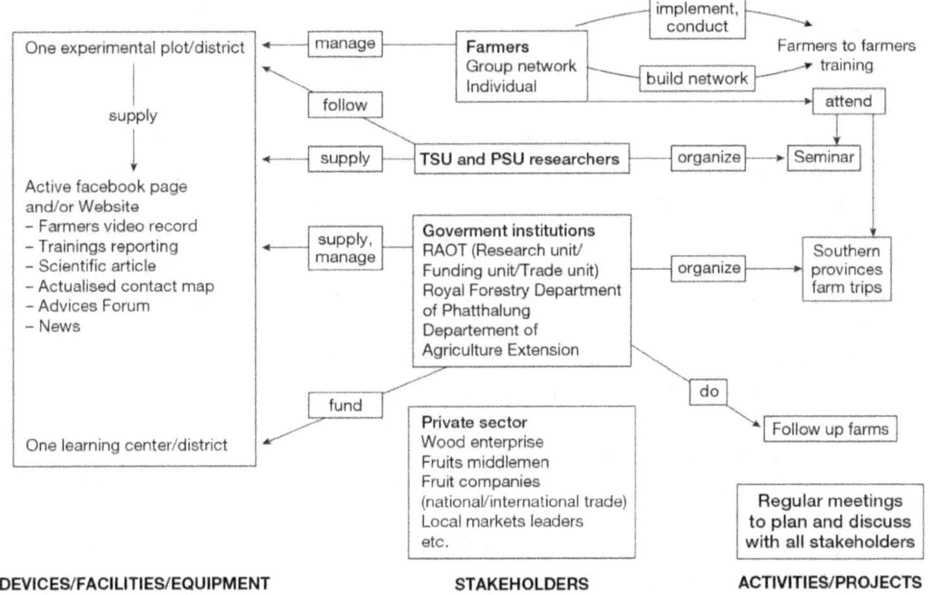

Figure 2.16. Proposed design for an innovation platform (Theriez et al., 2017)

The main axis of the proposed innovation platform is farmer-to-farmer training courses in agroforestry practices. Existing AFS plantations can be selected as "demonstration plots" for training purposes. New plantations could be monitored by researchers like farm trials for the purpose of comparison (two farmers already lead experimental plots), in particular to test double spacing systems which prioritize associating crops and rubber, with fruits that already sell well such as mangosteen and durian. The wide variety of fruits that can be associated with rubber (as well as timber) requires looking for local markets for each product, but at the same time, the variety of products also diversifies the market, which obviously reduces the risk of over production. This study showed that there is real potential for the emergence of an innovation platform for RAS in the Phatthalung area as well as a real demand from local RAOT offices, whose staff are relatively close to farmers via replanting programmes. It is thus possible to offer a range of training courses on the transfer of RAS technology from existing RAS farmers and groups to other farmers. However, the creation of an innovation platform in Southern Thailand requires a concrete political decision by top institutions like RAOT, while forestry institutions also seem to be very keen to contribute.

An innovation platform is a social tool designed to implement new ideas, promote new systems in a rubber-growing world where monoculture still dominates and to innovate rapidly. In Phatthalung, the aim would be to co-design RAS with producers, researchers, development and funding agencies (RAOT, Ministry of Forestry, agricultural extension/DOAE, local banks) and the private sector (fruit sector). Producers' groups organised around neighbourhood networks have existed for more than 20 years and are ready to train other farmers and to innovate agroforestry practices. This study has shown that an important inter-producer interaction network already exists; what is needed now is a detailed analysis of ongoing dialogue between producers in networks which still lack AFS advisers, the frequency of meetings between peers, and the geographical distances at which people interact. In other words, a complete overview of the social and geographical dynamics of the flow of information, knowledge and techniques in agroforestry in Phatthalung is needed.

For farmers, joining a group is a way to join a big network, share knowledge and access government support. The group makes it possible for farmers to participate in many local activities and to share knowledge. The group could be monitored by government agencies including RAOT, DOAE, the natural resources department, the health care department, the livestock department, among others. All local stakeholders already lead actions and organise events to help farmers promote their systems and sell products. Some also provide specific funding. Universities, in particular PSU and TSU, contribute to the dynamics of the sector. Teachers/researchers visit farmers and conduct farm trials. In association with government agencies, they exchange with other groups during official government activities, which concern more than 300 groups in 11 districts in Phatthalung. RAOT should play a central role in the organisation of meetings and the transfer of knowledge and know-how between farmers' groups in a future innovation platform. Meetings, farm visits and training are the most important activities to be promoted by the platform, but access to knowledge via the Internet would also be efficient.

A study of the attendance rate and the type of audience of the "Ko-so-no" (alternative education centre) where computers are accessible free of charge would make it possible to determine whether this infrastructure can offset farmers' current lack of equipment. A strong interactive network of identifiable reference farmers is essential to enable a regional innovation platform involving the main rubber development institutions, such as RAOT, researchers, rubber collectors and buyers, timber, vegetable and fruit collectors, to become immediately operational.

There is a need for further studies of local fruit value chains and for an analysis of the fruit and timber markets, in particular to judge the potential for expansion of the legal timber trade since the market has evolved in the last 10 years. Concerning the fruit sector, it is indispensable to know the saturation levels of the current market and its potential expansion if AFS expands. In 2016, how the timber industry functions is still not known due to illegal trade on the national and global timber markets. The whole sector needs to be reorganised, including the establishment of sawmills and of local timber industries. Finally, the pre-existence of such AFS networks, the sum of immediately exploitable knowledge and know-how and the goodwill of local AFS producers, research and regional institutions have created a climate that is particularly favourable for the establishment of a rubber agroforestry innovation platform.

Some institutions, including the Rubber Authority of Thailand (RAOT), the forestry department under the supervision of the Ministry of Environment and Natural Resources, and agricultural extension departments (DOAE) are aware of the need to support agroforestry. These stakeholders have converging interests, are positively engaged in the promotion of AFS, and are therefore potential partners for the emergence of an innovation platform for rubber agroforestry innovation. A complementary sociological survey at regional level is needed in countries where AFS groups and networks are poorly known.

Conclusions on Thailand

Although long recommended by Thai rubber authorities, intercropping during the immature period of the plantation has not been systematically adopted either by all farmers or in all the plots belonging to one farmer. In addition, intercropping during the immature period has not automatically involved maintaining RAS during the mature period of the plantations.

Despite their advantages, permanent RAS are still rarely adopted by Thai rubber farmers. Although there was formerly a tradition of agroforestry in Southern Thailand, it was almost totally replaced by rubber monoculture with the implementation of the rubber replanting scheme. Nevertheless, changes in the farmers' environment (government measures, rubber prices, research interest) in the last 20 years have created a more favourable context for the development of RAS. Farmers' initiatives have opened the way for changes to the rubber cropping system which need to be supported by research and extension services. There is also a global trend to promote RAS with strong involvement of environmental NGOs (for instance, see Penot et al., 2024) notably in the framework of the Global Platform for Natural Sustainable Rubber (GPSNR).

▸▸ Rubber versus other alternatives: what role for RAS?

The need for a specific economic analysis

This section has been originally published in Torquebiau and Penot (2006)[58].

We argue that there is an economic rationale behind the importance of agroforests worldwide, but that this rationale is complex to identify and measure. In the first instance, direct sales of agroforestry products (timber, fruits, vegetables, resin, nuts, rattan, medicinal products, etc.) and self-consumption, which enables significant savings in daily household expenses, are complementary. Beyond this aspect, it has been shown that long-term patrimonial strategies are of utmost importance to the farmers who do practice agroforestry. However, conventional economic analyses based on discounting rates are not ideal for these perennial, multi-component and multi-cycle systems, where future discounted values of tree products are difficult to predict and as such, are seldom taken into account by farmers in their planting choices (Torquebiau et al., 2002), unless the harvested products are easily marketable and generate a net margin which covers replanting costs (e.g. clonal rubber). Finally, farmers also plant and tend agroforests because of their social functions (land tenure, social status, living environment). So, while scientists have continually argued that agroforests are environmentally sound, this is probably not a major incentive for farmers. In Indonesia, the high biodiversity provided by jungle rubber is threatened by oil palm plantations. If a comprehensive economic analysis of agroforests is to be undertaken, it is legitimate that their environmental attributes be taken into account.

---

58. Chapter "Ecology vs Economics in Tropical Agroforests".

Thus, the objective here is to try to show that the reason behind the "enigma of tropical home gardens" (Kumar and Nair, 2004) lies in elements of positive externalities that are not accounted for in standard economic analyses, yet matter to farmers and perhaps also to other stakeholders (e.g. timber for sawmills). If agroforestry scientists want to convince farmers and policy makers that agroforests are worth considering as land-use options, and not only as relics of the past, appropriate economic analyses of agroforests need to be conducted that include ecological services (e.g. watershed protection, nutrient cycling, functioning as a carbon sink, as a bio-habitat, conserving biodiversity) as well as social, cultural, and aesthetic values.

Following Coase (in Cooter, 1982) and his analysis of social costs, we distinguish between "giving a value to a service" (potentially but not automatically tradable) and "paying for a service" (which leads to "who is going to pay?"). Taking into account (giving a value to a service) or internalising positive externalities (paying for a service) relates to resources or services that cannot be included in private accounting because they are public goods (e.g. landscape beauty, pollinating insects) or because they are preserved for future generations (e.g. biodiversity, soil resources). We argue that such global goods, considered as services to the community, should not only be taken into account in international negotiations on climate change or biodiversity, but also in agricultural policies, incentives and, as a result, in farmers' day-to-day decisions.

One of the services that was likely to be taken into account in the 2000s was the carbon sink function of the Clean Development Mechanism (CDM), which should have been applied in 2012 as scheduled in the Rio and Kyoto rounds. In fact, this mechanism never really worked and was removed in 2016 and replaced by another carbon programme. As rubber was (at the time) the only tree crop FAO considered eligible for CDM (beside timber trees), rubber based (and timber based) agroforests were theoretically eligible. If this is the case, their carbon sink service can be valued and considered in the trade or exchange of pollution rights (Cacho and Hean, 2001). In the 2020s, this mechanism might take first place in a context of climate change and of a "zero deforestation" policy.

Farming system level approach

A first pragmatic approach would be to conduct a household level analysis of the cost saved by products provided by agroforests that consequently do not need to be purchased (e.g. building and fencing materials, food, medicines, raw materials for handicrafts). Next, accounting for environmental benefits could also take place at household level if the surveys are sufficiently detailed and use data compiled over at least a year. Modelling farming systems (e.g. with software like *Olympe*), is a useful way to process data on production, value, cost of production and labour, to be able to compare return to labour and the gross margin of each cropping system at the farm level.

Several case studies have been conducted in Indonesia using *Olympe*, which was specifically designed to obtain an easy, dynamic yet detailed view of the main economic features of a farm, such as margin/activity or crop(/ha/year), return to labour, as well as all sources of income from on-farm and off-farm activities. The simulation is based on a 10-year period. Scenarios can be built according to hypotheses on price volatility (inputs or outputs), yield, or the impact of climatic events. The software provides a dynamic view of the farm trajectory and changes to it resulting from the decisions taken by the farmer, as well as external factors such as prices, risks, or yield (see Penot, 2007).

Economic analysis: a social-ecological perspective

While a farming system approach can enable a better understanding of the different roles of agroforests, there is also a need for a new approach to agroforest analysis to deal with higher levels of complexity and translate their social-ecological[59] performance into economic performance. One apparently irrational behaviour that has been observed in Indonesia is maintaining old rubber agroforests rather than planting economically highly profitable oil palm. One hypothesis was that agroforests would gradually pave the way for oil palm plantations: the social value of agroforests (control over land), the possibility to improve agroforest (with clonal rubber) and diversification strategies that would eventually lead to new self-development of improved rubber agroforests that are within the financial capacity of local farmers who have no access to credit or insufficient capital building capability. In the meantime, despite the enormous gain in return to labour and net margin provided by oil palm, agroforests have never completely disappeared, proof of the value of such systems when they are analysed in a farming system framework and from a social perspective. As a "reserve land factor" or a "long-term land control factor", agroforests may not represent a direct value but they do have an indirect value as a capital reproduction factor or as a potential expanding factor.

Patrimonial analysis, i.e., of changes in ways of building capital and transmitting assets, could be used, as agroforests are considered as land reserves that can be traded, and because trees represent a strategy for building capital for future investment. Long-term multi-cycle analyses should provide a frame to understand farmers' behaviour and long-term trends in farmers' strategies. Economic analysis of mixtures of plants with different length life cycles is also possible through farming system modelling. Smoothing long-term and patrimonial strategies (Torquebiau and Penot, 2006) may help account for the time factor and the historical perspective (e.g. accumulating capital, capacity building).

Subsistence versus cash

The merits of agroforests in providing subsistence food for families, enabling flexible crop production or reducing the need for external inputs also need to be taken into account. The comparison of farms with and without agroforests could reveal the real savings and their impact on household income. However, not all agroforests are based on food crops. Some agroforests are completely cash-oriented e.g. rubber (jungle rubber), resin (Damar agroforest), spices and timber (e.g. cinnamon-durian-timber based agroforest). Home gardens can be labour intensive, and require considerable quantities of inputs.

In 2023, markets are the main driving factor of RAS and most products are sold rather than self-consumed.

---

59. The term "social-ecological" implies an interactive system with social and ecological components of equal importance, while the conventional meaning of "socio-ecological" is simply an ecological system with some social aspects (Sayer and Campbell, 2004).

Landscape amenity and social conviviality

The potential of agroforests for the provision of values such as landscape beauty or conviviality for rural societies also needs to be incorporated in their assessment. It seems clear that in many situations, agroforests, and in particular agroforests managed by local communities, and hence considered as a public good with limited but shared access to local resources (fruit, timber), play an important social role. The *Tembawang* of the Dayak people in Kalimantan (Indonesia) is a typical example. In addition to being a reserve of forest products through "extractivism" when original forests have disappeared, these agroforests have important social dimensions as graveyards for Dayak populations in Kalimantan, or may play a role in village protection through the maintenance of a green belt around the village, or by forming concentric layers around the village. Even if there is no economic value (even as a service), its social value will generally prevent its destruction.

Economics in the context of agroforests: rehabilitating the micro-economic approach

Clearly, many particular features of agroforests cannot only be valued as goods. Social value, or the long-term strategic value of a piece of land, are relevant justifications for the existence of agroforests. Risk buffering may be one of the most powerful incentives to maintain or expand agroforests. Modelling farming systems and a prospective approach make it possible to assess the effect of such buffering on risks. Prospective analysis using scenarios will enable the identification of economic thresholds and boundaries and the definition of a domain of economic feasibility.

If the benefits to be had from agroforests (e.g. providing free fuelwood, meeting some nutrient needs, spreading income, contributing to nutritional security, to integrated pest management, to crop pollination, reducing crop failure, acting as a carbon sink) are appropriate for market value analysis, then neo-classical environmental economics can be used and externalities can be included (or re-internalized) in the process of income generation. The cost of pollution and delayed growth can be accounted for as negative externalities or as constraints to further development. Environmental services can be valued according to a "system of values" that is recognised locally as being relevant at a higher level (community or provincial level). The upper level could be the CDM (see, e.g. Cacho et al., 2002 or Albrecht and Kandji, 2003), for an analysis of accounting for carbon sequestration in agroforests in Indonesia (see Hamel and Eschbach, 2001, for the potential impact of CDM on natural rubber).

The real problem is therefore to see if farmers really do — or possibly could — benefit from externalities or from the advantages of agroforestry. In some cases, the answer is yes in terms of savings on the cost of building a home, food in the case of self-consumption, medical treatment thanks to the use of medicinal plants. The answer is less clear, or at least there is no direct profit to be obtained from long-term externalities such as the "sustainability" of land (and hence of production), but which is obviously taken into account by farmers in their decision on whether to invest in perennial crops or tree-based agroforestry systems. Patrimonial transmission as a result of capital building is also an indirect advantage as it provides the next generation with a sustainable and valuable production system. Some benefits, such as social benefits, are traditionally not adequately accounted for in the analysis simply because they are so difficult to assess. Lastly, some benefits are potential: in 2024, biodiversity could be a source of income tomorrow, biodiversity conservation may be considered as a "global service" in which case farmers would be entitled to payment by the international community for its provision, as suggested by the ICRAF/RUPES project (Shared Investment in Pro-poor Environmental Services). It was expected that the carbon sink value would lead to indirect profit through project implementation after 2012 according to Kyoto protocol (for rubber and timber trees in particular, see Hamel and Eschbach, 2001). In reality, the carbon market was not successful.

Only a detailed economic analysis of farming systems will enable correct identification of both direct and indirect benefits when considered in terms of farmers' risk prevention strategies, long-term investments and production sustainability. Research on rubber agroforestry systems is currently underway in Indonesia to identify the potential and real benefits of agroforestry practices compared to monoculture or other alternatives (Penot, 1996b, 1997, 2001, 2002, 2006, 2021-2024; Werner, 1997; Joshi et al., 2000; Lawrence, 1996; Rubis project, Dwi Sninta Agustina, 2022, personal communication).

The context of most developing countries means there are huge income gaps due to strong social stratification, information asymmetry, high transaction costs and institutional failures that have major implications for local economies — particularly when the time factor is important — in identifying and understanding farmers' strategies. Micro economics makes it possible to account for environmental assets, complexity, uncertainty, and implies stakeholder participation. When dealing with agroforests, benefits linked to public goods or goods that cannot be given a market value because they are intended for future generations (e.g. biodiversity, landscape amenity, carbon sink, cultural and aesthetic values) need to be apprehended from a different perspective. We have seen that a multi-functional approach, inspired by that developed by the Common Agricultural Policy (CAP) for European farmers, can be a source of ideas on how to take these externalities into account (but not necessarily the accompanying subsidies policy!). New mechanisms such as those developed for the CDM could be explored, in particular for global issues such as biodiversity conservation.

Agroforest attributes should also be accounted for in national accounting. Policy makers should acknowledge the fact that, if resource depletion was taken into account in an environmental economics approach, agroforests would rank very high amongst land-use options because they generate an "agroforest rent"[60] (Ruf, 1995; Ruf et al., 1999) which is much higher than the rent (i.e. income) obtained from conventional agriculture or other forms of resource exploitation (e.g. logging, mining, depleting the soil through excessive harvests). Agroforestry rent is similar to Ruf's theory of forest rent but generated by an agroforest (less disease, better soil, better productivity, less need for fertilizer, etc.).

Farmers who contribute to this resource rent could receive direct payment or, even better, indirect incentives (e.g. tax exemption) to stimulate land-use options that contribute to such public goods or to the provision of such goods for future generations.

---

60. The term "agroforest rent" is used here according to the definition of "forest rent" provided by F. Ruf in "Booms et crises du cacao", Karthala, Paris, 1995.

To achieve this status, agroforests need to be recommended along with other land-use options, they require a reference framework that accounts for these alternative economic analyses. Otherwise, they will always be rejected or marginalised as not fitting conventional economics and hence not matching development objectives. Whether for commercially oriented agroforests or subsistence-oriented home gardens, a long-term perspective must be part of a farmer's strategy when dealing with multi-strata agroforestry. However, there is obviously a biased debate between short term (economics) and long term (ecology). In both cases, farmers have developed long-term farming practices as a result of a long innovation process that ultimately accounts for long-term economics through the risk buffering capacity of agroforests. In most cases, social organisation is tightly linked with the technical constraints involved in the production and reliance on food reliance, securing income and potentially, control over land. There is a strong link between technical systems (technical pathways) and social systems (Penot, 2004a). Customary laws take this important point into account and are generally able to adapt to changes. There is an economic strategy behind maintaining agroforestry practices that have proved to be able to secure production and maintain control over land. In other words, long-term economics is fully associated with ecology in terms of sustainability as already well documented for traditional agroforestry systems in, for instance, West Sumatra (Michon and de Foresta, 1991). An appropriate economic analysis should fully account for the long term. A major challenge for the very near future is resolving the dilemma between the internalisation of externalities, by giving a value to "services" through a multifunctional approach, and by giving ecological criteria a real value added.

Conclusion

Rubber farmers have developed a series of innovations to adapt rubber to their extensive agroforestry practices (jungle rubber) or the estate model (SRDP in Indonesia) by associating rubber with perennial or annual crops. However, they have now reached a stage where options for further innovations are limited and productivity cannot be increased without using rubber clones, which require different management. SRAP wishes to respond to this demand. RAS based on clones are the best alternatives for farmers. Technical change is driven by economic necessity, in particular since the Indonesian crisis. RAS are the expression of the recombination of indigenous knowledge (agroforestry practices) and external knowledge based on intensification (clones and chemical inputs). Such technical change leads to more affordable rubber cropping systems that are better suited to the range of different local situations. In parallel, the positive externalities of RAS including biodiversity conservation and environmental sustainability are appealing for future large-scale developments.

Smallholders need reliable information, access to credit, good quality planting material, and recognition of the relevance of complex agroforestry systems by all actors, including by civil servants involved in agricultural development.

Alongside strategies aimed at diversification through the adoption of oil palm, rubber has still an important role to play for local farmers who do not wish to rely on one export crop alone.

➡ Changes in RAS patterns in West Kalimantan from 1994 to 2019

The focus of RAS/SRAP trials in 1994 was to move on from jungle rubber to clonal based rubber systems in agroforestry. In 1995, more than 80% of rubber was produced in jungle rubber systems. Agroforestry was one the technical options to increase the rubber gross margin per hectare. Most farmers wanted to change to clonal rubber to improve rubber productivity (from 500 kg/ha/yea in jungle rubber to 1,400/1,800 kg/ha/year using clones). However, at the same time, oil palm was undergoing colossal expansion in private estates and associated smallholder development schemes. From representing a good potential opportunity to diversify farmers' income, oil palm became a major competitor for rubber because of its better economic performance, and, in the meantime, the situation has changed, the objective for most local rubber farmers was no longer to replace jungle rubber by clonal rubber but to replace jungle rubber with oil palm. The three most important changes in the period 1995-2023 were (i) oil palm has become the most frequently planted crop and now accounts for 50-75% of the land formerly under jungle rubber in Sumatra and Kalimantan, (ii) most jungle rubber has completely disappeared. From the original 3 million ha in 1995, only 500,000 to 1 million ha of old jungle rubber probably remains in 2023, and this land is considered by local farmers as reserved for the future, irrespective of which crop, but generally oil palm, and (iii) part of the area under old jungle rubber has been replanted with clonal rubber, so that in 2023, most of the rubber produced comes from clonal rubber.

Given this trend, interest in agroforestry has evaporated because most farmers have already integrated both clonal rubber and oil palm in their farming systems. In this context, this is the perfect time to review the results of the 1994/1997 RAS/SRAP trial up to 2019.

The situation of RAS in 2019

In surveys conducted in 1997, RAS 1 was found to perform best in terms of maintaining soil fertility, preventing erosion and low cost of establishment during the immature period, and a survey conducted in 2007 showed that, in the long run, more than 80% of farmers had continued to maintain their RAS plots. This was the case of most smallholders who were reluctant to invest US$2,000 per ha to create a new clonal rubber plantation using their own savings (in contrast to planting oil palm by local oil palm estates with a dedicated credit). In 1997, the cost of establishment and maintenance during the first 3 years were estimated to be US$700 per ha (Boutin et al., 2000).

RAS 2 was the most widely adopted system, thanks to the associated trees (fruit trees and more recently, timber species) despite the fact that poor markets for fruits and timber are real constraints for further development.

RAS 3 "did the job" in areas infested by alang-alang (*Imperata cylindrica*), as control of the weed was very good thanks to the shade provided by associated trees and a cover crop (*Flemingia congesta*). Excellent results were obtained without the use of the herbicide Roundup in transmigration areas and in some villages like Pana (Boutin et al., 2000).

The changes observed in different trial plots were the following:
– conversion to oil palm in 20% of SRAP plots, or to clonal rubber monoculture in 20% of SRAP plots, mainly those located in Trimulia, with RAS 1 or 2 agroforestry systems in 50% of the SRAP plots and *tembawang* in 10% of the SRAP plots;
– specifically in Trimulia village (transmigration area): 100% of rubber plots were under monoculture due to poor sandy soils, lack of water for associated trees and the priority given to rubber trees;
– in Kopar: 80% of rubber plots were under RAS 1, where continuing access to forest products is still important for the local population;
– in Engkayu: 60% of rubber plots were under RAS 2, where total productivity through fruit production is important to ensure a stable agricultural income;
– in Embaong: 30% of rubber plots were under RAS 2, the rest was a mix of RAS 1 and monoculture;
– in Pana: 90% of rubber plots were under RAS 2;
– in Sanjan (former SRDP where no SRAP trials were performed): 50% of the area was still under clonal rubber and 25% of the rubber plantations under agroforestry;
– less than 10% of the plots changed to *tembawang*, a local fruit/timber-based agroforest.

Most trials took place between 1994 and 1996 in the villages of Kopar, Engkayu, Embaong, Trimulia, Pana (Sangau area) and Pariban baru (Sintang area). Another set of trial plots in the village of Pana were established between 2000 and 2005. The trial plots were visited regularly between 1994 and 2007. The photos show the situation in 1994/1997, then in 2005/2007, and most recently, in 2019. Today, all forests and most jungle rubber has been replaced by oil palm in roughly 2/3 of the area and of clonal rubber, either in monoculture or agroforestry in1/3 of the area.

The biggest change in land use and in farmers' strategies in our study area has clearly been the expansion of oil palm which rapidly became the number one priority for local smallholders. At the same time, local estates took over most of available land for their own oil palm plantations while the low rubber price killed any interest in cultivating rubber. Nevertheless, smallholders did not want to completely and permanently abandon rubber. In 2023, rubber continues to be planted, as it makes better use of available family labour, complementary to that used for oil palm production and as a way of diversifying income (mainly monoculture and RAS 2).

Lessons learned from changes in RAS

In 2023, farmers are in the same situation and face the same problems as in 1994: poor access to clonal planting material, no training in tapping frequency or practices, but they do have some knowledge about clones and agroforestry. Rubber agroforestry cultivation techniques no longer appear to be passed on by farmers to their sons or to other young farmers, but the two biggest differences are that (i) oil palm accounts for 2/3 of the land and is now the main source of income, and (ii) jungle rubber disappeared rapidly and all farmers now have plantations comprised of clonal rubber that produce yields, while old jungle rubber is considered as land reserved for future plantations.

All the trial plantations have now reached the end of their life span, which was reduced to 20-25 years due to diseases and poor tapping practices. Agroforestry was considered by most farmers to be very useful (i) during the immature period of rubber trees, because it enabled a better return to land with intercropping, or because of the reduced establishment costs depending on the type of RAS, and (ii) thanks to income diversification (through different kinds of fruit and timber species, either for self-consumption or for sale), improved farm resilience in the face of commodity price volatility.

The lessons learned are the following (i) rubber agroforestry trials were conducted at the right time (in 1994), when there was a strong demand from farmers for systems with low establishment costs that ensured income diversification: at the right time and in the right place, but… (ii) oil palm arrived in 1997 and its adoption was encouraged by very strong pressure from private companies (thanks to the concessions policy) and was a lucrative alternative to rubber cultivation with full credit (but loss of land) and better return to labour, (iii) in 2024, interest in agroforestry practices remains high among old men but evidence for interest in agroforestry among members of the younger generation is lacking, (iv) now is the time to replant rubber because the trees are old, but the same problems persist: access to planting material is difficult, (v) there is still no way for farmers to learn good tapping practices (e.g. through specific training, access to technical information on panel management, upward tapping, etc.), which are essential to prolong the life span of the rubber trees to 35 years, (vi) the severe impact of white root and other root diseases in areas previously under forest or old jungle rubber, and finally (vii) low rubber prices especially compared to palm oil, all of which discourage farmers from cultivating rubber.

As mentioned above, due to the impact of diseases and poor tapping practices, most trial plots are now at the end of their life span. It is thus the ideal time to conduct an in-depth socio-economic survey of all SRAP farmers to assess the current situation in terms of farmers' income (from oil palm/rubber and all other sources), their current and planned long-term strategies, and to explore the reasons for their continued interest in clonal rubber and agroforestry systems. A historical and prospective analysis could assess the impact of oil palm and rubber price volatility. The survey could be implemented in the following villages: Kopar, Engkayu, Embaong and Pana in Dayak area, Trimulia and Pariban Baru in transmigration areas, as well as in Sanjan for former SRDP farmers and include up to 80 farmers.

Three major questions are obviously part of the research agenda:
– Under agroforestry systems, what is the impact of fruit production on food security and on the quality of the diet of local families?
– What is the impact of timber production, both for household use and for sale?
– To what extent can agroforestry systems provide better climatic resilience for both rubber and intercropped varieties?

Future research should include (i) a perception analysis of agroforestry practices as a way to reduce the cost of rubber establishment and provide more income diversification at farm level (for improved resilience to price volatility), and (ii) a study on existing markets (Durian, Gaharu, Duku, etc.) and newly emerging markets for associated trees in RAS (Pekawai, Petai, Jengkol, timber trees).

Chapter 3

RAS in the rubber world: current agroforestry practices in various countries

Éric Penot, Noé Biatry

▸▸ Origin and development of RAS

Before 1900, natural rubber was mainly produced by tapping *Hevea brasiliensis* rubber trees growing in natural forests in Amazonia, the homeland of rubber (Dean, 2002; Stanfield, 1998), and in West Africa, by exploiting common latex lianas, *Landolphia* spp. (Danthu et al., 2016). In Amazonia, rubber trees often grow at very low densities with only one or two tappable trees per hectare (Dean, 1987). In both cases, the practice is classified as "extractivism" from natural forests.

After rubber was introduced in South East Asia in the form of monoculture in large estates by colonial companies, it was rapidly and extensively adopted by local smallholders in the form of "jungle rubber", a diversified extensive system derived from swidden cultivation (Feintrenie and Levang, 2009; Gouyon et al., 1993). In the 1930s, some examples of smallholder intercropping experiments with coffee and tobacco were reported in Indonesia (Dickman, 1951) as well as some intercropping during the immature period (year 1 or 2), i.e., before natural forest regrowth makes intercropping impossible, resulting in jungle rubber. In the 1950s and 1960s in Malaysia and Thailand, and a little later, in the 1970s, in Indonesia, most farmers started using clonal rubber, originally as monoculture, because it was more productive than jungle rubber and because monoculture was the main technical practice promoted by extension services at the time. The next step was the development of rubber agroforestry systems based on clonal rubber (see chapter 2).

Intercropping with annual crops during the immature period boomed in Thailand and Indonesia between1970 and the 2000s with the expansion of clonal plantations. To a lesser extent, the same applied to associating rubber with tree crops during the mature period, which accounted for 10-15% of the cropped area in Thailand, for 40% in rubber development projects in Indonesia in the 1980s and 1990s (Chambon, 2001; Stroesser et al., 2018). On Hainan Island (China), in the 1970s and 1980s, intercropping was strongly promoted by the government during both the immature and mature periods to diversify the portfolio of agricultural products to reduce the impact of typhoon damage and hence to stabilise farmers' income (Langenberger, 2015, 2017). In contrast, in Thailand and Indonesia, agroforestry was not officially promoted but was developed by farmers themselves. It is clear that most RAS developed in this way, relying on the farmers' own experience and know-how.

In remote areas, crops including upland rice, maize, vegetables, peanuts and chilli peppers were primarily cultivated for home consumption. Where access to markets was good, for example in Sri Lanka and in Indonesia, cash crops like pepper (the spice, *Piper Nigrum*), banana and pineapple were grown in the immediate vicinity of big cities of Sumatra (Palembang, Pekanbaru, Medan) (Bagnall-Oakeley et al., 1996; Penot, 2001). In 2024, the associated crops used in RAS are market driven.

Smallholder use of cover crops has always been very limited, and when they are used, the aim is to prevent erosion or to control weeds. There are two major reasons why cover crops were not widely adopted by smallholders: (i) growing a non-productive plant was not considered favourably in the smallholder sector and (ii) successful management of cover crops is labour intensive.

Agroforestry during the mature stage of the rubber trees includes cash crops, e.g. tea (Sri Lanka), coffee, cocoa (with very limited success at normal planting density) but mainly fruit and timber trees (in Thailand and Indonesia) or spices (in India). Fruit/timber-based RAS have been developed in Columbia and Brazil with local specifically Amazonian fruits including pupuna, acai, castanhera, and copoadzu (*Theobroma grandiflorum*). See the list of Amazonian species in appendices (Table S3).

Some farmers began to cultivate perennial crops when the rubber trees were big enough to provide shade, i.e., after 5 or 6 years, and continued growing the crop until the end of the rubber cycle, either on their own initiative (Thailand) or after government promotion (Sri Lanka, mainly cacao).

In some countries, RAS underwent a significant decline in the 2000s due to a more market-oriented policy (for instance in China or in some places in Indonesia, e.g. Jambi province), competition with oil palm (Indonesia), and in other countries, due to the diversification strategies of local farmers. In China, promising intercropping species such as medicinal ginger *Alpinia oxyphylla*, were rapidly abandoned due to low market prices (Zeng et al., 2012), the same was true of another ginger plant (*Amomum villosum)* in Yunnan, despite the fact it was strongly promoted for years (Zhou, 1993).

Incorporating livestock (sheep, chickens or cattle), in rubber plantations was also tested in North Sumatra, but poorly adopted by growers (Ng et al., 1997; Payne, 1985; Shelton and Stur, 1991; Waidyunatha et al., 1982). Some local livestock-based RAS developed in Vietnam (Penot et al., 2022) and eastern Sri Lanka (Penot et al., 2023a).

⫸ The key impact of shade

When clone rubber is grown at the normal density, in normal conditions, and in the absence of leaf diseases, shade generally reaches between 80% and 90%. Some clones such as RRIM 600, which is widely used in Thailand, provide limited shade, i.e., between 60% and 80%. Leaf diseases can reduce shade by as much as 50% (South Sumatra, Indonesia) depending on local conditions.

The results of the many experiments conducted since the 1990s indicate that cacao and coffee do not produce good yields under more than 80% shade in addition to strong competition with rubber roots in the upper soil layer. According to Wood (1985b), experiments using rubber at normal planting density as shade trees for cacao failed economically: the cacao trees survived in the shade but produced no yield, a fact that was confirmed in the Sri Lanka. Wintgens and Descroix (2009) claimed that rubber "has never been a successful shade tree for coffee". We can make the same claim for cacao, at least at normal rubber planting density. Timber trees are usually more suitable, as most species (for instance *Dipterocarpaces* species) are shade tolerant and may even require shade in their young stage.

A number of trials and surveys have shown that most fruit trees produce between 30% and 60% of their normal yield in the full sun (Penot, 2001; Stroesser et al., 2018; Penot et al., 2022). In sub-optimal and marginal growing conditions, shade may benefit certain plant associations, for instance, it can prevent overheating, which is a major advantage in the context of global warming, and in some species including coffee, fruit quality may even be enhanced if the shade is not too dense (Chaudhuri et al., 2013; Descroix and Snoeck, 2009; Muschler, 2009; Wintgens, 2009). However, shade normally considerably reduces the yield of associated plants. The solution is finding the right balance that will enable associated plants to be productive under a certain degree of shade. The shade threshold can differ from one location to another and more research is needed on that particular point. Some rubber clones are more suited to agroforestry with a canopy considered as "light" as it provides on average 70% shade (such as RRIM 600). The expansion of leaf diseases in many areas, in particular in Indonesia, also creates better conditions for agroforestry.

An alternative way to deal with shade constraints is to use double-row spacing, with a 15-25 metre inter-row, to guarantee sufficient light for the associated crops. Both coffee and cocoa succeed in such systems and we explore that possibility in a dedicated paragraph.

❯❯ The situation in South and South East Asia

The two main countries with original or widely practiced agroforestry systems are Indonesia and Thailand (the latter was covered in chapter 2). Here we review other countries with rubber agroforestry systems: Indonesia, China, Malaysia, and Sri Lanka.

Indonesia

After having explored past experience with RAS in Indonesia in the 1990s, here we explore recent developments. In September 2022 and October 2023, two meetings were hold in Jambi and North Sumatra (GPSNR) on agroforestry practices gave us the opportunity to talk to local rubber smallholders about the current situation facing rubber, the serious impact of low rubber prices in the long run, the impact of the wide adoption of oil palm, and the need for income diversification, but also to discuss current agroforestry practices in these conditions. Oil palm has become the priority thanks its high margin/ha and excellent return to labour. In 2022, it is true that most farmers have already diversified with oil palm and do not need another way to diversify based on agroforestry with rubber. Agroforestry enabled the introduction of rubber in Indonesia through jungle rubber, and then through RAS in the 1980s and 1990s. But competition with oil palm and the long series of low rubber prices since 2012 seriously reduced interest in growing rubber and hence in any kind of agroforestry, particularly in Indonesia. Nevertheless, some farmers continue to practice agroforestry for economic or even social reasons. On the occasion of the GPSNR meeting, farmers who were still practising rubber monoculture, expressed their interest in the following aspects:

Fruit/timber systems based on the normal rubber planting density
– Local fruits such as Jengkol (*Archidendron pauciflorum*) and petai/stink bean (*Parkia speciosa*), which are easy to plant and grow (adapted to shade and already present in jungle rubber systems) and for which there are local markets and high demand for Jakarta. The same trend has been observed in West Kalimantan province in Borneo (Penot et al., 2019).
– "Alam" or local/traditional/robusta coffee grown from local seeds and sold on local markets, with also high demand from Jakarta as well as arabica coffee in North Sumatra.
– Mangosteen ("Manggis") which starts yielding four years after planting and produces a reasonable yield in the shade provided by rubber trees planted at their normal density.
– Local markets for sugar palm and lemongrass for distillation (North Sumatra).
– There is a strong demand for the development of associated trees nurseries to provide local farmers with high-quality plantlets of associated tree species.
– The main problems with grafted fruit trees like durian is finding good quality nurseries and the cost.
– There is a possibility of transition from the rubber-based system to durian-based system at the end of the rubber lifespan (currently between 25 and 35 years depending on tapping quality). According to the strategy used by local farmers, this is a good alternative following rubber. This idea is very similar to that of *tembawang*, the "fruit and timber tree" based agroforestry systems found in West Kalimantan.

Normal spacing vs. double spacing
– 3 x 7 m or 2.5 × 6 m is best for local fruit and timber species which can grow in the shade or the current situation of Indonesia with leaf diseases leading to a shade level of 70%.
– The choice depends on the farmers' strategies and level of intensification.
– There is a need for demonstration plots with different types of agroforestry systems based on demand from local markets.

RAS in Sumatra in 2023

A recent farming system survey of a sample of 100 farmers conducted by the Indonesian Rubber Research Institute (IRRI) as part of the Rubis project, reported that farmers still show a real interest in agroforestry practices with rubber in both immature and mature periods, particularly in South and North Sumatra provinces. In North Sumatra, 94% of farmers use RAS, 54% in South Sumatra and 41% in Jambi, confirming that RAS still interests most farmers despite the local interest in oil palm (source: Rubis project, Dr Dwi Shinta Agustina/IRRI, unpublished data, personal communication).

The boom in "Amorphophallus" in Indonesia

In Indonesia, we recently observed a boom in intercropping using "*Amorphophallus*": a tuberous plant that takes its name from its giant phallus-shaped inflorescence, and is called "porang" in Indonesia, named after the species *Amorphophallus muelleri*. It was ignored for a long time and then rediscovered, cultivated and massively exported for its glucomannan content, which is used for weight reduction. In 2020, Indonesia exported 32,000 tons of porang to Japan, China, Vietnam, and Australia, for a value of US$100,000. The plant is also widely used in industrial, food, and health products.

Agarwood

Agarwood oil, *or Gaharu*[61], often referred to as oud oil, eagle wood oil, or aloe wood oil is a resinous, fragrant high value heartwood. The essential oil is derived from heartwood infected by a fungus that produces a dark aromatic resin. The most popular species are *Aquilaria malaccensis, Aquilaria agallocha* and *Aquilaria crassna*. Agarwood oil is native to India and several countries in South East Asia including Vietnam, Philippines and Indonesia. Its extreme rarity in local forests makes it very expensive. In 2024, agarwood oil is obtained from cultivated trees and artificial infection and requires considerable investment per hectare. The essential oil is extracted by distillation in water and can be found in several quality grades that depend on the original grade of the wood and the length of the distillation period. Typically, the longer the distillation time, the higher the grade.

Agarwood oil is extensively used as a medicine (aid to digestion, to repair damaged skin, relieve allergies, treat insomnia and acne, relieve joint pain, and pain during and after birth, nausea and vomiting) as well as for spiritual purposes. The anticancer properties of agarwood oil have been investigated. It is also used in cosmetics, perfume and aromatherapy. The market for therapeutic and cosmetic applications is expected to increase in the very near future. Agarwood oil has long been used by Ayurvedic practitioners for spiritual and emotional applications. India, Indonesia, Singapore, Malaysia, and Sri Lanka are top exporters/manufacturers of agarwood oil followed by North America and Europe. The main importers ranked in decreasing order of importance are U.S.A., Germany, U.K., China, Saudi Arabia, and France.

China

Agroforestry has a long history in certain parts of China. According to Chen Yung's (1943) "History of Forestry in China and Administration of the Republic", agroforestry was already practised in Shanyang County 1,700 years ago. Records show that 300 years ago, some forest farmers planted agricultural intercrops under young plantations of Chinese timber that very closely resembles the Taungya system[62] (Sin et al., 1998) in Myanmar with teak. A civil officer, Zheng Hui, suggested planting a hedgerow of timber to overcome local timber shortages. The development of rubber in China had negative environmental effects due to extensive deforestation it caused. A controversy emerged on the sustainability of rubber farming in Xishuangbanna and other locations in South East Asia (Ziegler et al., 2009). Intercropping is one of the alternative ways of reaching both ecological and economic goals (Wu et al., 2001; Ziegler et al., 2009)

Hainan

Rubber with medicinal plants

Medicinal plants are very widely cultivated in China (Ghobarni et al., 2011).

Rubber plus *Amomum villosum* and *Clerodendranthus spicatus*. In the Xishuangbanna area of south-west China, intercropping rubber with medicinal herbs is regarded as

---

61. https://www.transparencymarketresearch.com

62. The Taunggya system was used in government teak plantations in Myanmar (still called Burma during the colonial era). The estate allowed local farmers to grow intercrops during immature period to reduce maintenance costs.

a promising way to reduce the negative hydrological effects of rubber monoculture and to improve the sustainability of old rubber plantations. Briefly, intercropping with shallow rooted medicinal herbs (*A. villosum and C. spicatus*) can help old rubber trees tap water in deeper soil layers, thereby hopefully helping the rubber tree and the intercrop create complementary water-absorbing patterns in the pronounced dry season. However, no real improvement has been observed to date, consequently intercropping rubber with medicinal herbs still needs more research, particularly soil water conservation and management.

Rubber plus Traditional Chinese Medicinal Models with *Alpinia/Oxyphylla* during the immature period. *Oxyphylla* c.f. and *A. villosum* (Zhou, 1993) belong to the ginger family (Zingiberaceae). This intercropping system was promoted in mountainous and hilly regions with fertile soils. Mountain slopes were terraced and rubber spaced 2 × 7 m or 2 × 8 m. Three years later, about 400 kg/ha of seeds were collected from each variety in the first harvest year and 400-500 kg/ha of the seeds were harvested in years 4 and 5 (Haishui and Kejun, 1991), which are reasonable yields.

*Rubber plus *Morinda officinalis* during the immature period. The environmental conditions needed to grow *M. officinalis* resemble those of *Alpinia/oxyphylla*. Cuttings of *M officinalis* were planted under rubber trees in three to five rows with 0.5 x 0.8 m spacing. A yield of 250-300 kg/ha was harvested five years after planting.

Rubber plus *Amomum longiligulare,* adapted to shade. The best time to intercrop *A. longiligulare* is three or four years after planting rubber. 0.6 × 0.7 m spacing was used in this example. *A. longiligulare* took up 50% of the inter-row. It began fruiting three or four years after planting, when rubber was almost ready for tapping. The annual yield of *A. longiligulare* was 80-120 kg/ha which seems relatively low.

As most of these trials appear to have been abandoned or seriously reduced in recent years, yields and prices were probably not sufficient to maintain farmers' interest in such cropping systems.

Rubber plus lemongrass during the immature period

Lemongrass cannot grow well in the shade, so it was planted in the inter-rows at the same time as rubber trees with 0.8 × 1.0 m or 0.5 × 0.7 m between the rows of rubber trees; and harvested five or six months after planting, then once every four or five months. At each harvest, 10-15 tons/ha of fresh leaves were collected and turned into 100-150 kg citronella oil.

Rubber with tea

In Hainan province, most rubber is intercropped with tea, which is recognised as an effective way to reduce soil erosion (Guo et al., 2006).

Guo et al. (2006) conducted an economic analysis of rubber monoculture and rubber intercropped with tea using a large state-owned farm in Hainan, China as a case study. Whilst 7 × 3 m spacing was the normal density used for rubber monoculture, in the intercropped system, 12 × 2 m spacing was used for the rubber trees to accommodate 14,400 tea plants/ha (with 1.6 × 0.3 m spacing between tea plants). The loss rate of the rubber plants was 15% in both systems due to typhoons and/or temperature changes. The production cycle of rubber monoculture was 33 years and that of rubber intercropped with tea 34 years, and the first tapping took place in the 8th and 7th year, respectively. Normally, tea can be harvested in year 2. Using an interest rate of 5.76% as discount rate and the company's own cost and price record, the authors calculated the net present value (NPV) of the two systems and found that the rubber-tea intercropping system was consistently more profitable than rubber monoculture, and that the optimal rotation age was 29 years for rubber monoculture and 26 years for rubber-tea intercropping system.

The same authors, Guo et al. (2006), also conducted some sensitivity tests using different discount rates and fluctuations in the prices of tea or rubber and found that rubber monoculture would only out-perform rubber-tea intercropping if the price of tea decreased by 30%. The rubber-tea combination was more profitable because tea is a high value secondary plant and grows well under 30–40% shade. Intercropping also benefited rubber growth so rubber tapping was able start a year earlier and end a year later than rubber monoculture.

Production and cost data were collected for another trial with rubber plus black tea (Guo et al., 2006). The tea species selected for the study, *Camellia sinensis var. assamica*, which was widely planted in the past, was the right species for the Black Powder tea sold on the Chinese market. Rubber-tea intercropping is more labour intensive for both rubber and tea and thus depends on local labour availability. Intercropping under rubber is not feasible at high altitudes since rubber trees can be damaged by the cold. Because markets have changed in the past decade, black tea has been largely replaced by green tea.

The environmental conditions of more than half the rubber plantations in Hainan are similar to conditions in the plantation used for the study by Guo et al. (2006). However, the annual yield of latex, as well as stumpage may not be as high in some plantations because of stronger typhoons, lower temperatures or poor topographical and soil conditions.

Rubber and Alocasia

The effects of interspecific competition for nutrients on calla lily growth, development, and nutrient uptake when calla lily (*Alocasia macrorrhizos L.* or songe Caraibe/Alocasia/giant taro) were intercropped with rubber, were recently investigated in rubber systems (Sun Li-juan et al., 2019). Calla lily grows in shady environments. It reproduces rapidly and its tubers have high starch and sugar content. Whole calla lily plants also have medicinal and ornamental applications. Rubber tree/calla lily intercropping is a new and promising intercropping system suitable for implementation in shaded mature rubber plantations where the main product of the calla lily is industrial starch.

China: Yunan

In Xishuangbanna, rubber is mainly grown as a monoculture (Liu et al., 2006), although intercropping was previously recommended (Wu et al., 2001; Ziegler et al., 2009). In a case study of smallholder rubber farmers in Daka village in Xishuangbanna, Fu et al. (2009) identified several intercrops in rubber plantations including upland rice, taro and pineapple grown at normal rubber planting density. Leshem et al. (2010) analysed rubber intercropping practices in Xishuangbanna based on interviews with 15 experts and in-depth interviews with 25 farmers in two villages. They found that, depending on altitude and on the choice of crop, intercropping had positive economic and ecologic effects, for example, rubber intercropped with tea reduced economic uncertainty and improved the income of farmers in high altitude areas.

The Xishuangbanna Tropical Botanical Garden also conducts trials incorporating different tree and shrub species. The search for economic improvement has also led to the identification of "miracle trees", such as eagle wood (Aquilaria spp.) which is promoted for short-term production of agarwood. Agarwood is a highly aromatic and highly priced hardwood produced in old trees as a result of a fungal infection and related wood tissue reactions (Persoon, 2007). Agarwood is used to make incense and perfume. It is currently well developed in Indonesia as detailed in the previous chapter, and the market appears to be promising.

Another trial was conducted with native (indigenous) tree species in mature rubber plantations at normal planting density (Langenberger et al., 2017). The criteria used to select the species were: (i) it must be adapted to the prevailing environmental conditions, (ii) must be shade tolerant, (iii) its vertical growth must not be affected by light, (iv) conservation value, (v) economic potential, (vi) easy to manage. Rubber trees are usually planted in rows with 6-8 m between rows, with 2.5 to 3 m between rubber trees. The following species were selected for the demonstration sites: *Parashorea chinensis*, a valuable timber tree; *Taxus mairei*, a multi-purpose tree, providing good timber but also an anti-cancer drug, taxol; and *Nyssa yunnanensis*, selected for its conservation value. At the end of the economic life span of the rubber trees (about 30 years) there will be several options, but the three main options are: (i) the rubber plantation can be replanted, although the harvest of the *Parashorea chinensis* trees would be premature, while the *Taxus mairei* trees could be maintained through a new plantation cycle, (ii) both the rubber and the intercropped trees could be maintained for future timber and taxol production, and (iii) the plantation could be transformed into a managed sustainable forest scheme.

In Xishuangbanna, a study used cross-section data on 600 rubber farmers in Xishuangbanna (Min et al., 2017b), as a basis to develop four empirical models to analyse the adoption of intercropping at farm and at plot level. The study showed that only a small proportion of rubber farmers have adopted intercropping, tea being the most frequent intercrop. However, other studies indicate that intercropping remains an important source of income for households in the lower income category. The adoption of intercropping is affected by ethnicity, household wealth and family labour. On average, intercrops contribute 16.5% of the total household income, suggesting that intercropping is an important source of income for smallholder rubber farmers.

Malaysia

Rubber and livestock

Integrating animals in rubber plantations was found to be more profitable than rubber monoculture by both Majid et al. (1990) and San and Deaton (1999). Based on data collected from 51 farms and FELDA[63], Majid et al. (1990) found that the system based

---

63. Federal Land Development Authority (FELDA) was established on July 1, 1956 in Malaysia under the Land Development Ordinance of 1956 for the development of land and relocation with the objective of poverty eradication through the cultivation of oil palm and rubber.

on 50 ewes and one ram grazing in a rubber plantation from year 3 to year 25 was more profitable than rubber monoculture and the net present value (NPV) was 11.9% higher than rubber monoculture with a pay-back period reduced by one year to 8–9 years. San and Deaton (1999) assessed the feasibility of integrating sheep and soybean in rubber plantations based on data collected from 85 farms participating in Nucleus Estate Smallholder Scheme (NES) Development projects. Using a linear programming model, they found that the optimal combination of rubber trees with 16 remaining productive years (i.e. 22 years old) and sheep for a smallholder rubber farmer was 593 rubber trees with eight years of annual soybean production. Adding soybean was more profitable than rubber monoculture or only including sheep in the rubber plantation. Although no details on rubber planting density and yield were provided, having sheep graze rubber plantations would benefit rubber growth both by adding nutrients to the soil and by reducing the cost of weeding. However, it should be noted that these studies were conducted in the 1990s and the results may no longer be valid in the current economic situation.

Rubber and timber

A study was conducted by Yahya et al. (2023) on associating plantation crops like oil palm (*Elaeis guineensis*) and rubber (*Hevea brasiliensis*) and timber species like sentang (*Azadirachta excelsa*), a species usually found in secondary forest, and teak (*Tectona grandis*). The study involved government and private agencies and 50 farmers who practiced this type of agroforestry. The most frequent type of combination was found to be rubber and sentang used by smallholder farmers to supplement the low income they obtained from rubber particularly during the development stages when productivity is limited and when rubber is being replanted.

Rubber agroforestry systems in India

Unfortunately, very few publications are available to illustrate agroforestry practices in India even though India probably has the highest rate of agroforestry in the world and 80-90% of rubber is probably grown in association with other plants. In fact, as most rubber plantations are located in areas with a relatively high population density (i.e. more than 500 inhabitants per ha) there is a need for integrated, more economically efficient and more intensive agroforestry practices.

A high proportion of home gardens in Kerala have been converted into rubber plantations (Kumar and Nair, 2004). The cropped area of most smallholdings is less than 0.5 ha which has led to intensification through a combination of agroforestry and cropping, an important feature of Indian smallholder rubber (Menon, 2002). Banana, pineapple, different vegetables, cassava, tuber crops, ginger and turmeric are the intercrops most commonly cultivated in young rubber plantations (Jessy et al., 2017) using a normal planting density (6.7 and 3.4 m spacing). The probability of adoption of intercropping was highest for three crops (Rajasekharan and Veeraputhran, 2002): banana (*Musa* spp.), cassava (*Manihot esculenta*) and pineapple (*Ananas comosus*).

Shade tolerant medicinal plants like *Strobilanthus haenianus* (Karimkurinji), *Adhatoda vasica* (Valiya Adalodakam) and *Plumbago rosea* (Chuvanna Koduveli) can also be grown in mature rubber plantations without adversely affecting rubber growth and yield. The local marketability of such crops should be checked before large-scale cultivation is launched. Siju et al. (2012) conducted a study in central Kerala that revealed the growing popularity of contract farming with pineapple as intercrop in the immature phase at normal planting density. The results of their analysis highlighted the growing divergence between the recommended agro-management practices, and those actually adopted, for example, pineapple under contract farming, and the potential challenges this implies.

Rubber associated with *Garcinia*, vanilla and medicinal plants

Three experiments were conducted by the Rubber Research Institute of India in the period 2001-2014 with the aim to produce additional income and to improve small-holder welfare by integrating diverse crops in rubber ecosystems (Jessy et al., 2017).

In one experiment, coffee, vanilla with *Gliricidia sepium* as support stands, *Garcinia* and nutmeg (*Myristica fragrans*) were cultivated along with rubber at normal planting density. Three experiments were implemented to test other crops. Experiment 1 was based on intercropping rubber (clone RRII 105) using coffee, *Garcinia* (445/ha), vanilla and nutmeg (175/ha). Experiment 2 evaluated nine shade tolerant medicinal plants intercropped with mature rubber: *Adathoda beddomei, Alpinia calcarata, Andrographis paniculata, Asparagus racemosus, Desmodium gangeticum, Piper longum, Pseudarthea viscida, Rauvolfia serpentina* and *Strobilanthes cuspidal.* Experiment 3 tested short cycle vegetables during the wintering period of mature rubber using cow pea and *Amaranthus.*

Light availability was measured and subsequently reduced to 88%, 53% and 4.8% after respectively, 3.5, 4.5 and 5.5 years. The rubber yield was between 1,000 and 1,300 kg/ha/year, which is rather low compared to normal standards. The growth of rubber was significantly improved with intercropping and rubber yield was comparable to local monoculture systems. Soil moisture status in summer and microbial populations were higher in mixed planting systems and soil nutrient status remained stable. Yields of all the intercrops were good up to year 4. As the shade provided by rubber became more dense, *Garcinia* perished but vanilla and coffee continued to produce reasonably good yields. All the medicinal plants established well and produced reasonable biomass, but among the nine, the performances of *Strobilanthes cuspida* and *Alpinia calcarata* were comparatively better. Experiment 2 showed that short duration vegetables like *Amaranthus* and salad cucumber can be cultivated during the annual leaf shedding period in mature rubber plantations to meet part of domestic needs. The yields of coffee, vanilla and *Garcinia* obtained in experiment 1 are listed in Table 3.1 below.

Table 3.1. Yields of crops associated with rubber in experiment 1

Cropping year	2005	2006	2007	2008	2009	2010	2011
Yield of coffee as intercrop kg/ha	890	596	347	645	236	616	346
Yield of fresh vanilla beans/pl	0.42	0.36	0.5	0.38	0.28	0	0
Yield of *Garcinia* as intercrop kg/ha	13	35	394	503	521	245	89

Rubber and elephant foot yam

Trials were conducted in 2012/2013 at the Regional Centre of ICAR-Central Tuber Crops Research Institute, Dumuduma, Bhubaneswar, Odisha (India) with elephant foot yam (*Amorphophallus paeoniifolius* Dennst.), which is shade tolerant and can consequently be cultivated during the rubber mature period (Jata et al., 2018). The aim of the experiment was to study the efficiency of cropping systems with different treatments (irrigation, nutrient use efficiency and quality of elephant foot yam). The experiment used a split plot design with elephant foot yam plus green gram (*Vigna radiata* L.); elephant foot yam as the sole crop in the main plots with surface irrigation, drip irrigation at 100% cumulative pan evaporation (CPE), drip irrigation at 80% CPE, and drip irrigation at 60% CPE in the sub-plots. Five replicates were made of each treatment. Elephant foot yam plus green gram produced a larger yield of corms than the other treatments. Drip irrigation at 100%, 80% and 60% CPE resulted in respectively, 16.7% and 14.9%, 16.4% and 14.6%, 12.3% and 11.5% higher yield than surface irrigation. The highest nutrient use efficiency was obtained by intercropping elephant foot yam plus green gram with drip irrigation at 100% CPE followed by drip irrigation at 80% CPE. The combination of elephant foot yam plus green gram produced higher concentrations of protein and sugar in the corms than sole cropping. Optimum yields of elephant foot yam were obtained by intercropping elephant foot yam plus green gram with drip irrigation at 80% CPE, with an average yield of elephant yam of 30 t/ha.

Other crop associations

Intercropping a variety of crops before the rubber plantation reaches maturity provides the farmers with additional income. In a rubber-based multi-strata system, when the plantation reaches maturity, the canopy will be closed. Apart from food crops, medicinal and aromatic plants, for instance, medicinal yam (*Dioscorea loribunda*) can also be intercropped with rubber (Singh et al., 2021). Rubber trees can be associated with food and beverage crops including maize, cassava, sweet potato, coffee, tea, pepper, and lemon grass, and with fruit crops like banana and pineapple (Zheng and He, 1991). Combinations of crops grown under the rubber trees at different stages of development enable optimal use of land and water resources (Jiang et al., 2020) and also results in higher soil organic matter content (Xiao et al., 2019).

Fruit tree species associated with rubber are mango, sapota (*Pouteria sapota*), guava, citrus, custard apple, areca nut tree, banana, papaya, coconut, and drum stick.

Intercropping helps maintain the nutrient and moisture balance on the farm, provides a permanent ground cover, and helps increase farm yields while simultaneously enhancing soil productivity and preventing erosion. The selection of crops for intercropping will vary depending on regional preferences, climate conditions and market mechanisms.

Depending on local climate conditions, some of the tried and tested rubber-based agroforestry systems recommended in India (Kropf and Hoang (2012) cited in Singh (2018); reported in Table 3.2):
– jackfruit, acacia or mahogany can be grown as windbreaks for rubber trees;
– if the farm is located on marginal soils, elephant grass can be associated with rubber to provide fodder for cattle, buffaloes, goats and rabbits;

– food crops including peanuts, beans, sweet potatoes, taro, maize, or cassava can be associated with rubber in the first four years after rubber trees are planted;
– growing cassava with rubber in the first three years after rubber establishment can yield 14–18 tons of root tubers per hectare per year. No information was available on potential impacts of combining cassava and rubber on rubber growth or diseases;
– coffee and pepper can be intercropped with rubber trees in the early years of the rubber plantation.

Table 3.2. Comparison of contribution of rubber based agroforestry systems to income diversification of the community

Type of RAS	Tripura India		Assam India		Meghalaya India		Songhla Thailand	
	Income	Rank	Income	Rank	Income	Rank	Income	Rank
Monoculture	54,292	7	44,427	7	45,519	7	29,027	7
Rubber + fruit + agriculture	57,057	5	47,672	5	49,837	4	44,811	1
Rubber + poultry	55,715	6	45,807	6	46,764	6	31,314	6
Rubber + livestock	60,325	1	50,288	1	21,316	2	42,948	2
Rubber + rice	58,080	4	49,412	3	49,595	5	32,775	5
Rubber + fishery	58,466	3	47,733	4	51,502	1	40,476	3
Rubber + piggery	59,398	2	50,193	2	51,030	3	37,187	4

Source: extracted from Vishwanathan, 2008. The currency is Indian Rupiah.

Sri Lanka

The impact of shade is one of the main factors to take into account along with suffi-cient water for both rubber and the associated crop(s). Sankalpa et al. (2020), Penot and Ilahang (2023), and Chambon and Wibawa. (2023) identified the most common inter-crops during the immature period to be peanut, maize, cowpea, chili and banana, and the most common intercrops during the mature period to be dairy cattle, pepper, cacao, Ceylon cinnamon (*Cinamomum* sp.*), passion fruit (*Passiflora)*, *Citronella* sp. (known locally as "pangiri"), and banana. Rodrigo et al. (2001) estimated that 25–50% of farmers in the traditional rubber growing areas in eastern Ski Lanka practice rubber-based agro-forestry mainly combined with tea, banana, cinnamon, pepper, and pineapple during the mature period. Harvesting cassava or peanuts alongside clearing or burning weeds increases the risk of soil erosion whereas growing *Citronella* sp. reduces soil erosion.

Banana is the most widely reported intercrop (Rodrigo et al., 2001; Rodrigo et al., 2001) in Sri Lanka in both the established eastern rubber growing areas and in Moneragala/Ampara districts, which are new rubber growing areas in the west (Iqbal et al., 2006; Rodrigo, 2001) at normal rubber planting density (3 × 7 m or 2.5 × 8 m). The original planting design recommended in Sri Lanka for intercropping with banana was a single row of banana planted between rows of rubber trees planted at normal density (Rodrigo, 1997) to reduce competition for rubber as far as possible. In practice, it has been shown that banana planting density can be increased three-fold with no detrimental effect on rubber growth (Rodrigo et al., 1995; Senevirathna et al., 2010).

Sugarcane is a traditional high value crop in the dry and intermediate zones of Sri Lanka, particularly in Uva province where the market has been good for many years (Moneragala/Ampara districts). Sugarcane is grown under contract, i.e., local sugarcane companies provide inputs and contract to purchase the harvest. Intercropping rubber with sugarcane with rubber at the normal planting density in the first 5 years of rubber tree growth has been shown to improve the growth of rubber during the first 6 years as well as to protect the young rubber plants from drying out, which was confirmed by all the farmers in the area. Both the trunk girth and the height of the rubber plants were comparable with those in the traditional eastern humid zone (Rodrigo et al., 1995; Rodrigo et al., 2001).

Pueraria phaseoloides and *Calopogonium muconoides* are the most commonly grown cover crops. Farmers do not distinguish between the two species and locally refer to them both as "awarana waga" or "pohorawel". Cover crops are apparently poorly adopted by local farmers.

The crops most often associated with rubber during the mature rubber period in the Moneragala/Ampara areas are cocoa, pepper and some timber trees at normal rubber planting density (Penot et al., 2023; Iqbal et al., 2006) and tea and cinnamon with double rubber spacing in traditional western rubber areas.

The association of rubber and tea with double spacing is quite common in the traditional western rubber growing area where tea is generally considered as moderately shade tolerant (Hajra and Kumar, 1999), but most of the tea which is grown as an associated crop is grown under full sun in systems with wide rubber inter-rows of between 18 and 25 m (Iqbal et al., 2016; Penot, 2004; Rodrigo et al., 2001).

In the eastern region, pepper is planted close to the rubber trees and uses the trunk of the rubber tree as a support, but according to recent survey conducted in 2023 with Laura Guilonnet (published in Penot et al., 2023), pepper is no longer popular because of severe local diseases particularly in Moneralaga, and because the local variety is not suitable for the local humid climate.

Some fruit and spice trees (soursop, mango, citrus, cinnamon) were tested with rubber in some demonstration plots at RRISL (Rubber Research Institute of Sri Lanka) but they were rarely adopted by the farmers as RRISL and DRD (Development of Rubber Division) had previously focussed on associating cocoa with rubber at normal rubber planting density. In practice, associating cacao with rubber at normal planting density does not provide good results because the shade provided by the rubber trees is far too dense after year 5. Timber trees, harvested before they are 10 years old, have been successfully associated with rubber in Uva province.

Countries where there has been no RAS in the mature period

Laos and Philippines have no experience of RAS as such, although some farmers may practice intercropping during the immature period. Cambodia and Myanmar have only used RAS in some trials conducted by researchers. In these countries RAS are not used either by smallholders or by estates.

In one province of Vietnam, there is some limited association of livestock with rubber in the immature period (Penot et al., 2022, FTA).

Cambodia

The rubber industry in Cambodia is currently expanding, with an increase in both smallholder plantations and private estates through "Economic Land Concessions" (ELCs), with almost 45% of plantations now owned by smallholders. This intensive development of rubber is a threat to the environment as most new plantations are created at the expense of local forests and cause soil degradation. ELCs and land ownership by smallholders are the main causes of deforestation in Cambodia (Diepart and Dupuis, 2012). Several tree species, timber and other species are under threat because of illegal logging and loss of forests. Ex-situ conservation of tree seeds began with Danida and Forestry Administration in the Cambodian Tree Seeds Project (CTSP, 2001-2006) with 21 species selected for planting in Kbal Chhay, Sihanoukville (Strange et al., 2007). The overall aim of the 2005/2008 study by Cirad and the Cambodian Rubber Research Institute (CRRI) was to reconcile associations of rubber and forest tree species and to expand the inclusion of timber trees in agroforestry. The specific goal of the study was to assess the long-term behaviour of simple associations of rubber and timber trees using double-row spacing but also to create a legal market for timber trees to compete with the illegal logging sector and combat deforestation, land grabbing, and soil erosion. The trial included rubber associated with 14 different tree species selected for their different characteristics and including high-value/endangered trees (particularly over-logged high-value species), well-known fast-growing trees in plantations plus trees that provide non-timber products. These kinds of plantations represent a long-term investment because many years pass before timber can be logged, and possible loss of rubber due to competition between species for water. High potential margins can be expected from timber trees, for instance, for *Dalbergia* sp., whose sale price can reach US$30,000/m3. The possibility of harvesting other non-timber products from the trees is also currently under study. Tree species that could be sources of complementary income are (i) *Dipterocarpus alatus* (oleoresin), but this species requires a long time before resin can be collected, (ii) *Moringa oleifera* (whose leaves and fruit are consumed as food supplements), and (iii) seeds of all the high-value trees that are collected to be sold. In 2023, the study is still underway on a 5.3 ha plot of the CRRI plantation at Chup (11° 57' N 105° 34' E), Kompong Cham Province, a historic rubber production centre in Cambodia. Maize was planted along with *Stylosantes guyanensis* in the first year after rubber was planted. *S. guyanensis* continued to be used as a regularly rolled cover crop from year 2 to year 5. Trees associated with rubber were planted in a 6.5 m × 6.5 m × 3 m pattern, while the rubber trees (clone GT 1) were planted using large spacing (16 m × 2.25 m) or rubber double-row spacing (14 m × 2 m × 2.5 m).

The tree species associated with rubber are (i) *Tectona grandis, Hopea odorata* using a simple (large) spacing pattern, and (ii) *Dipterocarpus alatus, Tectona grandis* using a double-spacing pattern. Other species used are *Adenanthera pavonine, Afzelia xylocarpa, Albizia lebbeck, Acacia auriculoformis, Coluta laccifera, Moringa oleifera, Sindora siamensis, troemia callyculata, Dalbergia bariensis, Pterocarpus macrocarpus, Dalbergia cochinchinensis* and *Xylia dolabriformis*.

A marked reduction was observed in the growth of rubber associated with *Acacia auriculoformis*, whereas when associated with teak plus a cover crop of *Stylosantes guyanensis*, rubber growth was only slightly affected. The observed effects of the growth of timber trees on young rubber trees differed: (i) small trees such as *Afzelia xylocarpa* and *Dalbergia bariensis*, had only slight effects on rubber tree growth, (ii) small trees such as *Hopea odorata*, had only slight effects on rubber growth, but their speed of growth increased when the rubber trees start to provide shade, (iii) medium size trees such as *Pavonina adenantera, Albizia lebbeck, Moringa oleifera, Dipterocarpus alatus, Sindora siamensis, Pterocarpus macrocarpus* had moderate effects on rubber growth, and (iv) fast growing trees, such as *Acacia auriculoformis*, had a major effect on rubber tree growth.

In the specific climatic conditions in Cambodia, where the "dry season" lasts 3 to 5 months, rubber production could be affected by the type of associated species. The length of the rotation remains an open question, since timber trees may have a longer lifespan than rubber. The economic perspective underlines (i) the extremely high performances of medium-term associations with high value timber species such as *Dalbergia* sp., which last 30 to 40 years, and (ii) the cost of leaving standing trees at the end of a concession. This possibility makes timber and rubber associations an attractive tool for afforestation, which could also be promoted by enforcement of a new law on ELCs allocations, requiring a minimum forest cover at the end of an ELC.

This trial has two probable main outcomes: (i) the need to carefully consider the planting design of rubber/timber trees in order to limit the risk of competition for water (and light) in a context of an increasingly constraining dry season under climate change, (ii) the possibility of creating a significant source of income at the end of the rubber lifespan to provide the capital required for replanting.

Myanmar

Some trials have been conducted in local research governmental stations. The rubber trees used in these trials are generally still young, i.e., in their immature period. The trials were designed as demonstration plots with the following combinations:
– coffee using normal and double spacing. In 2022, the coffee was not yet ready to be harvested;
– durian, banana and pineapple using double spacing;
– amarante (*Amarantgus* spp.), long beans, maize and food crops using normal spacing.

Conducting trials that include demonstration plots is a very promising way to convince local farmers of the possibility of developing RAS in Myanmar in the long run.

A study of rubber production in Tanintharyi region (Vagneron et al., 2017) reported some intercropping with rice when rubber was planted, followed in years 2 to 5 by cassava and pineapple. More rarely, the authors found areca nut (*Areca catechu*) and durian (*Durio zibethus*), during the mature period included in extensive agroforestry systems in the villages of Pyin Thar Taw and Thaung Thon Lon. In the moderately intensive systems in Thet Kal Kwat village, most farmers intercropped rubber with perennial cash crops such as areca and cashew nuts, with timber trees such as agar wood, and with fruits such as rambutan, banana and pineapple. Common intercrops in the intensive system in Pa Kar Yi village included rice at planting, followed by cassava and pineapple, or more rarely, by areca nut and durian.

Vietnam

Some preliminary trials were conducted in the 1990s in the area of Pleiku with rubber and tea but no monitoring was done. In the 2000s, a trial was conducted at the Rubber Research Institute of Vietnam (RRIV) in southern Vietnam using double spacing with 2 lines of rubber with 2.5 m between rubber trees × 4 m (between rubber rows) with an inter-row of 16 metres and one row of timber-rubber trees in between. The term timber-rubber tree refers to rubber trees selected for their better growth and capacity to also be used as timber at the end of their lifespan. No published data are available concerning that trial. Similar trials were conducted in Cambodia with other timber species.

▸▸ The situation in Africa

Like in South East Asia, in Africa, rubber was introduced by colonial private estates primarily in the form of rubber monoculture. Smallholder rubber production developed later in the 1950s also based on monoculture, which was considered to be the best system for rubber. Farmers did not develop their own local agroforestry practices because extension services usually only permitted monoculture.

Côte d'Ivoire

Preliminary trials in the 1990s

The RAS trials in the 1990s at the National Rubber Research Centre of Côte d'Ivoire, (*Centre national de recherche Hévéa de la Côte d'Ivoire, CNRH*; Snoeck et al., 2013) included a comparative study of rubber as a monocrop with four rubber-based possibilities for diversification: rubber intercropped with coffee or cacao or cola or lemon, for which real data from a 17-year field trial in south-western Côte d'Ivoire were used. Rubber tree density for rubber monoculture was 510 trees/ha whilst that used in the intercropped system was 420 trees/ha. The density of the intercrops was 682 trees/ha for coffee and cacao (small shade tolerant trees), 55 trees/ha for cola and lemon (the trees are bigger and need more sunlight). Rubber tapping started in the 7th year. When Snoeck et al. (2013) calculated year-on-year cumulative return, defined as the sum of each year's gross margin, they found that, from year 3 to year 12, three diversified systems (rubber intercropped with coffee, cacao and cola) were statistically significantly more profitable than rubber monoculture from year 3 to year 12. The biggest difference was in year 10 when the cumulative return of rubber-coffee, rubber-cacao and rubber-cola was respectively, 98%, 65% and 22% higher than that of rubber monoculture.

From the 11th year on, the difference between the intercropping systems and rubber monoculture became progressively smaller and from the 13th year on, the rubber-cola combination became less profitable, but the rubber-coffee and rubber-cacao combinations remained more profitable using the double-spacing system than rubber monoculture although the difference was small. Intercropping rubber with lemon was less profitable from the 8th year on due to the sharp drop in lemon yield starting in year 6. Cola can be harvested between year 7 and 13 with peak yield occurring in year 10. This study (like other studies) showed that rubber latex yield actually benefited from the presence of intercrops, as the yield per tree was slightly higher than the yield obtained with rubber monoculture. However, the 17.6% reduction in rubber trees in the diversified systems needs to be offset by sufficient profit from secondary products. In the same RAS study, it was found that the growth of lemon and cola (both species need sunlight) was most adversely affected by shade from rubber and stopped producing in year 13, whereas coffee and cacao continued producing until year 17 although their yields peaked in year 7 and year 8, respectively. This explains why cola and lemon were less profitable than rubber monoculture, rubber-coffee or rubber-cacao combinations. However, it is possible that, if one considers the complete lifespan of rubber, the cumulative return of rubber monoculture is higher than intercropping with coffee and cacao.

Keli et al. (2005) studied both short and long-term intercropping. Data from 20 years of experimentation showed that intercropping food and industrial crops in the first three or four years after rubber planting was profitable. Food crops including upland rice, yam, groundnut, plantain banana, maize, and industrial crops like coffee, cocoa, oil palm, pineapple, coca and lemon trees, were found to be suitable intercrops. Rubber tree growth was improved by 29% under intercropping compared with growth in the control plot (rubber with a legume cover crop). The yields of the intercrops were moderate, but still profitable for the farmers. Some rubber-based crop rotation systems were thus recommended to small-scale farmers. The recommended three-year successions were yam – rice/peanut – maize/peanut, yam – rice/peanut – plantain, yam – yam – yam, plantain – plantain – peanut. The results confirmed the beneficial effect of food crops, in particular rice, peanut and plantain, on the growth of rubber trees, compared to the standard control rubber tree-*Pueraria phaseoloids* legume covercrop (Keli and De La Serve, 1988; Keli et al., 2005) Concerning light and shade, in some trials, measurements of the width of the crown of rubber trees indicated that five years after the rubber trees were established in permanent associations, 29% of the initial area reserved for crops (16 m) was shaded by the rubber tree canopy with the clone GT 1, while 38% was shaded by the PB 260 clone (Kouadio et al., 2021).

In this trial, the yields of associated crops were (i) low to mediocre for peanut (200 to 600 kg/ha) and yam (200 to 5000 kg/ha), (ii) average for maize (2,000 kg/ha) and (iii) satisfactory for plantain (7 to 21 t/ha), rice (800 to 3000 kg/ha) and cassava (10 to 30 t/ha). Then, after 6 years, the yields of both temporary and permanent crops grown in association with rubber dropped significantly. The authors concluded that increasing the width of the line spacing (16 m) was not sufficient to maintain the yields at a level that justify the cultivation of food crops in association with rubber (Kouadio et al., 2021).

Example of modelling rubber agroforestry

In Côte d'Ivoire in 2022, deforestation was almost complete. The role of cocoa as the main driver of deforestation and consumer of the "forest rent"[64] (Ruf, 1992) since the 1980s is well established. Surveys conducted since the 2000s found that rubber trees have mainly been planted on low-lying shrubby or fallow land, on land previously used for food crops, or in old cocoa and coffee plantations. The main constraint is

---

64. The forest rent is based on the fact that a new cocoa plantation will profit from net positive forest advantages in terms of soil structure and fertility and no diseases.

replanting cocoa once the forest rent disappeared, in other words once there is no more forest. Most rubber smallholders were cocoa farmers whose aim was to diversify by planting rubber.

Production of rubber by smallholders in Côte d'Ivoire proved to be enormously successful, approaching one million tons/year, with a boom in the last 15 years. A valid question is thus how diversifying to rubber affected the management of the "cocoa farms" and the changes this involved for their families.

As the economic results of growing rubber depend on rubber prices and the price of rubber has been very low since 2012, in December 2019, the authors conducted a survey of farming systems used by 150 cocoa/rubber smallholders in all the cocoa producing regions of Côte d'Ivoire, from Abengourou in the east to San Pedro in the west, with a focus on smallholders who lived in villages. Using the data gathered during this survey, several scenarios were designed to explore possible ways of improving the prevailing situation, including the adoption of rubber agroforestry (Biatry, 2021) through diversification with the aim of increasing the farmers' income. One limitation of the scenarios is the fact that it was only possible to interview a few local "absentee owners", especially the local elite, such as high-ranking civil servants, or lawyers.

The preliminary result of the survey confirmed that a major driver of diversification was the difficulty involved in replanting cacao on old cocoa cacao. Another result was that 25/35 years of rubber would make it possible to replant cocoa in better conditions than replanting cacao without a break. In this sense, choosing rubber can be interpreted not only as a mainstream strategy of income diversification but also a long-term strategy to make it possible to replant cacao in better conditions. In retrospect, if cocoa farmers had diversified with rubber, a "rubber cocoa complex" would have replaced the old "coffee-cocoa complex" that prevailed in recent decades.

Our scenarios show that diversifying as a way of facing the volatility of the price of natural rubber is the main advantage of the "RAS fruit tree" scenario, primarily aimed at improving sustainability. In 2023, the rubber industry in Côte d'Ivoire is facing several challenges: (i) the need for accurate estimation of the quantity of raw rubber material to be processed by local factories (because part of the rubber yield is exported as raw material), (ii) guaranteed access to Apromac[65] prices (the official price paid by local rubber factories), and (iii) promotion of further diversification through agroforestry practices in order to increase the gross margin of rubber plots and to maintain local farmers' interest in rubber despite lower prices.

Eight model farms representative of our sample of 150 farmers were therefore created together with their economic models using *Olympe* software (Penot, 2007). Different scenarios were tested to evaluate the impacts of price fluctuations, of changes in the type of labour used (family or hired), and of the implementation of rubber-based agroforestry systems. The results of the scenarios showed that the agronomic and economic performances of the farms varied, but that they still had many reservoirs of productivity. Payment at the official price, the use of family labour, and agroforestry systems are strategies farmers could adopt to stabilise and improve their agricultural income. The survey revealed that farmers frequently intercropped with food crops

---

65. Apromac : Association des Professionnels du Caoutchouc Naturel de Côte d'Ivoire.

(cassava, yam, peanuts, plantain, maize) in the first 1 to 4 years after establishment of rubber trees, which were planted at a normal density of 6 × 3 m.

Several model hypotheses were formulated involving different levels of certainty: (i) rubber cultivation occupies an increasingly essential place in farmers' income in Côte d'Ivoire and is a viable alternative to cocoa monoculture, (ii) the rubber marketing channel is one of the key factors that influences farmers' agricultural income, and (iii) rubber-based agroforestry systems are more profitable than monoculture and have positive environmental externalities.

Here we use the A2 farm type to illustrate the impact of all the scenarios. The A2 farm is representative of local farms in the area (see Biatry, 2021).

The RAS scenario with fruit trees

This scenario is based on the association of rubber and fruit trees:
– orange trees: 40 trees/ha producing a yield of 1,250 kg sold at 115 FCFA/kg (1 $US is 600 FCFA);
– cola trees: 30 trees/ha producing a yield of 300 kg sold at 350 FCFA/kg;
– mango trees: 30 trees/ha producing a yield of 900 kg sold at 110 FCFA/kg;
– avocado trees: 40 trees/ha producing a yield of 1,000 kg sold at 95 FCFA/kg.

To simulate the effect of competition with rubber for light, the fruit tree yields were reduced by 50% compared to those reported in the literature. Despite the relatively high production costs due to the use of inputs and hired labour, the adoption of RAS has a positive impact on agricultural income (Figure 3.1). Fruit trees represent a new source of diversification for rubber/cocoa planters and allow them to reach the threshold of 2,500,000 FCFA net agricultural income.

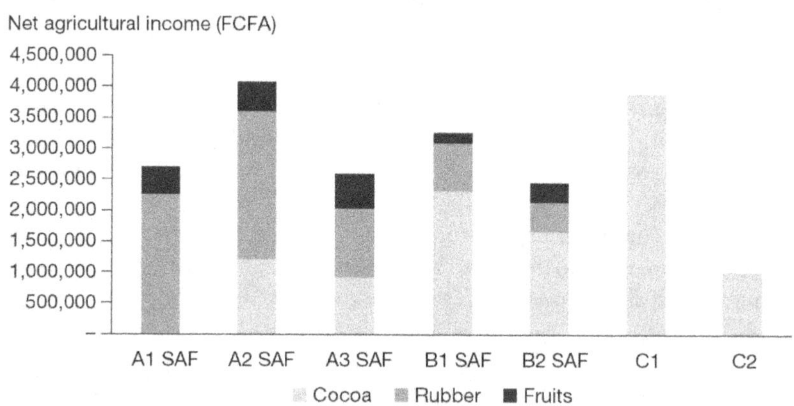

Figure 3.1. Net agricultural income from the different RAS cropping systems

The sale of fruits (oranges, mangoes, avocados, cola nuts) provides an extra gross margin/ha of 154,000 FCFA per hectare, thereby increasing the gross margin of the farm. The proportion of income represented by each type of fruit ranges from 21% to 33%: 33% for oranges, 24% for cola nuts, 22% for mangoes and 21% for avocados. The sub-types of farms (A3 and B2) which sell their raw rubber at the prices offered by "collectors" have a relatively low gross margin per hectare, and adding fruit trees increases their margin by respectively, 52% and 68% (Table 3.3). However, it is certain that the farm types based on rubber trees, and which are already present in the official Apromac sales channels (A1 and A2) are those most likely to be established (Table 3.3).

Table 3.3 Comparative analysis of the economic performance of each type of RAS farm

Farm Type	GM/ha rubber monoculture	GM/ha fruit based RAS	Difference (%)	Current agricultural income	Agricultural income Fruit based RAS	Difference (%)
A1	775,840	929,840	19.8	2,249,936	2,696,536	19.8
A2	768,085	922,085	20.0	3,596,064	4,073,464	13.3
A3	298,900	452,900	51.5	2,021,931	2,591,731	28.2
B1	642,830	796,830	24.0	3,086,910	3,271,710	6.0
B2	227,800	381,800	67.6	2,133,009	2,456,409	15.2

From the rubber tree's point of view, agroforestry is a "neutral" crop association: unlike agroforestry systems based on cacao, where shade can sometimes reduce the production potential of the cacao tree, rubber is not penalised by the presence of fruit trees, as when a maximum of 250 trees are associated with rubber per ha, it continues to dominate the canopy. In fact, it is the yield of the fruit trees that is negatively affected by the shade provided by the rubber trees compared with the yields of fruit trees growing in full sun.

The usual RAS scenario is based on the following items:
– rubber with 1 associated tree species (fruit tree, timber tree, coffee, cacoa) which produces a yield during the tapping period of the rubber tree (Snoeck et al., 2013). In this case, rubber is planted at the normal density (6 × 3 m);
– annual food crops can be planted between the rows of young rubber trees during the immature period, i.e., during the first 3 or 4 years.

The choice of associated crop focusses on the "best-bet" alternative between rubber trees and perennial crops. This is how the "AF/fruit tree" scenario was chosen, with fruit trees in the inter-row of rubber trees, with rubber planted at normal density. For this scenario, the species of fruit trees were selected because in a previous study on agroforestry practices in Côte d'Ivoire (Sanial, 2018), they had been observed growing in the cocoa plots. Cola, orange, mango and avocado trees were selected in addition because there was already a market (even if limited) for these products. Data on the yields and costs of the crops were taken from the *Agronomist's memento* (Cirad and GRET, 2006). To account for the effect of the shade provided by the rubber canopy on the fruit tree, yields of fruit were reduced by 50% compared to yields obtained in a conventional plantation.

A variant of this scenario was created that incorporates teak timber trees in the plantation. Teak requires little maintenance, and has the advantage that it can be exploited when the rubber trees themselves reach the end of their lifespan. The data were taken from a teak planting guide published by the National Agronomic Research Centre (CNRA) of Côte d'Ivoire (N'guessan et al., 2023).

The hypotheses in this agroforestry scenario are agronomic, economic and social:
– the inclusion of other trees has no impact on rubber yield as long as the density of other trees is less than 200 trees/ha and there is no canopy above that of rubber

(Penot, 2001). The neutral effect on rubber tree yield has been reported in other countries (Indonesia, Thailand) but remains to be confirmed in the case of the fruit trees selected in the scenario described here;
– structural market conditions may change in response to the supply of fruit. With the population of Côte d'Ivoire predicted to double by 2050, and a strong trend towards urbanisation, the market for fruit will probably increase;
– assimilation of the techniques used by planters and tappers in agroforestry cropping systems. In the past, companies did not recommend associating other crops with rubber, consequently industry players will need to play an active role in setting up this type of system.

Case study based on the A2 farm

The reason we chose to use the study of the A2 planter as an example was because it is the most "representative and balanced" farm, in the sense that the areas under rubber and cacao are almost identical. Above all, it represents a possible future for a large number of farms, with agricultural income primarily coming from rubber trees (raw rubber sold at the official price), and cacao representing diversification to face the risk of volatile rubber prices. This is the type of farm on which the most scenarios could plausibly be tested, with different variations (Figure 3.2).

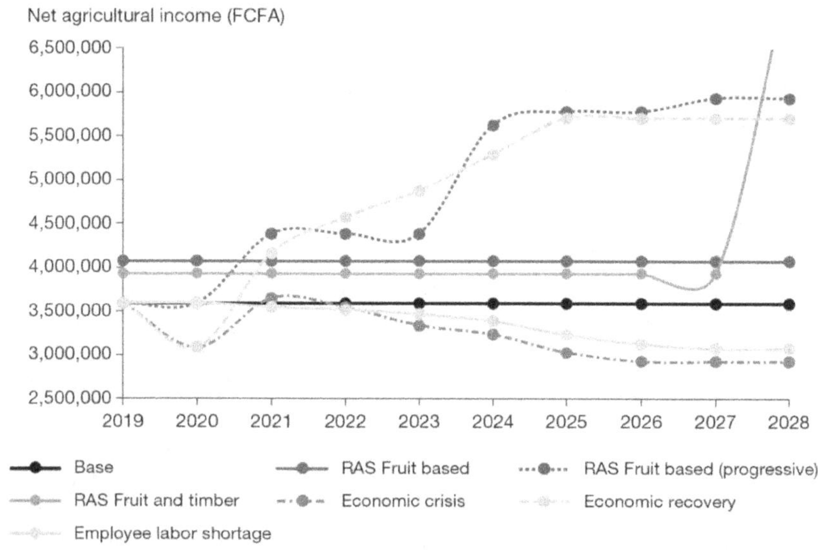

Figure 3.2. An example of one type of A2 farm: net agricultural income varies with the RAS scenario tested

The diversification of farm income for better resilience against fluctuating prices for natural rubber is the main asset of the "RAS Fruit tree" scenario, and this type of system could improve the sustainability of the rubber value chain in Côte d'Ivoire. The impact of the lack of a structured rubber industry is highlighted by the "collectors" price offered by sales intermediaries, which is the main factor explaining the different economic performances observed. This could happen again if rubber prices rise again and trigger a wave of new rubber plantations.

In 2024, the Ivorian rubber industry faces new challenges, in particular, (i) the real accurate estimation of the quantities of raw rubber material in order to adapt the potential quantity that could be processed by local factories, (ii) the guarantee all producers are going to be paid with the Apromac price, and (iii) promotion of more diversification through agroforestry practices.

Ghana

Many Ghanaian smallholders intercrop young rubber trees with shorter duration cash crops to increase their income (Tetteh et al., 2019; Malézieux et al., 2009; Langenberger et al., 2017). A promising species is *Thaumatococcus daniellii* which provides an extremely efficient sweetener with zero calories. It has been tested in African rubber plantations in Ghana with good results (Waliszewski, 2010).

In the 2010s, a trial was conducted with plantain banana at two different locations in the Western Region of Ghana. The first trial was conducted at the location of the Council for Scientific and Industrial Research-Crops Research Institute (CSIR-CRI) in Ellembelle district. The second on-farm farmer participatory trial, Tikobo No. 2 – Ehiamadwen, was conducted in Jomoro district. The conditions used were plantain monoculture (P) and three others, one with one row of plantain, one with two rows of plantain and one with three rows of plantain intercropped. The results showed that the system with three rows of plantain and two rows of rubber produces the highest yield for banana

Nigeria

In the 1990s, there were around 300,000 ha of old more or less abandoned rubber plantations (clonal or not), similar to jungle rubber. Today (2023) nobody can say how many of these old plantations still exist.

The main intercrops used during the immature period of rubber planted at normal density that have been recorded since then are cassava, maize, millet, plantain, yam, pineapple and leaf and fruit vegetables. The most common fruit crops grown with rubber in Nigeria are cherry, cola nut, bitter kola, pear, and *Irvingia*, as well as the rubber-livestock system, i.e., raising rabbits, snails, goats, sheep, pigs, and bees in mature rubber plantations.

In 2006, a study was conducted by Timothy U. Esekhade on rubber intercropped with melon/maize/cassava in year 1, melon/maize/yam/cassava in year 2, and melon/pineapple in year 3. This trial was an extension of an experiment started in 1993 (Esekhade et al., 2003, 2012). At the time the smallholders lacked extension services and access to technical information.

Later, another trial was conducted with the following crops: cherry (*Chrysophyllum albidum)* and avinger (*Irvingia gabonensis)* in the inter-rows of rubber (Wudpecker, 2014). Among the intercropping systems tested, rubber intercropped with cassava and plantain reached maturity earlier than rubber planted as sole crop or with tree crops (avinger and cherry) as intercrops.

Another experiment conducted by Esekhade (published in 2003) intercropped one, two, three and four crops selected among cowpea, soy bean, melon, maize and cassava with young rubber. The most robust girth of rubber samplings was observed in the rubber/soybean/melon and rubber/melon/maize systems, in the 1998 and 1999 seasons. Young rubber in the rubber/cow-pea system had the highest height increment rate. The area harvest equivalent ratio (AHER) revealed the comparative advantage of multiple cropping over monocropping. Rubber with melon had the highest AHER of 2.41 while rubber with soybean had the lowest AHER (1.20) of the intercrops.

A study by Mesike et al. in 2019 based on a survey of 200 smallholders on the adoption of rubber agroforestry in Nigeria reported positive differences in income for intercrops and mini-livestock but not for fruit trees in the rubber mature period.

Other countries in Africa

Except for temporary intercropping with annual crops in the first 3 years of rubber, no RAS is reported in Cameroon or Republic of Congo.

▸▸ The situation in South and Central America

No RAS is reported in Mexico, but RAS is known to exist in Brazil and Colombia.

Brazil

Michelin conducted trials with palmito (*Bactris gassipaes*) in Brazil in the 1990s, but the results were not published. A similar system was tested in Guatemala on a private estate, apparently sucessfull but not published. Some rubber systems were transitioning to fruit based agroforestry systems in Tome-Acu in the 2000s, where rubber was associated with pupunha/palmito/chontaduro *(Bactris gasipaes)*, at a density of 150 vines/ha including acai palm trees *(Euterpe olearacea)* and andiroba timber (*Carapa* spp.).

Another interesting association is rubber with acai/cupuaçu. These two fruit trees can be grown together perfectly as they occupy a different stratum and the acai palm provides a little shade for the cupuaçu (*Theobroma grandiflorum*). Other fruits can be associated with them: banana during the immature period of the cupuaçu, and pineapple during the first two years after rubber is planted. A third stratum could be occupied by slow-growing timber trees such as mahogany, teak, or ipe.

Other fruit trees mentioned by local farmers were Roocou/Achiote/Urucu (*Bixa Orellana*), guarana (*Paullinia cupana*), araçá/guave/cherry guava of China (*Psidium cattleianum)* — a different variety from the one observed in Colombia, which is called araçá for which there is a large juice and pulp market, comparable to the market for cupuacu in the Brazilian Amazon.

Columbia

Diversifying agricultural crops, including rubber, as alternatives to growing coca was a priority of the 1998-2005 Plante project in the Colombian Amazon, a pioneer area in the uninterrupted war waged by the government on drug lords and local guerrilla (FARC) until 2016. Plante was a presidential project launched in 1998 to offer local smallholders reliable and appropriate alternative crops to replace illicit crops and to put a halt to the associated violence. Rubber was identified as one of the most attractive and competitive crops for the purpose, as it is a reliable source of income since a good local market already exists for the product. It enables farmers to accumulate capital while simultaneously allowing them to invest in other diversified cropping or livestock systems (in particular fish ponds). In the 2000s, there was a fundamental change in societal demand in areas where colonisation is already past history and land-use has stabilised. The social cost of coca was considered unacceptable by local communities which led to a shift to less risky and less politically sensitive, but of course also potentially less profitable crops. Efforts have focused on competitive rubber monoculture or rubber-based agroforestry systems in partnership with the local rubber growers' association. The approach produced real social mobilisation and innovation capacities to cope with the social violence resulting from the cultivation of illegal crops and led to the adoption of more diversified and sustainable farming systems in the province of Caquetá.

Although cultivating coca originally provided a significant income, in the 2000s, this income no longer seemed sufficient for many producers located near cities who have other options, which is the case of rubber growers in Caquetá. The price of coca fluctuated considerably, an indirect effect of fumigation (undertaken to destroy the coca crop) and the social cost of coca. Most producers were looking for the return of social peace with guaranteed sales of their products (mainly fruit, milk and meat), fish farming, fruit crops, palmito, rubber, and to a lesser extent palm oil. These crops are profitable alternatives for agricultural development, and an elegant and sustainable solution for disengagement from — and ultimately the disappearance of — prohibited crops. Colombia has a long experience of what is historically and sociologically referred to as "Violencia".

The advantage of perennial crops, and in particular of fruit trees, is that a market already exists, particularly for "palmito/chontaduro" (depending on its use as a fruit or for heart of palm), and for the fruit trees that are typical of the region. For example, although *Bactris gasipaes* is successfully cultivated in Brazil for the production of palmito (Michelin perfectly mastered its cultivation in the Mato Grosso trials), this species requires careful management (between 3 and 15 years of age) for the production of chontaduro in Colombia, because *Bactris* grows much faster than rubber, overtakes it and can then quite simply prevent rubber growth from the 4th year on. Identifying the right species and the optimal densities can be achieved by setting up a full-scale experiment with small planters, with varying planting densities and different types of associations.

A participatory workshop was organised in 1999 with 4 groups of equal size in which the participants made suggestions for technical reference systems and associations of crops/intercrops with rubber for future experiments to test the technical feasibility of their proposals (Penot et al., 2012). Data from previous trials implemented by CORPOICA (the Colombian Corporation for Agricultural Research) were also used. In a plenary session, each group presented its results, which were then discussed by the group as a whole. The plenary session concluded that the interest of agroforestry practices, in particular the combination of rubber and Amazonian fruit trees, depends on the following criteria: (i) diversification of sources of income, (ii) flexibility of the system as a whole in the event of an excessive drop in the price of one or the other associated crops (in particular rubber), (iii) optimisation of work productivity, (iv) reduction in the overall cost of maintaining the agroforestry system (compared to the cost of running two monocultures), (v) the beneficial effect of the appropriate intercrop on rubber tree growth during the immature period, (vi) fighting *Brachiaria* spp. at the lowest cost (growing *Bracharia* spp. as an intercrop delays opening of the trees and tapping by 2 to 4 years), and (vii) priority in terms of growth in the immature period for rubber. The species tested in the above-mentioned trial are listed in Table 3.4.

Fast-growing timber trees (that are cut between 7 and 15 years of age) can be included in the system, but care must be taken to limit competition with rubber. Fast-growing timber species can be very aggressive, and it may be necessary to put off planting them until the second or third year after planting rubber. Non-climbing cover legumes, such as *Flemingia congesta* or similar plants, can be used to protect the soil, in anti-erosion lines, or along contour lines, or to fill the "gaps" between fruit trees.

The local shrubby legumes (*Albizzia falcataria*, cambullo (*Erythrina fusca*) including *Flemingia macrophylla*, which were tested by CORPOICA, were clearly too aggressive. Likewise, trees of the *Acacia mangium, Gmelina arborea* type should be avoided because they grow too fast. Although these tree species have advantages, for example, as a way to rehabilitate soils compacted by livestock farming, they are not right for associating with rubber, as we saw during our visits to the trial at the Macagual station in Caquetá in 1999 and 2001.

The association of rubber producers and the efforts to improve the technical itineraries of rubber-based agroforestry systems made in the early 2000s in order to compete with coca are evidence for this dynamism, for the strong capacity for social mobilisation and for the innovative processes underway in this region. The resumption of the war in 2003 put a temporary stop to the process. A peace agreement was reached in 2016. We have been unable to check if the agroforestry trend is still underway, but it remains an important source for successful diversification and an opportunity to break the vicious circle of coca cultivation, at least for local rubber growers.

▸▸ Double-spacing systems: an alternative system for full-sun species associated with rubber

The double-spacing system could be a way to limit excessive shading of associated plants. Double spacing is based on a larger inter-row to optimise plant access to light for longer than the original 4-5 years with normal planting density. The system is based on double, or triple rows of rubber separated by 8, 10, 15, 20 or even 25-metre inter-rows. The wider the inter-row, the longer the associated crop can benefit from light and will produce more when subsequently shaded by the rubber trees. Several inter-row spaces have been tested (i) 12-15 m for cacao or 25 m for tea in Sri Lanka, (ii) 16 m with a double row of cacao with good yields for the first 12 years in Côte d'Ivoire (Snoeck et al., 2013), (iii) 20 to 25 m for 15 to 20 years of light for timber, fruit trees (Cambodia, Vietnam) or rice (Indonesia).

Double spacing systems generally contain fewer rubber trees per hectare (350 to 500 trees/ha) which, in some conditions, may reduce the rubber yield. Double spacing systems were not widely promoted by extension services until very recently (since 2020) when the search started for systems with a better gross margin per ha than standard monoculture.

Double spacing in Indonesia

All the double-spacing systems in Indonesia have only been tested in trials in research stations and have not been adopted by smallholders so far.

Since 2011, there have been significant changes due to low rubber prices, which seriously affected smallholder farmers' incomes and triggered the need for diversification. Some smallholders converted rubber plantations into oil palm or coffee plantations or grew annual crops such as maize and upland rice. Intensification through agroforestry is a potential way of preventing further conversion away from rubber to other crops. A study including a demonstration plot with double-row spacing was conducted in Tanah laut District, in South Kalimantan Province (700 ha) and in the Musi Rawas District, South Sumatra (400 ha). The study began in 2009 and ended in 2016 (Sahuri, 2016). The double-spacing system used wide row spacing (18 m × 2 m) × 2.5 m and a density of 400 trees/ha. Smallholders were thus able to cultivate intercrops such as maize, paddy, or cassava in wide inter-rows with sufficient light to obtain normal yields. The planting distance of maize as an intercrop was 40 cm × 10 cm, for paddy it was 75 cm × 20 cm, and for cassava 100 cm × 100 cm. The rubber clone used was PB 260. A survey of 50 smallholders showed that the rubber trees grew well. The trees could be tapped at the age of 55 months whereas in Indonesia, tapping usually starts at 60-70 months. The average yield of rubber was 1,500 kg/ha/year, which is quite satisfactory in Indonesian conditions. The yield of hybrid maize in the third year reached 5,000 – 5,500 kg/ha, that of Dayang Rindu upland rice varieties 2,000 – 2,250 kg/ha, peanut yield was 2,000 – 2,200 kg/ha, and cassava yield 16 to 19 t/ha. From an economic point of view, double spaced RAS with annual crops such as rice, maize or cassava is profitable as long as there is a strong demand for the crop concerned and, in the case of upland rice, if it can compete with irrigated rice.

The results of the study showed that with single-row spacing, the growth in the first year of tapping was slightly better than with double-row spacing but that the yield of latex from individual trees was the same in the two. The yield/ha of rubber was thus higher with standard single-row spacing because the planting density was higher.

Double spacing in China

Rubber and medicinal plants

A field experiment was conducted at the Experiment Farm belonging to the Chinese Academy of Tropical Agricultural Sciences, in Danzhou, Hainan, China, using the Reyan 7-20-59 rubber clone and two planting patterns: (i) single-row (SR) avenue planting 3 m × 7 m, and (ii) double-row (DR) avenue planting pattern (2 m × 4 m) × 20 m (Lin Weifoo, 1991). Although relatively fewer rubber trees were being tapped per unit area in the DR system, with 98% of SR, the yield per hectare was not significantly affected due to ithe higher yield per tree. In addition, the double-row system allowed more light to penetrate. When the overall performances of the two planting patterns were compared, the double-row system proved to be more suitable for long-term intercropping in rubber plantations. Due to the lack of markets, sales of some intercrops including *Alphinia* and *Oxyphylla* gradually decreased (Lin et al., 1999).

Rubber with tea, coffee and other crops

Combining rubber and tea was successful with large double spacing. Limited light penetration had previously been a major constraint to maximising land use using intercropping in conventional single-row rubber plantations. In 2002, a long-term field experiment based on a double-row pattern using the CATAS 7-20-59 rubber tree clone was set up to investigate whether land use efficiency could be increased by using double spaced intercropping. At the end of the experiment, the results showed that with the improved double-row system, 42-50% of the total land area could be used for intercropping with more crops (Huang et al., 2020).

Both annual and perennial crops were tested in the experiment: yam bean (*Pachyrhizus erosus* [L.] Urb.), peanut (*Arachis hypogaea* L.), maize (*Zea mays* L.), soybean (*Glycine max* [L.] Merr.), elephant grass (*Pennisetum sinese* Roxb.), ginger (*Zingiber officinale* Rosc.), common bean (*Phaseolus vulgaris* Linn.), arrowroot (*Maranta arundinacea* L.), coffee (*Coffea arabica*), cinnamon (*Cinnamomum cassia*), and cacao (*Theobroma cacao* L.) were planted when the rubber trees were mature. In China, 99% of coffee beans are arabica, which are mainly grown in high-altitude regions (700–1,840 m above sea level) throughout Yunnan Province. Robusta coffee is mainly planted in low-altitude regions (normally < 200 m) on Hainan Island (Liang et al., 2014; Zhang et al., 2014; Yang et al., 2021).

Among the annual intercrops tested, yam produced 74% of the yield it produces in monoculture. Peanut produced the lowest yield at only 38% of its yield in monoculture. Yields of intercropped arabica coffee (Catimor variety), which was introduced from Yunnan province, and local robusta coffee (Reyan variety) were recorded. Robusta coffee produced 1,226 kg/ha, equivalent to 35% of its yield in monoculture. Arabica coffee produced 1,319 kg/ha, close to its yield in monoculture. Double-row planting increases the light available to annual crops and other crops which, like coffee, require at least 70% light. Under the double-row system, Catimor arabica coffee is more suitable than robusta for intercropping with rubber.

The rubber/yam bean and rubber/Arabica coffee schemes were the two most promising intercropping patterns.

Impact of double spacing on rubber yield

In terms of rubber yield potential, Rodrigo et al. (2004) and Snoeck et al. (2013) reported that the rubber tree yields obtained using the double-row system were 77.1% and 87.7% of those obtained using the conventional single-row system. Similarly, from 2010 to 2018, the yields obtained in the double-row plot were 89% of those obtained in the single-row plot, due to lower rubber tree planting density. Although, from 2010 to 2018, the double-row system produced a significantly lower total yield of rubber (1,296 kg/ha) than the single-row system (1,445 kg/ha), there was no difference in rubber yield per tree between the double-row and single-row systems, confirming the results of other similar experiments.

⟫ Timber species

Timber as a challenge

The main challenge in timber RAS is including more trees in the different types of AFS in order to produce more timber and fuelwood at marginal cost because timber species actually profit from maintenance of both the main crop and of other associated trees. Most valuable timber species require the shade that is provided by AFS. In other words, it is far easier to include 5 to 10 additional timber trees in existing AFS that taken together add up to millions of ha (and will produce millions of albeit scattered individual timber trees) than to create new productive forests to grow valuable timber species which require high initial investment and continuous maintenance for 40/50 years.

AFS therefore offer a real opportunity. It should also be noted that farmers need trees in their production model, because even at a small scale, growing trees is advantageous because trees and agriculture are complementary, and because of the robustness of AFS in terms of technical itineraries. Indeed, at the scale of the planet, more trees are needed because of their role in mitigating climate change. Indeed, in some countries in West Africa, for instance in Côte d'Ivoire, the medium-term consequences of climate change for cocoa production are a cause for concern. Finally, farmers also have a financial need for timber trees as a source of income which can be exploited for the purpose of investment at the end of the lifespan of the main crop or when the farmer needs cash (with trees cut at 5, 8, 12 or 15 years like teak). But farmers can also use timber trees themselves, notably for building and construction.

AFS farmers could thus be the next timber producers: in terms of resources (low cost, low maintenance, good integration in AFS), they are the best placed to take up the challenge. Government policies should be aware of this opportunity and create favourable contexts and regulations to boost timber production. An appropriate regulatory framework is essential to secure farmers' investments and to guarantee the wood produced in tree plantations will continue to be used in the future.

During the period when the timber represented by rubber trees at the end of their lifespan was changing from being considered a waste product to becoming an economically important component of rubber plantation management (Killmann and Hong, 2000; Shigematsu et al., 2011, 2013), there were also reports of timber trees being included in RAS (Jongrungrot and Thungwa, 2014; Jongrungrot et al., 2014; Somboonsuke et al., 2011; Penot, 1997). In Thailand, the main tree species cited in the reports (Somboonsuke et al., 2011) were teak (*Tectona grandis*) and neem (*Azadirachta indica*). Since teak requires light, it needs to be incorporated during the early establishment stage of rubber and requires an appropriate light regime throughout the rotation. Many different timber species are associated with rubber in Indonesia. A wide range of spontaneous timber tree species associated with rubber were identified in the 1997 Cirad/ICRAF/SRAP trials (Table 3.5).

Since that 1997 survey, some of these species have been recently re-introduced in agroforests in the 2020s (Table 3.6), in particular in *tembawang* systems, or are protected when they emerge in natural regrowth in jungle rubber and RAS systems.

Table 3.5. Spontaneous timber species maintained in local agroforests and their uses

Local names	Latin names	Uses
Leban	*Vitex pinnata*	Timber, wood, spice, medicinal
Medang	*Litsea elliptica*	Timber, latex
Ramboutan	*Nephelium lappaceum*	Fruits, timber
Jengkol	*Pithecellobium jiringa*	Fruits, vegetable, timber, medicinal
Durian	*Durio zibethinus*	Fruits, timber
Pingam	*Artocarpus* sp.	Fruits, timber, vegetable
Cempedak	*Artocarpus integra*	Fruits, medicinal, vegetable
Lengsat	*Lansium domesticum*	Fruits, medicinal, handicrafts
Pekawai	*Durio* c.f. *dulcis*	Fruits
Mentawa	*Artocarpus* c.f. *anisophyllus*	Fruits
Nyatuh	*Palaquium* spp.	Timber, latex
Owan		Timber, handicrafts
Bungkang	*Polyalthia rumpfii*	Timber, spice
Belian	*Eusideroxylon zwageri*	Timber
Ubah	*Glochidion* sp.	Timber
Kemenyan	*Styrax benzoin*	Timber, latex, livestock feed
Tantang	*Buchania sessifolia*	Timber
Bidara	*Nephelium maingayi*	Fruits

Table 3.6. Useful spontaneous vegetation in rubber gardens in West Sumatra and Jambi that is not cleared by the farmers (1997 survey)

	Fruit tree species		Medicinal plants
Durian	*Durio zibethinus*	Sicerek	*Clausena* c.f. *excavata*
Nangka	*Artocarpus heterophyllus*	Sidingin	*Kalanchoe pinnata*
Rambutan	*Nephelium lappaceum*	Jirak	*Eurya acuminata*
Macang	*Mangifera foetida*	Sitawa	*Costus speciosa*
Mango	*Mangifera indica*	Bidaro	*Eurycoma longifolia*
Langsat and Duku	*Lansium domesticum*	Daun kasai	*Pometia pinnata*
Jambu	*Eugenia aquea*	Sikarau	*Cyrtandra* sp.
Petai	*Parkia speciosa*	Kunyit	*Curcuma domestica*
Mangosteen	*Garcinia mangostana*	Kunyit balai	*Zingiber purpurteum*
Jengkol	*Pithecellobium jiringa*	Sikumpai	Not determined
Kabau	*Pithecellobium bubalinum*		
	Timber species		**Plants with other uses**
Sungkai	*Peronema canescens*	Rimbang	*Solanum torvum*
Meranti	various genera and families, but esp. *Shorea* spp.	Daun kayu sibuk	

	Timber species		Plants with other uses
Kulim	*Scorodocarpus borneensis*	Damar	*Dipterocarpaceae*
Petaling	*Ochanostachys amentacea*	Kopi	*Coffea robusta*
Kumpabok	Indet.	Jambu monyet	.
Maraneh	*Elaeocarpus palembanicus*	Sitarak	*Macaranga* c.f. *nicopina*
Tamalun	Indet.	Dalo	*Macaranga javanica*
Kawang	Indet.		
Madang	Various genera and families but esp. *Lauraceae*		
Surian	*Toona sureni*		

In a survey conducted in 2022, as part of the Rubis project/RRII, other species were identified: kernang, a rattan like specie, gaharu (agarwood), various types of meranti, pulai (*Alstonia scholaris*), coconut tree and durian.

Potential timber trees in South East Asia

Malaysia

According to the Malaysian Timber Council (MTC, 2015), the eight recommended fast growing multipurpose trees species in Malaysia are *Acacia mangium* (acacia), *Khaya ivorensis* (African mahagony), *Tectona grandis* (teak), *Neolamarckia cadamba* (kelampayan), *Azadirachta excelsa* (sentang), *Octomeles sumatrana* (binuang) and *Paraserianthes falcataria* (batai).

Indonesia

Several timber species were cited by Jambi farmers during a GPSNR agroforestry workshop held in Jambi in September 2022 (Penot et al., 2022 GPSNR):

– Pulai *Alstonia angustiflia* (with FSC certification)

Similar to *Acacia mangium* in terms of uses. The wood is creamy white to pale yellow in colour, and slightly lustrous. The grain is straight and the texture medium. The grain is sometimes irregular or oblique. Latex canals are present. Density at 12% moisture content is 0.45 g/cm3. The blunting effect is normal; peeling is reported to be good but slicing is not recommended or is of no interest. Tools need to be kept sharp to avoid fuzzy surfaces. Filling is recommended for a good finish. Nailing is not good but gluing is satisfactory, although the glue dries rapidly and there is a risk of a blue stain appearing during drying. Pulai is not resistant to fungi and is susceptible to dry wood borers. Pulai can be used for several applications e.g. boxes and crates, matches, veneer, panelling, furniture, joinery, moulding.

– Medang *Litsea leytensis* Merr.

Good quality: but now rare since most forests have disappeared (Source ITTO).

Other names: Theptharo (Thailand); Medang serai (Indonesia); Medang lesah (Indonesia); Medang busok (Indonesia); Laso; Keplah wangi (Malaysia); Kepaleh;

Kayu gadis; Kajoe lada; Kajoe gadis; Gadis; Chintamula hitam; Safrol laurel (United Kingdom); Re huong (Vietnam); Thep tharo (Thailand); Karawa (Myanmar); Ki sereh (Indonesia); Safrol laurel (United States of America); Bunsod (Sabah); Keplah wangi (Sarawak); Medang kemangi (Malaysia); Rawali (Borneo); Selasihan (Indonesia); Teja (Sarawak); Teja (Malaysia); Huru (Indonesia); Medang (Indonesia). The Scientific Name Synonyms: *Cinnamomum sumatranum* (Miq.) Meissner; *Cinnamomum porrectum* (Roxb.); *Cinnamomum glanduliferum* C. Nees (Roxb.). The uses are: general housing, panelling, furniture and cabinets, luxury furniture, plywood and veneer, turning, tools, agricultural tools, containers, truck bodies, naval construction, other and musical instruments, handicrafts, shoes, coffin, moulding. Essential oils are obtained by steam distillation of medang. (Source ITTO).

– Mahang *Macaranga* spp.

A light density wood of poor quality. *Macaranga* spp. is a small tree which can grow up to 25 m in height and has a diameter at breast height (DBH) of 30 cm. This is an early successional tree that grows mainly in swamps up to 100 m above sea level. *Macaranga* spp. comprise 250 species, of which 30 grow in tropical Africa and Madagascar, and the rest in tropical Asia (from India to Indo-China, China, Taiwan and Ryukyu Island), throughout the Malaysian region, northern Australia and the east Pacific up to Fiji. Most of its diversity is found in Malaysia, where some 160 species grow, with an exceptionally high number of endemic species in Borneo and New Guinea (Sosef et al., 1998). Mahang wood is traditionally used for temporary constructions, especially for the parts of native houses which are not in contact with the ground. It can also be used for light frames, interior trim, moulding, shingles, packing cases and match splints. In the Philippines, mahang wood is a favourite material for wooden shoes. *Macaranga* produces high quality pulps and particle boards, cement-bonded boards and wood-wool boards. It is also suitable for the production of plywood, and is known to make good fuel wood.

– Berumbung/*Adina minutiflora*

Used for housing/doors and windows, etc.

– Nyatoh

Nyatoh is the trade name used for wood of a number of hardwood species of the genera *Palaquium* and *Payena* growing in rainforest environments in South East Asia, particularly in Indonesia and the Philippines. Nyatoh wood is reddish and most species are easy to work with and stain and polish well. It has a tight straight grain that resembles cherry wood. The surface is dark brown/red in colour. Nyatoh is generally considered to be a sustainable resource, but several species of related genera *Palaquium* and *Payena* are on the IUCN Red List due to overexploitation and an alarming reduction in their habitats. Rated as non-durable and as susceptible to insect attack. Common uses: furniture, plywood, interior joinery, and recently building solid-body electric guitars (Source Wikipedia).

– Mahoni

There are two types of mahoni: (i) *Swietenia macrophylla*, commonly known as Honduras mahogany, or big-leaf mahogany, is a species belonging to the Meliaceae family. It is one of three species that yield genuine mahogany timber (*Swietenia*), the others being *Swietenia mahagoni* and *Swietenia humilis*. Mahoni is native to South

America, Mexico and Central America, but naturalised in the Philippines, Singapore, Malaysia and Hawaii, and cultivated in plantations and as wind-breaks elsewhere. Unlike mahogany sourced from its native locations, trade in plantation mahogany grown in Asia is not restricted, and in 2024, the mahogany timber grown in these Asian plantations is the main source of international trade in genuine mahogany. The Asian countries in which the majority of *Swietenia macrophylla* grow are India, Indonesia, Malaysia, Bangladesh, Fiji, Philippines, and Singapore; and (ii) *Swietenia mahagoni*, commonly known as American mahogany, Cuban mahogany, small-leaved mahogany, or West Indian mahogany, is a species of *Swietenia* native to South Florida in the United States and islands in the Caribbean including the Bahamas, Cuba, Jamaica, and Hispaniola (this species is not grown in Asia).

Among the very durable heavy hardwoods are: balau (*Shorea* spp.), belian (*Eusideroxylon zwageri*), traditional in old jungle rubber, giam (*Hopea* spp.), malagangai (*Eusideroxylon malagangai*).

The majority of the light hardwood falls in either the moderately durable or the non-durable category.

Among the non-durable commercial timber species whose applications are limited to indoors or to environments where they will not be in contact with the soil or moisture are: light red meranti (*Shorea* spp.), jelutong (*Dyera costulata*), sesenduk (*Endospermum diadenum*) and mahang (*Macaranga* sp.).

List of Non-durable plantation species (10-year-lifespan):
– Acacia mangium,
– Acacia crassicarpa,
– *Acacia auriculiformis*,
– *Gmelina arborea* (yamane),
– *Azdirachta excelsa* (sentang) or neem tree.

Conclusions concerning the role of timber species in RAS

The main benefits of timber species in RAS are: (i) beneficial effect on the environment, better adaptation to climate change with more trees/ha that might mimic forest, improve animal biodiversity, (ii) good water conservation, prevent erosion, retain soil moisture, (iii) create capital for replanting at the end of the rubber lifespan, (iv) easy to integrate, and (v) local markets exist for use in housing, boat building.

There is a need for a more organised timber marketing chain: farmers do not receive their fair share of the margin along the value chain. It is important to promote optimisation of the timber chain such that farmers receive more of the profits. Some farmers prefer fruit trees, others may choose timber depending on their strategies. Tables S4 and S5 (Appendices) are a complete list of timber species that can be included in RAS for example, in Indonesia.

Incorporating native trees in rubber monoculture is another option that has already been tested. The aim is to mitigate negative environmental impacts and to provide alternative sources of income for farmers.

▸▸ Conclusion

Farmers almost everywhere practice intercropping during the immature rubber period to ensure an income during the first 5-7 years when rubber is not yet productive.

It can be concluded that to be adopted at a large scale, and to permanently include other plants, intercropping must either to be very profitable or at least require little labour, as labour availability is becoming a real problem in many rubber producing countries. For rubber-based intercropping systems, profitability is closely linked to the following biophysical and economic factors (Langenberger, 2017; Penot, 2020):

– Biophysical interactions between rubber and inter-crops

Some species benefit from the shade provided by rubber trees (Guo et al., 2006). This is true of certain timber species and medicinal plants, but for most associated plants, shade is a real constraint. Other shade tolerant secondary species are (i) coffee and cacao (Snoeck et al., 2013) but with at the most, 30-40% shade, which is not compatible with the degree of shade provided by rubber planted at normal density, (ii) bamboo (Charernjiratragul et al., 2014) but which could compete too strongly for water, (iii) the leaf legume *Gnetum*/pak Lieng (Simien and Penot, 2011), and (iv) species such as camphor reported in Winarni et al. (2018), and cardamom. Species to avoid in RAS are those which require a lot of sunlight (citrus, some fruit trees (Snoeck et al., 2013), as well as species which grow taller than rubber.

– Lifespan of secondary species

The best-bet species are those whose lifespan resembles that of rubber. Some species live much longer than rubber (i.e. more than 35 years) including durian, petai, most tropical fruit trees and many timber trees. This may lead to a different strategy at the end of the rubber lifespan, i.e., the decision to change from rubber to long-term fruit/timber agroforestry systems. Examples of profitable combinations of rubber with durian and/or petai are reported in studies conducted by Somboonsuke (2001), Simien and Penot (2011), Stroesser et al. (2018), Winarni et al. (2018) and Wulan et al. (2006). The same applies to timber trees. Fast growing timber trees were generally found to be less profitable, as they have to be cut and replanted two or three times during the course of one rubber lifecycle (Wulan et al., 2006).

– Regularity of production of intercrops

The number of harvests ranges from almost daily (pak Lieng), two or more times a year (annual crops), once a year (most fruits and nuts) to one-off harvesting at the end of the lifespan (timber trees).

Regular harvesting implies higher labour costs but can improve the cash balance particularly for smallholder farmers with limited land resources. Timber trees harvested at the end of lifespan provide sufficient capital to cover the cost of replanting. But harvesting after 35 years is not a solution for rubber farmers who require a regular income (Stroesser et al., 2018).

– Planting density for rubber and secondary species

The standard planting density for rubber monoculture (7 × 3 m spacing), has tradi-tionally been used by smallholders because most crop associations that use this density did not adversely affect rubber yield (Charernjiratragul et al., 2014; Penot,

2001). The optimal density of different species in a rubber agroforestry system needs to account for individual farming households' multiple goals (Gosling et al., 2020) as well as for the wider socio-economic-ecological environment. Double spacing systems enable higher productivity of associated crops in large inter-rows.

– Market value of rubber and associated species

To combine rubber with high value timber and fruit trees with good market stability is probably the best way to cope with fluctuating rubber prices. Durian (Simien and Penot, 2011) and tea (Guo et al., 2006) have been found to withstand price fluctuations. Durian has been identified as a desirable plant to complement rubber in both Thailand (Simien and Penot, 2011) and Indonesia (Winarni et al., 2018) as there are many stable markets for durian. However, what works in one region or country may not work in another area for environmental, political and socio-economic reasons. Despite the positive ecological benefits (Drescher et al., 2016), rubber agroforestry is still not widely used, particularly after the end of the immature period (Langenberger et al., 2017).

Additional income from intercropping, cash availability and improved return to labour are key requirements to increase the adoption of agroforestry (Penot, 2001; Gosling et al., 2020). Most current rubber agroforestry systems include both indigenous trees, as well as a range of livestock species but more rarely than plant production. Constraints reported in the literature are the additional labour requirements and local labour shortage (Guo et al., 2006; Snoeck et al., 2013; Stroesser et al., 2018), the necessary agroforestry knowledge and related skills (Somboonsuke, 2001; Penot, 2001), government policies (Penot et al., 2019), potential pests and diseases associated with the intercrop (Somboonsuke, 2001; Langenberger et al., 2017). Resilience is becoming a major concern for producers. RAS can contribute to economic and environmental resilience. Rubber-based agroforestry clearly has the potential to reduce most smallholders' vulnerability to rubber price volatility.

To promote RAS requires a reasonable understanding of the diversification processes used by smallholder farmers (Barrett et al., 2001) and of producers' strategies concerning tropical tree crops (Schrott and Ruf, 2014). More research is required on optimal density and density patterns, the proportion of different species of secondary crops combined with rubber linked with existing markets (Zeng et al., 2012; Zhou, 2000) and soil/climate conditions. Intercropping is currently purely market driven. Economic development linked with the proximity of cities and industrial centres also increases the availability of off-farm income options to the detriment of RAS. Table S6 (Appendices) summarizes the current situation of RAS in rubber producing regions worldwide. And Table S7 (Appendices) displays the total species encountered in RAS.

Chapter 4

Expectations of RAS, impacts and contribution to current's main challenges in 2024

Éric Penot, Bénédicte Chambon, Alexis Thoumazeau, Phillipe Thaler

Like all other commodities, rubber plantations have less rich biodiversity than natural forests. Both forest degradation and deforestation cause loss of biodiversity (Jongrungrot et al., 2014; Fern, 2018; Peerawat et al., 2018). According to Orozco and Salber (2019), the expansion of industrial plantations also increases pressure on animal biodiversity in non-degraded forest located on the margins of plantations and can disturb wildlife corridors used by primates or elephants.

However, the impact of the expansion of rubber plantations differs depending on (i) the previous land use, and (ii) the current cropping system. For instance, agroforestry rubber plantations can have a positive effect on biodiversity when planted following monoculture (Feintrenie and Levang, 2009; Penot and Ollivier, 2009; Jongrungrot et al., 2014; Penot and Feintrenie, 2014; Fern, 2018).

▸▸ Agroforest cropping systems provide miscellaneous goods and services

Multiple roles

Farmers worldwide, but especially in developing countries, do not only focus on agricultural production. While they are seldom sensitive to global issues such as biodiversity conservation or carbon sequestration, as opposed to their own family priorities, they nevertheless contribute a series of "goods and services" that are not always marketed or even recognised. The multi-functional role of agriculture is now acknowledged and promoted in some parts of the world (e.g. in Europe) in reaction to "productivist" agriculture, and has enabled the reduction of direct subsidies for production in favour of paying subsidies for the environmental functions of farms.

Agroforests can fulfil this multi-functional role better than other cropping systems because they have more positive externalities than land-use options based on mono-crops. Agroforests, consequently, merit tailored economic analyses to account for both goods and environmental services as well as short- and long-term issues.

Agroforests, particularly home gardens, generally combine a long-term strategy for the production of resin, nuts, fruits and timber, for instance and annual or bi-annual food crops such as legumes, cassava, banana, with short-term products for immediate consumption. Farming systems models can include components on externalities or services to analyse this multifunctionality, but some components including biodiversity conservation may be easier to treat at regional or macro level. While so far, priority has been given to plant biodiversity, some studies have pointed to the role of agroforests as buffer zones for game (Nyhus and Tilson, 2004).

Another important role of agroforests is the production of a "forest rent" as defined by Ruf (1995a), i.e., a reduction in the cost of – and risks involved in – establishing a perennial plantation thanks to the positive externalities provided by forests, such as preserving or improving soil quality, controlling weeds and pests. The "forest rent" concept was extended to agroforests by Penot (2001), based on the fact that agroforests have similar attributes. Among other functions, agroforests do maintain (and sometimes improve) the forest rent whereas conventional monoculture plantation crops (cacao, coffee, oil palm), generally –at least partially– consume it. Therefore, when the time comes to renew plantation crops in an agroforest, economic sustainability is favoured because the cost of establishment is similar to that of replanting after a forest.

On the other hand, agroforests also have certain constraints. Crop mixtures being the rule, some crops are favoured while others are not. Agroforests sometimes provide such small yields of a particular crop that it can only be sold locally. The period before associated trees may produce a yield will increase the length of the wait before the farmers receive a return on their investment, and this is even more true for timber. Most smallholders use unimproved genetic planting material which may be of questionable quality, particularly in the case of fruits that may be not suitable for export. Most agroforests are extensive, and rely only on family labour, but some –including RAS– could be made more intensive by applying fertilisers during the immature period to improve growth and reduce the length of the immature period of the main crop, as well as by using improved planting materiel (coconut trees, rubber clones, selected grafted durian and grafted fruit trees, for instance).

Agroforests are particular cropping systems with a range of specifications which makes them more difficult to analyse than monocropping systems or even than multiple cropping systems comprised of associations of annual crops. It is hypothesised that this lack of analysis has made it difficult for agronomists and extension agents to promote agroforests and has prevented research on agroforests going beyonddescriptive studies to become truly analytical (Penot, 2001; Kumar and Nair, 2004).

Plant biodiversity in RAS

Planted biodiversity

By essence, RAS host higher planted biodiversity than monoculture rubber plantations due to the large number of associated trees and plants. Moreover, monocrop plantations mostly contain very few clones. Where estates tend to use up to a dozen different clones in their large plantations, smallholder plant only a few clones, usually only one (Clément-Demange et al., 2007). Moreover, one clone tends to dominate in any given country, in which case, intra-rubber biodiversity may also be very reduced.

For example, according to RAOT, in Thailand, 95% of the total area under rubber is planted with clone RRIM600. Rubber tree intra-diversity is extremely limited in such systems. In this case, any association with another planted species increases total plant biodiversity. The most diverse systems are traditional "jungle rubber" systems (see details in chapter 1). In their basic form, they are indeed secondary forests enriched in rubber seeds. Hence, in Indonesia, to give an example, their plant biodiversity is almost 60% of that of natural forests. In other words, only natural forests can be richer. However, jungle rubber systems have tended to be replaced by other land uses, including rubber monoculture, because of their low productivity. Clonal RAS have intermediate plant biodiversity. Although some RAS systems (e.g. RAS 1) could theoretically host the same plant biodiversity as traditional jungle rubber at least in the inter-rows because they consist in rows of monoclonal rubber trees planted in secondary forest, in practice, they probably host fewer plant species than jungle rubber, as most species are partially weeded out at least once during their life span.

As explained in chapter 2, the composition of RAS in Indonesia varies widely. Some RAS are extremely complex, and may include more than 20 cultivated plant species, but some include only one. Thus, RAS in itself does not ensure plant species richness, which depends on the pattern chosen by each farmer. The type of plant that can be inter-cropped in RAS depends on the rubber planting design and density. If, as in most cases in Thailand, the standard row pattern is used (6-7 m between rows, 2.5-3.5 m between the trees in a row, i.e., a density of 450-600 trees/ha), only shade tolerant species can be permanently intercropped. These are usually perennial fruit trees, shrubs that provide leafy vegetables (*Gnetum*) or fruit trees, or alternatively, semi-perennial species like banana and bamboo (Jongrungrot et al., 2014; Stroesser et al., 2018). Tall timber trees can also be included in this type of design. These trees are usually shade tolerant species that are planted in the rubber inter-row some years after the rubber trees to avoid too much competition with the rubber trees they will eventually overtop (Wu et al. (2016) on below-ground interspecific competition for water; Yang et al. (2020) on intercrops and surface water availability improvement; Zhao et al., 2023; Zhu et al., 2019).

An interesting option was proposed by Langenberger et al. (2017): planting endangered forest tree species, preferably of local/regional origin, in normally spaced rubber tree plantations, with financial support from conservation agencies. A similar idea was developed in Sabah, Malaysia, with on-farm trials with oil palm and local riparian forest species to create biodiversity corridors for endangered local animal species. This project, called the "Trails" project[66], was implemented on a private estate. Such schemes would have a remarkable impact on high-value tree biodiversity. Double-row spacing is also possible in RAS to provide more space (up to 25 m) between the rows of rubber trees, which themselves are often planted in double or triple rows. In such cases, because more light is available, more species can be grown, including annual crops such as maize or rice, or perennials such as coffee, cocoa, or tea (as is already the case in Sri Lanka, Penot et al., 2023a).

Niche effect and management

Beyond the direct increase in biodiversity obtained by planting other species with rubber, the complexification of the ecosystem can host additional plant species such

---

66. Trails: climaTe Resilient lAndscapes for wIldLife conservation.

as epiphytes that can grow on the associated species. Palm trees in particular often harbour ferns and are known to provide favourable conditions for epiphytes (Bekuma et al., 2007). Associating shrubs like pakliang (*Gnetum*) with rubber, like in Thailand, also probably favours the development of understorey vegetation as does monoculture with other non-productive shrubby vegetation. A network of plants of different heights would ensure a connection between the ground and the rubber tree canopy. In turn, this would provide habitats for other organisms (microorganisms, mesofauna, etc.) paving the way for restoring richer ecosystems on rubber farms. These aspects require further study. Conversely, wider spacing between rubber rows creates heterogeneous environmental conditions in a given plot, which also favours biodiversity, with plant species adapted to varying degrees of shade (Jongrungrot et al., 2014).

Whatever the type of plantation, the way it is managed has a fundamental impact on plant biodiversity. In practice, in RAS, some crops are intensively managed including pesticide use and tillage, but rubber monocrops can be lightly managed, particularly after canopy closure. When crops are permanently associated with rubber, typically in alley systems, the natural understorey can be seriously limited by the agricultural practices applied to the associated crops. Manual or chemical weeding can affect a more extensive area in RAS than in monocrops, where only the vegetation growing along the rubber tree rows is controlled. It is not rare to see the whole surface of the plot "cleared" of natural vegetation in RAS. In such a system, plant diversity is obviously very low. Conversely, many farmers, notably those living in traditional areas, use very limited weed control, and let natural vegetation grow almost freely in the inter-rows. They only "clear" a strip of land along the row of rubber trees to facilitate the task of the tappers. In this kind of monoculture system, plant biodiversity can be quite high. Indeed, Panklang et al. (2022a) found either land covered by profuse natural vegetation or almost bare soil in the 4 rubber systems they compared: monoculture rubber, shrub RAS, fruit-tree RAS and timber RAS. This confirms that management (i.e. the farmers' practices), rather than the system per se (RAS vs. monoculture) limits or adds to plant biodiversity in rubber plantations.

Animal biodiversity

Association with cattle and poultry

Livestock is seldom included in RAS. The presence of a large number of big animals (cows, buffalos, or even pigs) hardly seems compatible with rubber cropping, as they pose a risk for the tapping equipment, if not for the trees themselves. This kind of husbandry is consequently usually limited to letting few animals graze close to homestead. Poultry (chicken or ducks) are more common in rubber plantations and may represent one of the main activities of some farmers. However, the surface area concerned remains limited. Creating fish ponds between the rubber rows is also possible, but rare. On the whole, animal husbandry cannot be considered to have a significant influence on animal biodiversity in RAS.

Habitat for insects, arthropods, birds, wild mammals and reptiles

Several studies have shown that the presence of birds in rubber monocrop plantations is very limited. In studies in which bird life is considered to be an indicator of natural biodiversity, rubber plantations score very poorly. The main reasons are that the fruits and seeds produced by rubber trees are basically inedible for birds (Bekuma et al., 2007) and the structure of rubber tree canopy is not suitable for nesting, at least when the standard planting design and density are used (Putri et al., 2020). In this context, including fruit trees or shrubs can have a very positive effect on insect and bird populations, both in terms of diversity and number. Warren-Thomas et al. (2020) found that agroforestry had a positive effect on butterflies, but not on birds in southern Thailand. For the same reasons, rubber plantations do not offer a very favourable habitat for mammals. However, as rubber plantations are often located on forest margins, roaming elephants are not rare and may damage the trees. In the 1990s, many trial plots included in the SRAP research project in Jambi were also damaged by local monkeys, as forests were still extensive in the area.

In regions where some forest remains or where there are national parks, the co-existence of wild elephants and agriculture is an increasing problem, as is the case for instance in the RLU (Royal Lestari Utara/Michelin) plantation in Jambi province in Indonesia because the plantation is located close the "Bukit Tigapuluh" national reserve. As food for elephants is freely accessible in RAS, vegetal biodiversity attracts these animals and RAS are considered to be more at risk than monoculture. In young RAS, elephants not only eat the associated crops, but sometimes uproot the rubber trees, for no obvious reason (Penot, personal observation in Gabon). Snakes, in particular arboreal species such as cobras or mambas are common in the rubber canopy. Like for birds, the association of rubber with fruit trees or shrubs is favourable for mammals and reptiles, with fruit eating species in turn becoming prey for other predators. The food web is improved when species of different height and size are associated and create a more complex and better connected habitat for the fauna. However, for the reasons explained above, RAS will not necessarily enrich wildlife when only a few plant species are associated and "harmful" practices are used such as spraying insecticides or herbicides. Rubber monocrops also provide an appropriate habitat for wildlife when the natural vegetation in the inter-row is only lightly controlled. According to the farmers, the presence of dangerous species such as venomous snakes or insects in dense and high understorey is the main reason for more intensive weeding in rubber plantations.

Considered as a whole, plantations provide connections with surrounding ecosystems. Deforestation affects wildlife not only directly through the destruction of habitats, but also indirectly due to fragmentation of the forest. When portions of forest are fragmented by a rubber plantation, the connectivity between the remaining patches of forest can be disrupted, since, as shown above, rubber monocrops offer few food resources and opportunites for shelter. RAS in general, and particularly the most complex ones, are thus considered to favour ecosystem connectivity.

▸▸ Impact on soils

Including crops/trees between the rows of rubber trees can improve soil health. Several multi-criteria studies focussed on soils have underlined the positive effects of associated crops on soil systems (Chen et al., 2017; Zhu et al., 2019a). The choice of associated species can also have different impacts on the soil, as reported by Chen et al. (2017). It is thus important to take the soil characteristics into account when designing RAS. Several soil ecosystem services can be affected by RAS, and several soil processes/components are discussed below.

Erosion

Rubber cultivation can also have negative effects on soil quality, particular when the soil is bare, and soil erosion can be severe in the rainy season. The main drivers of soil loss in plantations are applying herbicides and removing understory vegetation (Liu et al., 2016). The efficiency of the increased soil cover under agroforestry has been demonstrated (Liu et al., 2017). Diversification combined with proper understorey management is thus an efficient way to limit soil losses. On the other hand, multiplying the number of cycles of rubber cultivation in monoculture has been shown to have negative effects on soil fertility – at least, this has been demonstrated in the third cycle (Panklang et al., 2022b). The serious disturbance of the soil that happens during planting also results in soil loss. Fewer disturbances after planting combined with protecting the soil with cover crops can help to reduce soil loss during the rubber immature stage (Hu et al., 2023).

Water cycle

Similar problems are frequently mentioned concerning water resources (IUCN, 2011; Guardiola-Claramonte et al., 2010; Hauser et al., 2015; Fern, 2018; Higonnet et al., 2019). In rubber monoculture (particularly when it replaces natural forest), disturbance of the hydrological cycle and use of agrichemicals are responsible for surface water pollution. There are obviously significant differences between cropping systems, particularly between monoculture and agroforestry. Firstly, the use of chemicals has a major impact, and as a matter of fact, the majority of rubber plantations (whether or not under agroforestry) do not really need fertilisers or pesticides during the mature period as nutrient exports are very low. Secondly, according to Penot and Ollivier (2009), Jongrungrot et al. (2014) and Fern (2018), complex agroforestry systems like jungle rubber have less impact on water and soil quality (and hence on erosion and fertility) than most simple agroforestry systems, in particular, than monocropping.

Nutrient availability

In addition to possible effects of crop associations on the water cycle, nutrient cycles may be affected by diversification. The nutrient requirements of each of the associated crops need to be evaluated to avoid nutrient deficiencies and hence reduced yields. Research only recently identified the exact nutrient requirements of rubber (Chotiphan et al., 2019; Vrignon-Brenas et al., 2019). However, associations with other crops under rubber agroforestry systems may alter the nutrient balance and, depending on which species of trees are associated with rubber, nutrient availability may also be affected (Wu et al., 2020), and can lead to nutrient deficiency in the main crop, in this case rubber. Zhao et al. (2023) reported that using several intercropped species had a negative impact on the soil nutrient status and resulted in a shortage of phosphorus. The design of the RAS should thus account for the number of species, their functional role and their needs in order to avoid depending on fertilisation to overcome competition for nutrients among plant species.

Carbon storage in soils

Rubber is the only crop recognised by the Kyoto protocol for its carbon storage capacity (Penot and Ollivier, 2009). Indeed, in certain conditions, for instance, when rubber trees are planted on cleared land or to replace another crop, rubber plantations can store carbon. However, according to Hauser et al. (2015), these cases are rare, and the balance is more often negative when monoculture rubber plantations replace primary forest, secondary forest or even swidden cultivation. Rubber agroforestry systems are more virtuous in terms of carbon storage because they contain more trees per ha (Penot and Ollivier, 2009; Jongrungrot et al., 2014; Penot and Feintrenie, 2014).

Real impact of rubber agroforestry on soils?

Soils can also be restored after 40 years of rubber by using good management practices (Perron et al., 2022; Brauman and Thoumazau, 2020) in the absence of agroforestry practices. The impact of the practices is what counts for soil conservation

Soils can store a large quantity of carbon thereby mitigating the effects of climate change. The effect of land-use change on soil organic carbon (de Blécourt et al., 2013) and the evolution of soil carbon stocks in rubber tree stands (Blagòdatsky et al., 2016; Sun et al., 2017) are widely described in the literature. Agroforestry practices are considered promising ways to increase soil carbon stocks (Albrecht and Kandji, 2003), but the results of experiments on rubber agroforestry systems are still limited. Esekhade and Okore (2012) reported an increase in soil organic carbon during the immature stage in a study in which rubber was associated with banana. Increased carbon sequestration has also been reported in the rubber mature stage when certain tree species are associated with rubber (Li et al., 2020; Wu et al., 2020). The increased carbon inputs thanks to litter and via roots which are an integral part of the RAS system can thus increase soil organic carbon in rubber plots. In addition to crop associations, proper management of the understorey vegetation can also help build soil carbon stocks (Ren et al., 2023).

Soil biodiversity

Soils are one of the main reservoirs of biological diversity. Soil organisms probably represent around 25% of all species described worldwide. Such diversity is critical to sustain soil health and related ecosystem functioning including nutrient cycling, the transformation of organic matter, provision of physical support for micro-organisms (Bardgett and van der Putten, 2014). Soil biodiversity not only depends on a large number of organisms, the roles played by the organisms are critical in maintaining soil function. Increasing aboveground diversity through agroforestry systems has been highlighted as a key to fostering soil biodiversity, particularly in agroforests compared to croplands (Marsden et al., 2019). In the case of rubber agroforestry, the effects of increasing the number of planted species in this perennial-based system have been less studied than other components such as soil water and nutrient availibility. In their study, which focussed on the abundance of bacteria and/or fungi, Tongkaemkaew et al. (2018) found no difference in soil macrofauna between monoculture and RAS in South Thailand.

In mature plantations, Wang et al. (2017, 2020) and Jessy et al. (2017) showed that the microbial communities increased under RAS with certain associated crops. This increase in microbial abundance builds a more resilient system and may help to counteract the negative effects of rubber monoculture, e.g., acidifiction and nutrient depletion (Liu et al., 2019). However, other factors have been identified as being more important in explaining changes in soil diversity than which crops are associated with rubber. Management practices such as the use of chemical inputs or the management of understorey vegetation explain the differences in soil biodiversity better than the the fact of associating a particular crop with rubber (Liu et al., 2021; Therumthanam et al., 2014).

▸▸ Adaptation to climate change

Agroforestry is considered to be one of the best alternative cropping systems for adaptation to climate change based on three main assumptions: (i) plant diversification within a plot can mitigate the risk of damage caused by climate change, as different species differ in their reaction to a given climatic event. Next, the likelihood of having one or several species that are tolerant or resistant to the given stress is higher than with a monocrop. Similarly, the probability for the ecosystem to recover (resilience) is considered to be higher in a diversified system; (ii) agroforestry can mitigate some of the effects of climate change, particularly the increase in temperature, at the micro-climate scale as trees provide shade for understorey crops; (iii) multilayer vegetation can protect the ecosystem from extreme events, which will be more frequent in future, because several layers of vegetation protect the soil from erosion more efficiently, while trees protect smaller plants from strong winds.

However, in RAS, such positive effects often remain theoretical, as literature on the topic is sparse and because certain effects may be complex and have unintended consequences.

Plant diversification for risk mitigation

The susceptibility of different species to different kinds of climate stress may improve tolerance to future climate change, although this is difficult to demonstrate. Future predicted events including higher mean and extreme temperatures, more frequent and intense drought events, irregular rainfall patterns including more frequent rainstorms, will certainly affect all the crops that are usually planted in rubber growing areas. One possible mitigation factor is that the different crops used in RAS display varying degrees sensitivity to a given form of climate stress. For example, one species could be more sensitive to direct heat and another to soil drought, meaning there is less risk that all the plants in a given plot are affected by a given event. This is linked to asynchronous development of different plants, as sensitivity to climate stress depends on the phenological stage. To give but one example, rubber trees would be strongly affected by a heat wave during the leaf growth period early in the season, whereas fruit trees would be more sensitive during the fruit growth period, which comes later.

Direct environmental effects

Micro-climates: rubber trees provide shade for other crops

Using trees to shade crops reduces direct sunlight and hence the reduced temperature beneath the canopy is a widely used strategy in several RAS, as already well documented in coffee or cocoa agroforestry systems. Up to now, the shade provided by rubber has been too strong and has had negative effects on coffee or cocoa yields and so the association has not been recommended unless the inter-row is enlarged. However, with the continuing increase in temperature due to climate change, the need to reduce it could increase in parallel. Breeding for such systems should aim at developing shade-tolerant varieties of the associated crops on one hand, and rubber clones adapted to the association in question, on the other hand. However at the time of writing (2023), the required breeding criteria have not yet been clearly defined. In RAS, the presence of dense, multilayer vegetation under the rubber trees can also increase air humidity and reduce wind speed. But although the dense multilayer vegetation can limit damage caused by wind, higher air humidity can increase tapping panel diseases caused by *Phytophtora.*

Synergies (tapping different sources of water) versus competition for water

Another possible positive effect of plant diversification is that, when associated with annual crops or shrubs, trees tend to deepen their roots to avoid competition. As the trees then tap deeper sources of water, they become less vulnerable to drought (Panklang et al. 2022; Thoumazeau et al., 2022), while the associated crops exploit the more superficial water resource. Some associated species, like bamboo, can retain water in their dense root system, thereby increasing soil humidity. Nonetheless, it is important to be aware of possible competition for water between rubber and associated plants. In North-East Thailand, Clermont-Dauphin et al. (2018) showed that a cover crop of *Pueraria* competed too strongly with young rubber trees located in the upper part of a plot as the trees could not reach the water table, leading to the death of these trees in the dry season.

Protecting the soil against extreme events

Better soil cover can mitigate the effects of heavy rainfall on erosion. Erosion is already a major concern in rubber plantations, particularly – but not only – on sloping land. Using covercrops such as *Pueraria phaseoloides* or *Mucuna pruriens* during the immature phase of the plantations when the canopy cover is insufficient, is an efficient way to protect the soil. However, this practice is rarely used by smallholders because of the cost and labour required. Associated crops (e.g. rice, cassava) or pluri-annual crops (e.g. pineapple, banana) can play the same role if they are correctly managed. After the canopy closes, the cover crop or the associated crop decays naturally, leaving the soil almost bare. It is widely believed that the rubber tree canopy (or the canopy of other trees used as cover), will efficiently protect the soil from erosion. However, recent studies showed the contrary with a much higher rate of soil detachment and erosion in rubber monoculture than with open-field crops such as maize. The explanation is that the drops flowing off the relatively large rubber tree leaves are much bigger and heavier than normal raindrops that fall directly on the soil. Combined with the height of the canopy, this means the drops hit the bare soil with high kinetic energy. In RAS, the understorey crops can significantly mitigate this erosion process by intercepting these heavy drops of water before they hit the soil.

Sustainability in agroforests

Sustainability can be explained at different levels. It is the simultaneity of attributes in different domains which makes it a powerful concept. As far as agroforests are concerned, ecological sustainability is usually measured in terms of biodiversity conservation, natural resources management (soil, water) and pollution control (use of few phytochemicals or none at all). Economic sustainability can be visualized via the provision of stable, long-term diverse sources of income and patrimonial assets. The risk-buffering capacity of agroforests contributes to both ecological and economic sustainability. Social sustainability could be achieved through secure land tenure, secured by the capability of agroforests to avoid conflict (through people's common law or regulations, like *adat* in Indonesia), socialisation in a protected environment and the preservation of community values. These are values shared by a group and concern sustainability (preserving resources for the next generation), living environment (a "forest-like" landscape), a balance between fruit and timber resources and specific locations for social activities (holy forests or graveyard forests for the Dayaks in Indonesia, for instance). A vision shared by the members of a community reduces potential conflicts or sources of tension. A sense of sharing also reduces social differenciation. Again, among the Dayak people, timber that grows in the common *tembawang* can be used to build houses but not sold.

Provision of income for individual members of the population is very often balanced by collective decisions concerning the use of resources, attention paid to resources depletion and more generally, to social uses of agroforests. Institutional sustainability might be measured based on the fact that agroforests can be managed individually or jointly. Table 4.1 lists some arguments that link agroforests with sustainability.

Kumar and Nair (2004) rightly pointed out that home gardens may be on the verge of extinction due to new trends in agrarian structure, high market orientation, population pressure, land fragmentation, and acculturation. In the face of such constraints, the ecological foundations of home gardens may not be sufficient to ensure their survival. However, home gardens in Java persist despite an average population density of more than 800 people/km^2, and strongly market-oriented agriculture. The presence of some very high value crops (e.g. durian fruit) in the home gardens could explain this phenomenon. Yet, Java is not the only place where a positive correlation has been observed between the number of trees and human population density. Other examples have been found in Kenya (Tiffen, 1995), in Kerala, India, and in Sri Lanka (IRRDB, 1996).

Other multi-strata agroforests are also being influenced by changing economic factors. Jungle rubber (*Hevea brasiliensis*) and damar (*Shorea javanica*) gardens in Indonesia have had to face international price crises[67]. Diversification of local farming activities may occur at the expense of traditional agroforests, for instance, due to massive investment in oil palm. The effect of globalisation depends on access to markets and on the type of marketing involved. In Asia, most export products (rubber, oil palm, coffee, cocoa) have long been linked with international prices. In Africa, the commodity boards established in the 1970s to protect farmers from price volatility failed to deliver expected results and are now being called into question. As a result, globalisation has a stronger impact on African farmers than on Asian farmers, as the latter are used to adapting to international markets and to price cycles. We hypothesise that agroforests play a role in this adaptability, but other effects may have more impact: new decentralisation and local governance policies, new rules for access to credit, to projects or to information. Will agroforests be able to adapt to such changes more efficiently than conventional monocropping?

---

67. Rubber prices fluctuated from 2 US$ per kg in 1996, to 0.6 US$ in 2001 and then back to 1.2 US$ in 2004. In 2024, rubber price is between 1.3 to 1.9 US$/kg.

For many years, jungle rubber represented a great opportunity for poor farmers in pioneer areas. Now, monoculture or RAS using clonal rubber is much more profitable: yields and labour productivity are three to four-fold that of jungle rubber. In some areas, traditional RAS may not be a good economic option compared to either rubber or oil palm monoculture, for instance; but jungle rubber has nevertheless been replaced by monoculture RAS in some cases.

▸▸ Environmental concerns and externalities

If an economic perspective with emphasis on the local and regional levels were used to incorporate positive externalities such as agrobiodiversity management, improved nutrient cycling, integrated pest management, ecological sustainability and services, decision makers would possibly be convinced that home gardens and agroforests are profitable ventures. If an "agro-forest rent" approach is applied, policy makers and development professionals will consider agroforests as a profitable investment in the long term. This would lead to better consideration of agroforests in research and development programmes worldwide. If agroforests are still a success for many farmers, it is clearly not only for the sake of biodiversity conservation. Other values such as security (risk management and sustainability), diversity (and diversification), land control and reserve ("rights" to land and trees with emphasis on tree tenure), and social values, are included in the perception of agroforests, which are considered by most farmers as one cropping pattern amongst others.

Most farmers who continue to maintain agroforests also include some monocrops in their farming system, and these vary depending on the local context. The reason why farmers maintain agroforests in some countries, for instance, in India (Kerala), Indonesia (jungle rubber, *pekarangan*, damar systems), Sri Lanka (Kandy agroforests), or West Africa (oil palm based agroforests), is probably because their strategy has internalised the advantages of agroforests. A micro-economic analysis at farming system level including all sources of income, the cost-benefit of each activity and return to labour could explain such long-term strategies, provided it accounted for the dynamics (time effect) of perennial crops in home gardens and other types of agroforest.

Methods of economic analysis that use farming systems modelling capable of incorporating the outputs of mixtures of plants with different cycles and that allow smoothing of long-term and patrimonial strategies are certainly required to explain precisely what farmers do and why. Although agroforests are not a "panacea", their positive externalities and advantages seem to offer an ideal compromise between sustainability and risk spreading.

Beyond the economic advantages of agroforestry systems: what is the role and place of externalities in farmers' strategies?

Most RAS result from adaptations of cropping patterns to the local climate, local soil conditions, the farmer's cropping system, family self-consumption needs and in the case of AFS systems based on export crops, market conditions. AFS usually combine specific crops with the aim of producing different types of products, thereby helping diversify farmers' sources of income. Some systems are simply the result of local demand, for example, coconut tree-based systems in South East Asia that focus on food for self-consumption, or home gardens like the *pekarangan* in Indonesia (Torquebiau and Penot, 2006). Other systems are based on a main cash crop, usually rubber, cocoa, coffee, clove/nutmeg, resins (damar) fruits or timber species linked with the opportunity to grow a crop for export that developed during the colonial era in the 19th century (Michon et al., 1991, 1997). Both systems account for millions of hectares, particularly in South East Asia (jungle rubber covered 3 million ha in the 1990s). In 2024, agroforestry systems are still definitely part of local cropping systems largely used by local farmers but interest is reduces facing other opportunities such as oil palm.

Income diversification is a key to better global resilience of local cropping patterns through the production of the main crops and different kinds of fruit, firewood, timber wood, resin or rattan combined with other species including medicinal plants (Penot, 2001). The plants usually have a wide range of uses, health, home construction, food, handicrafts and furniture making (Momberg, 1993). Some products are sold and some self-consumed, usually depending on access to local markets. Some systems are based on the largest number of crops that can be combined while others focus on achieving the effect of associating one particular crop with the main crop, e.g., shading in the case of coffee and cocoa (Ruf, 1994). Most agroforestry systems result from marketing opportunities, which is the case of coffee, cocoa, rubber, cloves, or associations that are appropriate in a particular context (limited land availability, suitable soil and climate for a given tree/crop/livestock association) such as home gardens, AFS based on coconut, AFS based on fruit and timber trees, etc. Agroforestry systems with associations of trees host more biodiversity than monoculture, and their positive impacts on the soil which generally, but not always, include positive externalities. Most of these externalities, or those that are considered as such by external observers, for example by researchers or developers, are already an integral part of farmers' strategies and might not be perceived as externalities by the smallholders themselves, quite the reverse, they are an integral component of their cropping strategy. In other words, from the smallholders' point of view, do externalities deserve their name?

We believe they do not, particularly due to the social effects and the resilience of these components, as, right from the start, they are incorporated by the farmers in their decisions concerning the appropriate cropping systems, therefore, from the farmers' perspective, they are not considered as externalities. In other words, our main hypothesis is that most agroforestry farmers have already internalised externalities – mainly positive externalities – in their strategy and in their decision making concerning their choice of cropping systems.

From the point of view of agroforestry, we do need to know whether what economists call "externalities" are perceived as such by local farmers (even if the farmers are not familiar with the concept per se) or are already an integral part of farmers' strategies and consequently leave room for other aspects in addition to productivity such as resilience, long-term stability, and environmental concerns. In any case, measuring externalities is difficult as most services can be attributed a value indirectly or because unexpected advantages only emerge in the long run. We aimed to identify the role these externalities play in the survival and expansion of agroforestry as well as in the farmers' strategies and perceptions. We aimed to distinguish between the type of externalities that can be re-internalised to calculate economic value, those that are no longer considered as externalities by the farmers themselves (since they are an integral part of the technological package) and hence, whether technical, environmental, economic and social externalities attribute value to products that cannot be traded directly.

The concept of externalities in economics

In economics, the "externality" concept characterises the fact that through his/her activity, an economic agent has an "external effect", or more exactly an "unexpected effect", with no monetary compensation, i.e., that has a value or produces a benefit –positive externality–, or on the contrary, is a nuisance, causes damage without compensation – negative externality (Meade, 1952). In this way, an economic agent may be in a position to consciously or unconsciously influence the situation of other agents, who are not necessarily part of the decision. Externalities can involve different modalities depending on the topic: (i) technical components – erosion, fertility, water system, etc., (ii) economic components –margin, risks, (iii) environmental components –biodiversity, etc., and (iv) social components –farmers' patterns of organisation, etc. (Archibald et al., 1988; Oluyede, 2012).

A technical externality in production occurs when the production function of an actor is modified by the action of a third party. An economic externality in production occurs when the utility the actor derives from a good depends on the usefulness other consumers derive from the same good, and particularly on the position of the actor with respect to the position of the other actors in possession of the good, which is typically an "economist's perspective". An adoption externality, or network effect, occurs when the fact that other people perform the same action increases its usefulness (or value). In that case, the value of the product depends on how many users it has, which is typically the case of rubber agroforestry in Thailand, which concerns fewer than 5% of rubber farmers in the southern rubber production area. But that relatively small number concerns farmers who are organised in groups or networks (Theriez et al., 2017) and who are deeply involved in agroforestry and the specific knowledge and behaviour associated with it. In such cases, productivity and a purely "economic vision" are far from reality and do not explain the real components of farmers' strategies. Although concern for the environment has become a priority for most people, it was already a key component of farmers' perception of agroforestry, particularly stable production, agricultural sustainability and respect for the social value of biodiversity. In other words, externalities can be considered from different angles and need to be explored through different typologies to understand which components can be taken into account and potentially re-internalised.

Typologies to explore the concept

A standard typology can be identified by the type of economic and/or environmental effects it has: (i) "positive externalities", i.e., when an actor provides an economic service to a third party without being rewarded, and (ii) "negative externalities" when an actor economically disadvantages a third party without being obliged to compensate the affected parties (including him/herself) for the damage he/she caused. This typology is very efficient for descriptions and is by nature mostly qualitative. In agriculture, this typology is often used for technical or environmental externalities (Gomiero et al., 2011a). From a practical point of view, it is very effective for agroforestry.

Another typology can be identified based on the nature of the economic act. The term (i) "flow externalities" describes situations in which the economic action is a flow (for example, a flow of pollution), and the term (ii) "externalities of stock" applies where the economic action is a stock (for example, a stock of pollution). This typology is probably easier to use to quantify externalities, as flows and stocks are usually well documented and exploited in economics. The typology could be very useful in agroforestry, for example, to measure the long-term effect of pollution on soils, or the long-term negative impact on yield (for instance) or the positive impact of the biodiversity of the associated crops on soil fertility. But although such calculations are theorically possible, such long time series are very rarely available.

The final typology (Figure 4.1) can be identified by the type of economic act: (i) a "production externality" is when an actor profits by preventing the deterioration of a service or a product caused by another actor and (ii) a "consumption externality" in the case of consumption by another actor. This approach is rarely used in agroforestry.

Assessing externalities is a real challenge

Analysing externalities is difficult because relevant local agronomic or environmental data are rarely available (Pretty et al., 2000). Some negative externalities, like erosion, can be calculated using an appropriate equation (for instance Vishmayer's equation) plus information on soils and rainfall. Some, for example, global biodiversity, are impossible to calculate or to be attributed a value. Some can be extrapolated and valued by comparing them with other samples with no externalities. For instance, the value of collecting medicinal plants can be evaluated by indirectly calculating the health costs of a similar group of people with traditional cropping systems, but no positive externality like associated biodiversity.

Externalities cannot be measured directly by the consumer or by any other actor, but some, particularly technical externalities, can be measured at the level of the producer (Gomeiro et al., 2011b). Negative externalities (Figure 4.1) can penalise certain categories of economic agents (for instance, the cost of pollution caused by agricultural inputs for the production of Vittel mineral water in France (Benoit et al., 1997), nuisance, effects on health, etc.). Concerning positive externalities (Figure 4.1), a value has to be attributed that is recognised by all the actors, but such a value is usually not included in cost benefit analyses as it has no immediate return. Concerning negative externalities, as most are long term, they are also difficult to incorporate in economic calculations. However, the concept of multi-functionality in agriculture enabled the EU to recognise and attribute a value to some externalities by incorporating agro-environmental measures in the "Second Development Pillar", dedicated to rural development, established by the European Union (1999 reform). Originally externalities were an economic concept applied to value chains, products, and economic impacts on actors.

However, it is now clear that the technical and environmental dimensions of externalities are also required to adequately account for the impact of any complex system, particularly in agriculture, where environmental concerns are now priorities, as well as for complex multi-layered agroforestry systems. Producers' perspectives and perceptions consistent with global social evolution towards a more responsive civil society and more concerned consumers better account for externalities. In addition, the social value of externalities now has to be taken into account as the externalities concerned may be of importance in farmers' strategies based on their own perception of how agriculture should be implemented. However, as underlined above, externalities are very difficult to assess and measure, even in the long term. They may be considered as externalities that are "internalised" by smallholders right from the outset, as they very often contribute significantly to a smallholder's choice of a particular cropping system; this is the case in many local societies, for example rubber agroforestry farmers in Thailand, or clove agroforestry farmers in Madagascar. This social dimension is challenging to evaluate in terms of economic output but is now considered as one of the main assets in the livelihood approach (Serrat, 2017). The positive externalities of agroforestry may not be entirely "calculated" but they are definitely part of farmers' strategies, particularly their contribution to stablising agricultural income.

From an economic perspective, externalities reveal that a market price may not include all costs, and that services may not have all the effects expected. In terms of income analysis, whenever possible, efforts should be made to assess the cost and/or economic advantages of externalities to be able to include them when calculating the cost or margin. Re-internalising externalities is a real challenge. The three main difficulties involved in attributing a value to externalities are (i) the information required on all topics is not available locally, (ii) externalities may play a role in the long run and assessing long-term economic impacts is not easy, and (iii) social values, which are very important in some rural societies, may be impossible to evaluate economically. Thus it is almost impossible to account for all the costs and services provided by and/or the advantages of certain societal features (Alliot, 2003). Even if some features are well known, e.g., the impact of biodiversity on soil fertility and water management, their long-term effects are rarely well documented,

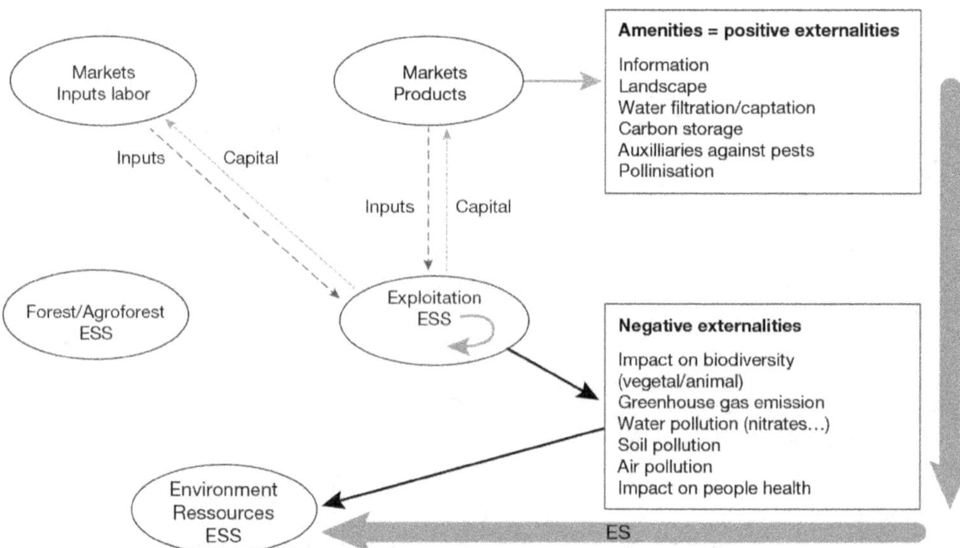

Figure 4.1. Schematic representation of positive and negative externalities of an agricultural production system

ESS: ecosystem services; ES: environmental services.

Most farmers who practice agroforestry have intuitive knowledge of positive externalities or have acquired genuine knowledge on their own and generally take these different types of knowledge into account in their strategies (Momberg, 1993). The same applies to other agricultural technologies such as conservation agriculture or permaculture, which are perceived as ecological intensification processes (FAO, 2001).

Most externalities are considered as environmental services. Aznar et al. (2007) offered three definitions of environmental services:
– Products of natural capital: environmental services are defined as services rendered by nature to man, not as products.
– Positive production externalities: the work of the Organisation for Economic Co-operation and Development (OECD) on rural amenities (1994) and on the multi-functionality of agriculture (1999) fit this definition. Environmental services are considered to be positive production externalities. The services provided are not planned, but are the result of practices in the same way as economic assessments measure the economics of environment (Baumol, 1988).
– The economy of services through the intentional character of the supply of a service: for instance, improving the environment is intentional from the point of view of a service provider (e.g. in the case of providing drinkable water).

Typical examples of environmental services are carbon sequestration by agroforests and the biodiversity of the associated crops, both of which are real externalities, but are not considered as such by farmers in their strategies.

For institutions, evaluating externalities at a macroeconomic level – national or international – allows them to demonstrate the merits of their proposals – advocacy, public policies, bills, etc. (de Foresta, 2013). This could be the best way to measure the global impact of externalities at regional or national level, given the previously mentioned difficulty in effectively assessing certain externalities and their real impact at the level of a territory. For private companies and/or estates, the question could be "what is the value of services the company renders indirectly, or the benefits obtained from using our products?"

Some technical and economic externalities can be included in the operating budget as costs and services for a particular product, but may be far more difficult to include in national accounts. The question is therefore how to transform services and problems into costs and advantages (Conway, 1993). Some attempts were made at the beginning of the 2000s in the European Union. In France, to give one example, with the Territorial Farming Contract (*Contrat Territoriaux d'Exploitation*, CTE; Aznar et al., 2005) in which subsidies were not only linked to the production of specific product (e.g. wheat) but also to some ecosystem services. Some methods of calculation are controversial, for example, contingent valuation, which consists in undertaking monetary valuations via surveys of the type "how much would you be willing to pay to preserve such a resource?"

It is also difficult to accurately measure the value of public goods. Any estimates will continue to be the subject of controversy by some actors (Pearce, 1998).

The objective: to re-internalise externalities in the case of agroforestry

All recent studies conclude that, with the exception of most agroforestry systems, the intensification of agricultural practices leads to loss of biodiversity and to degradation of some ecosystem services, e.g., pollination of flora by bees (Warren-Thomas, 2020).

Some important questions concerning agriculture in 2024 are thus:
– How can global ecosystem services be incorporated at the scale of a territory?
– How can nitrate pollution of an underground aquifer by a river be incorporated?
– How to account for the silting up of irrigated rice fields linked to the erosion of the surrounding hills? (Cacho and Hean, 2001)
– Does agroforestry limit greenhouse gas emissions thanks to its carbon sequestration potential?
– Do agroforests contribute as much as most forests?

Agroforestry should not only be assessed from the point of view of carbon storage. Agroforestry has other ecological advantages including (i) protecting drinking water resources (reducing nitrate and pesticide leaching), (ii) creating ecological corridors and more attractive landscapes, (iii) reducing soil erosion thanks to better control of runoff, (iv) improving the fertility of agricultural soils (Boonkird et al., 1984). It is difficult to attribute an economic value to such externalities, however, life cycle analysis (LCA) is probably a useful tool to give a value to some externalities, although to our knowledge, as yet, it has not been used for that particular purpose and further research is needed (Hendrickson et al., 2006). The exploitation of timber in agroforestry systems does contribute to a very long use of agroforestry products and long-term carbon storage, which could be demonstrated using LCA.

The existence of externalities partially explains the gap between the potential and actual adoption of sustainable land-use practices. There is substantial evidence for the advantages of biologically diversified agroforestry systems in conserving biodiversity, controlling certain pests, weeds and diseases, enabling pollination, maintaining soil quality, increasing energy-use efficiency and mitigating the effects of global warming by reducing the temperature, or an example of the other extreme, by protecting arabica coffee from freezing at 1,000 meters above sea level, a high altitude for coffee, in northern Vietnam. AFS may even allow rubber production when the maximum temperature (28 °C) needs to be reduced to enable the lactiferous system to continue to function. Agroforestry systems increase resistance and resilience in the face of extreme weather events. AFS enhances carbon sequestration and the water-holding capacity of surface soils (Gomiero et al., 2011a; Kremen and Miles, 2012). The real challenge is to re-internalise externalities in the economic analysis of agroforestry systems either as costs or as economic advantages. To this end, we need to:
– attribute a value to both positive and negative externalities, which is possible at plot level, difficult at land use/territory level;
– re-incorporate valued externalities as margin/ha or margin/farm account and whenever possible, when calculating income;
– evaluate certain externalities that are currently not fully or accurately evaluated, for example, maintaining the original soil fertility at plot level, or the global impact of ensuring the survival of pollinating bees.

Farmers are considered as key components of sustainability when they adopt agroforestry as the main component of their strategy with the aim of reducing vulnerability and increasing resilience, both at farm and environmental level. If the data are available, the technical and economic externalities can then be included in an operating budget.

Managing externalities in agroforestry

The environmental costs of agriculture have limited the ecosystem services on which we depend. To ensure the global sustainability of agroforestry systems, it is essential to include the costs (externalities) of agriculture which were originally "invisible", in particular in smallholder decision-making processes, in order to identify the "true" cost of using certain agricultural inputs and practices, both negative effects (chemical pesticides, pollution, etc.) and positive effects (soil fertility, biodiversity conservation, providing a natural habitat for birds and insects, etc). Externalities can be identified at different levels in agroforestry systems ranging from (i) the externalities of a particular cropping system to those of the plot itself, (ii) externalities to other actors (at landscape/farming systems/land-use level). The positive externalities of AFS are (i) biodiversity conservation (as a sanctuary, niche, or reservoir), (ii) protection against erosion, (iii) provision of ecosystem services, and (iv) social value (for instance, the religious offerings and traditional gifts of fruits in Thailand). The negative externalities of AFS are (i) their impact on and reduction in the yields of associated crops, (ii) labour requirements.

A typical example of an externality is the "biodiversity concept" with: (i) "Useful" biodiversity (timber, fuelwood, wild fruits, resin, etc.) is widely known and its components are combined to increase the resilience of AFS, (ii) crop diversification, which depends not only on a single product but provides several sources of income in the short and/or medium term, and also includes (iii) biodiversity with no marketable value as a way of providing long-term ecological services thereby making a significant contribution to long-term sustainability. Then, the following questions arise: (i) What is the role of these externalities in agroforestry development and in the associated farmers' strategies? (ii) Even if most income analyses find it difficult to attribute a value to these externalities, they may nevertheless play a key role in farmers' choices and preferences for agroforestry over monoculture –if they have the choice. Biodiversity is also a major component of landscape management (Schroth et al., 2013).

At land-use level, the main externalities are (i) mitigating greenhouse gas emissions due to "forest-like" carbon sequestration, (ii) protection of drinking water resources (by reducing nitrate and pesticide pollution), (iii) creating ecological corridors as sources of biodiversity and enabling the passage of a wide range of organisms including large animals, (iv) controlling soil erosion through better control of runoff, (v) improving the quality and fertility of agricultural soils, (vi) reducing negative externalities on health caused by pesticides (see Tables S8-S10 in appendices). Table S9 lists positive and negative economic externalities, while Table S10 lists externalities at landscape level. In the end, some externalities are not perceived as "externalities" by most smallholders as they are an integral part of their strategy and can be considered as major factors behind their decision-making process. Where is exactly the frontier?

Some examples of incorporating externalities in RAS in Thailand and Indonesia

In southern Thailand, rubber agroforestry is considered by some farmers as a specific cropping design because, since the 1960s, most farms have been based on monoculture through the actions of ORRAF/RAOT[68]. The main features of rubber based agroforestry systems are the following: (i) initially, the social value of fruits in fulfilling the tradition of welcoming any visitor and even family members with fruit, and later on, faced with the low rubber prices since 2012-2013, as a source of income diversification to cope with rubber price volatility, (ii) following the late sovereign's recommendations: respecting nature and re-introducing trees and forest-like environments (social value), (iii) attributing a value to medicinal plants[69] and (iv) a process of land intensification in areas where land availability is limited or scarce (like in South Thailand where all land is already being cropped, or in transmigration areas in Indonesia). In 2015, a survey of 34 local rubber farmers was conducted in 4 districts in Thailand (See Figure 3 in Stroesser et al., 2018) and showed that agroforestry systems were more resilient to rubber price volatility thanks to income diversification. Figure 2 (in Stroesser et al., 2018) shows the net gross margin/ha of 5 types of AFS. The 6th type is an RAS involving early tapping with low rubber production. All of them are compared with the current rubber monoculture system and an average international price for rubber of US\$1.3/kg which remained stable over the period 2015/2020. The study by Stroesser et al. (2018) showed that in 2015, rubber agroforestry systems provided an average income equivalent to that obtained from rubber monoculture at a rubber price of US\$3.2/3.7/kg. In other words, RAS do help maintain a stable agricultural income, reduce vulnerability and contribute to global resilience. Income stability is a key priority in farmers' strategies in which AFS play a key role in sustainability.

Such AFS have other externalities that are currently not taken into account to calculate the gross margin. The most important ones that are easy to calculate are:
– The value of stored carbon: a rubber plantation can produce up to 200 m3 of wood to be used to build furniture. If we add the 50 m3 of associated timber trees: the total quantity of wood produced is 200×500 kg $+ 50 \times 600$ kg $= 130$ tons of wood (with both forest equivalent carbon sequestration for 30/35 years and substitution through the use of the wood produced)[70]. The CO_2 value is therefore equivalent to $130/2= 65$ tons of C/ha, valued at 1,625 euros (i.e. 25 euros/ton) in 2019. If the carbon market were efficient, this value would largely contribute to the cost of replantation (the average cost of replanting in AFS is between 2,000 and 4,000 US\$/ha for the first five years).
– The presence of weeds and/or associated shrubs, small trees and/or small crops help maintain soil fertility, or even enable improvement, but real data are unfortunately rare (Neyrey et al., 2018). In 2019, Liu showed that using herbicides in rubber plots had a negative impact on erosion. Thoumazeau et al. (2018) demonstrated that the presence of weeds and shrubs maintains and even increases soil fertility in rubber plots. In the long run, the final output is environmentally sustainable agricultural production, which can be measured in terms of income stability, for instance. The difficulty in visualising a long-term impact is that it requires including 2 or 3 complete cycles of the perennial concerned.

---

68. RAOT (Rubber Authority of Thailand) is in the process of replacing ORRAF (Office of Rubber Replanting Aid Fund).

69. That point is also very important for Dayak farmers in Kalimantan for access to fuelwood and fruits in transmigration areas where forests have disappeared (from trees growing outside the forest).

70. The density of rubber wood is 450/650 kg/m^3 (average 500), that of teak wood (often present in AFS) is 480/850 kg/m^3 (average 600). One cubic meter of wood contains 1 ton of CO_2 and 1 ton of wood contains 0.5 tons of CO_2.

– It would be possible to measure the effect of AFS on the soil by comparing it with total yield and the income to be obtained from a second rubber cycle if soil fertility were not maintained and the result was consequently a decrease in yield. However, the result would be questionable, as climate and diseases also have a major influence on yield.

– As most actors mentioned, the social value of AFS is real but probably cannot be measured economically, which is true of both non-marketable plant and animal biodiversity.

– In Indonesia, farmers who have no easy access to health facilities, but instead use the medicinal plants that grow in the jungle rubber version of RAS, save between 5% and 8% on their normal family expenditure (Courbet et al., 1997). According to a survey conducted in Indonesia in 1997, using timber from AFS plots enabled farmers to build houses for their children at 30% of the normal cost.

– Re-internalising such costs and advantages will not fundamentally alter the analysis of farmers' strategies, as a value is attributed to services and factors that does not affect their immediate income, but does make it possible to compare different types of systems and to assess their sustainability over time, for instance oil palm AFS vs. RAS. Including the cost of fertiliser or of the pollution of water by pesticides, loss of soil fertility after several cycles and of the impact on diseases would enable a better assessment of the long-term sustainability of the different AFS cropping patterns.

One of the most important features we observed in RAS in South Thailand as well as among Dayak farmers in Kalimantan (Indonesia) with similar systems (Penot, 2001), is the social value attributed to AFS by local farmers (Stroesser et al., 2018; Theriez et al., 2017; Penot, 2001) before it also acquired economic value when there was a sharp fall in the prices of rubber.

Conserving biodiversity and the "original social value" of biodiversity (for instance of medicinal plants, but not exclusively) have been explored in Indonesia (Werner, 1997; Diaz-Novellon et al., 2002), and in Thailand (Warren-Thomas et al., 2020) where farmers referred to the need to protect local biodiversity reflecting the late king's recommendations concerning South Thailand (*"protect the environment and keep trees in the landscape"*). From a practical point of view, the use of medicinal plants either as a general health treatment or to treat a specific disease allows farmers to save on what they would normally spend on doctors and medicines in a context where most smallholders do not have a social security safety-net. Externalities that provide stability (erosion control, etc.) became more qualitative over time, and were perceived by local smallholders as inherent to a cropping system.

Conclusion

All potential components that render agricultural production more resilient and sustainable are already an integral part of local farmers' long-term strategies. In other words, some externalities (or which are considered as such by socio-economists), may not be considered as such by local farmers who incorporate the expected output right from the outset even if they cannot give it a monetary value. They do have a clear perception of environmental ecosystem services to which they attribute a social value that is locally recognised in the same way as its economic value. Among the expected outputs that are no longer externalities for farmers are fire prevention, protection against erosion, preservation of the water catchment and of soil fertility, and biodiversity conservation to ensure continued access to fuelwood and other products. To these can be added all factors that indirectly contribute to income stability, better return to labour and the overall long-term sustainability of cropping systems. This explains why – when they have the choice – so many farmers prefer agroforestry systems, particulary because it is usually easy to incorporate local plants (fruit and timber tree species, local non-timber forest products, etc.).

It would be useful to calculate the real value of externalities to be able to compare the economic and environmental efficiency of different types of cropping patterns as well as their impact on the landscape, for instance at watershed level. However, this might not even be necessary given that local farmers consider the "social value" of agricultural sustainability to be its most important aspect, whatever its real economic value. It seems more logical for local communities to respect social values by, for instance, referring to the late king's philosophy concerning forest and trees in Thailand than to a hypothetical calculated economic value that in fact is only of interest to researchers who need to compare situations.

Whenever possible – and when data were available – most values have been attributed and calculations made at plot level and rarely at landscape or land-use level (village, region, watershed, etc.). The real impact of externalities at landscape/land-use level is still largely under-estimated in 2024. The positive externalities of AFS have become key components in the adoption of AFS and in local farmers' strategies. We also observed that negative externalities are very limited in AFS (Table 4.4).

Other multi-strata agroforests are also under the influence of changing economic factors. Jungle rubber (*Hevea brasiliensis*) and damar (*Shorea javanica*) gardens in Indonesia have also had to cope with international price crises[71]. Diversification of local farming activities may take place at the expense of traditional agroforests, for instance due to massive investments in oil palm. Agroforests are hypothesised to play a role in this adaptability. Other effects might have a bigger impact: new decentralisation and local governance policies, new rules for access to credit, projects or information. Will agroforests be able to react to such changes more efficiently than conventional monocropping?

Natural rubber is a renewable resource when rubber plantations are well managed, unlike petroleum, which is used to make synthetic rubber. What is more, as a large proportion of rubber comes from village plantations, which has many positive social effects (Hauser et al., 2015; Pirard et al., 2017). As summarised by Gitz (2019) and developed in this work, the potential impacts of rubber expansion depend on three main factors: (i) the land use or land cover that rubber replaces (natural ecosystems or cultivated or degraded areas); for example, in a context of climate change, a rubber plantation will store more carbon than an oil palm plantation and rubber wood can be harvested at the end of the tree's life cycle; (ii) the type of production system (monoculture or agroforestry and its overall efficiency); for example, an agroforestry system has a less negative impact on biodiversity or water than industrial monoculture, and (iii) benefits smallholders and local populations by contributing to their economic and social resilience.

---

71. Rubber prices dropped from US$2/kg in 1996, to 0.6 in 2001 and then back to 1.2 in 2004.

Table 4.4. Summary of sustainability attributes of agroforests

Ecological	Economic	Social and institutional
Reduces soil erosion,	Significant use of endogenous resources,	Reduced and flexible labour requirements,
Increases soil organic matter content,	High safety factor against marketing and seasonality hazards,	Contributes to nutritional security,
Buffers soil moisture and temperature,	Reduces need for cash, thanks to the many diverse bio-physical outputs (plant and animal food, medicines, fibres, etc.),	Contributes to community socialisation,
Closes the nutrient cycle		Preserves traditional knowledge,
Improved soil physico-chemical properties,	Socio-economic outputs are diversified and distributed over time,	Biodiversity linked to traditions and practices,
Efficient use of light and water,		Gives women a key role,
High wild plant and animal biodiversity,	Balance between subsistence and cash income,	Enables equitable distribution of products,
Use of endogenous resources,	Possibility to build capital,	Functions as a land reserve (for alternative land uses),
Contribution to on-farm production of wood and fuel wood,	boosts rural industries and employment,	Maintains right of access to common goods (e.g., fruits),
High soil biotic activity,	Can adjust to different contexts,	Ensures flexibility of ownership (private vs. communal).
Better scope for evolution and diversification of plants with an economic value,	Stabilises yields,	
Differentiated vertical and horizontal management zones and related ecological niches,	Offers management flexibility (intensive vs. extensive),	
Potential for organically grown products.	Economic resilience (value as "land reserve").	

Source: Adapted from Torquebiau (1992), Penot (2003), and Kumar and Nair (2004).

These three factors will play a determining role in the sustainability of rubber production in the future. According to Gitz (2019), the rubber sector needs measures to connect downstream and upstream that involve different stakeholders, that build on science and knowledge and that promote transfer from one system to another in a practical way.

Evaluating the real economic impact of agroforests – a challenge for both agronomists and social scientists

Part of this section has been originally published as a keynote paper[72]. The advantages of sustainable agroforests originate from a trade-off between ecological and socio-economic attributes. Conventional economic approaches might have a hard time combining the two series of attributes in a comprehensive manner. Compared to a simple yield analysis, which is possible in conventional monocropping agriculture, the array of attributes of agroforests are a challenge to agronomists and social scientists alike.

---

72. See Penot (2016), Xishuangbanna Tropical Botanical Garden by the Chinese Academy of Sciences, Yunnan Province, China.

Among other facts, (1) the products are varied, their production is spread out over time, the plants have life cycles of different lengths, and combine subsistence, cash, capital and patrimonial objectives, (2) ecological benefits are crucial but are not internalised in analyses, and (3) some ecological attributes have no market value.

If neoclassical economics is used to assess the performance of agroforests, yield criteria, cost-benefit analysis and net present value may disqualify agroforests as opposed to conventional monocropping, because the analysis will exclude a series of agroforest outputs which are not traded on the market or adequately accounted for in farm economics. A value can be attributed to a good, whatever its final use (including savings made thanks to self-consumption), but services and positive externalities are far more difficult to assess. Risk buffering by agroforests needs to be measured, e.g., in the case of a drought, in an El Niño year, or in the face of commodity price volatility. Farm system models can perform this task and will produce different comparative scenarios.

The overarching question is simply: How can we measure the agricultural sustainability of agroforests?

Farming systems approach

The flexibility of tree and crop production in agroforests is tied to the mature and immature stages of the trees or crops involved. It is thus essential to take the life cycles of the different plants into account in long-term economic analyses. Specific discounting rates may be necessary as cycles can last up to 40 or 50 years. Different scenarios will be needed, as bias can occur in valuing products depending on the discounting rates chosen. For instance, in agroforests based on tree crops, rubber or resin is produced for more than 30 years, whereas annual and bi-annual crops are usually only cultivated during the first 3 to 6 years of the life of a plantation. Timber can only be harvested at the end of the life span of the agroforest. Thus, if detailed data are available to reliably assess real income (including self-consumption), comparing systems will be more valuable than absolute data (Penot, 2001).

When the benefits of agroforests can be analysed using the market value of their products and services, then neo-classical environmental economics can be used and externalities can be internalised (or re-internalised) in the process of income generation. The increase in pollution and its cost can be taken into account as negative externalities or constraints to further development. Environmental services (e.g. carbon sequestration; Albrecht and Kandji, 2003; Montagnini and Nair, 2004) can be valued according to a "system of values" recognised locally as being relevant at a higher level, that of the community or province. The problem is knowing whether farmers really do benefit from externalities and from the advantages of agroforestry, or at least have the potential to benefit.

Whether agroforests are commercially or subsistence oriented, a long-term perspective must be part of each farmer's strategy. However, there is obviously a bias in the debate between the short term (economics) and the long term (ecology). In both cases, farmers have developed their own long-term farming practices through a long-haul innovation process which may or may not account for economics through the risk buffering capacity of agroforests. In most cases, social organisation is tightly linked with technical production constraints, food security, reliable income and control over land. There is a strong coherence between technical systems (technical pathways) and social systems (Penot and Chambon, 2003).

⇥ Interest of certification

Today, rubber production is facing both environmental and social challenges. Additionally, acording to Fern (2018), compared to other agroindustries like palm oil, the rubber industry has been lagging behind in terms of sustainable and responsible production, and has been slow to act on deforestation, labour and human rights issues. Tire manufacturers are threatened by potential public scandals concerning the sources of their supplies (deforestation, exploitation of the poor for labour, etc.) and the pollution caused by the production of tires. Tire manufacturers may be using certification to partially protect themselves against such scandals. If a media scandal were to break out over deforestation or other environmental or social damage, companies that have not yet begun certifying their rubber as sustainable will be consigned to the sidelines.

Stakeholders in the rubber industry began to focus on the sustainability of the natural rubber industry in 2012, by creating a think tank in the IRSG (International Rubber Study Group), which resulted in an SNR (Sustainable Natural Rubber) initiative (iSNR). The main weakness of the IRSG is that it includes neither the main producers (e.g. Thailand) nor the main consumers (e.g. China). The working group produced a first draft of specifications, entitled iSNR, to promote sustainable rubber. So far, the project has led to the identification of 5 major axes: (i) productivity, (ii) quality, (iii) child labour, (iv) treatment and use of water, and (v) deforestation.

The limited involvement of producing countries led the tire manufacturers to launch a global platform for sustainable natural rubber (GPSNR) via the World Business Council for Sustainable Development (WBCSD) Tire Industry Project (TIP) at the end of 2017. The goal is to establish a fair, equitable and environmentally sound natural rubber value chain. Based on 12 criteria, the mission of GPSNR is to lead improvements in the socioeconomic and environmental performance of the natural rubber value chain.

⇥ Rubber and oil palm

Role and place of oil palm

Local smallholders rapidly included rubber in agroforestry systems when it was first introduced in Indonesia, in Bogor (Java), North Sumatra and then West Kalimantan Province. Since the 1970s, many government projects have been implemented with the aim of replanting using more productive rubber clones in monoculture systems (SRDP, TCSDP[73]).

During the same period, transmigration centres[74] were created to enable the settlement of Javanese immigrants in Kalimantan, based either on food crops (which was a dismal failure) or on tree crops, first with the rubber and then with oil palm NES[75] which was a relative success (Levang et al., 1997). The Indonesian government's transmigration projects were intended to relocate Javanese transmigrants from overcrowded Java to the outer, less populated islands. The presence of the official Javanese trans-migrant

---

73. SRDP: Smallholder Rubber Development Project. TCSDP: Tree Crop Smallholder Development Project.
74. The policy of moving surplus populations from Java island to the outer islands began in 1905. It was re-launched in 1950 following independence, and peaked in the 1980s. The Ministry for Transmigration was established in 1984.
75. NES: Nucleus Estates Smallholder Scheme (PIR in Indonesian).

populations almost never led to either social or land ownership conflicts with the local Dayaks. The conflicts that did break out in the province in 1998 and again in 2001 involved the Dayak and Madurese communities. Occupation of land by Madurese farmers in the absence of prior negotiation with local communities may well have been one source of conflict, but it was not the only one. Cultural differences and behaviours also triggered tensions between the two communities, and the process accelerated from 1985 to 2010.

This section has been partly originally published in Penot and Geissler (2004). Land occupation in West Kalimantan Province (Borneo), particularly in Sanggau district, changed considerably between the 1980s and 2020. Since the introduction of rubber in agroforests at the turn of 20th century, the Dayak's original slash-and-burn agriculture shifted to jungle rubber, and then, in the 1990s to oil palm. Oil palm and *Acacia mangium* estates were established at a very large scale between 1990 and 2015 thanks to the Indonesian government concession policy. Since the 1980s, the different actors (the State, private companies, local Dayak communities and Javanese transmigrant populations) have adopted land-use strategies that have caused a considerable modification of overall land use. The government policy of issuing concessions for oil palm and *Acacia mangium* led to a new legal redistribution of land to the detriment of local populations whose rights were based on "customary rights" (*adat* in Indonesian). This situation is a potential source of conflict between concession holders and local communities. In 1985, local communities had legal control over 52% of the district, but only over 29% in 1998. Protected forests accounted for 7% and transmigration projects for 3% of the total area. In reality, the situation is less alarming, because only a portion of the land under concessions was actually planted (20% on average in oil palm concessions and 10% in *Acacia mangium* concessions). According to a 1999/2000 survey (Geisser and Penot, 2000), at the end of the last century, in fact 54% of the area was still available for use by local communities. The alarming "legal situation", a source of potential conflict, was therefore tempered by the real rate of land occupation. What has changed in the meantime?

The new legal redistribution of land in the 1990s from the government lend to oil palm and *Acacia mangium* concession policy that took place to the detriment of local populations who, in addition, had not received any clear information on the subject[76]. If the situation continues and land becomes increasingly scarce, it might generate conflict between concession holders and local communities.

Oil palm is now the main crop grown both by local farmers in the area (in 2024 oil palm accounts for 72% of the cropped area) and by the estates, although rubber remains important for those local farmers who want to keep a certain level of crop diversification. In 2020 we found that most of the former jungle rubber area (90% of the whole rubber area in 1994) had been converted to oil palm and/or to a lesser extent, to clonal rubber. In other words, although rubber production continues, the majority of jungle rubber has disappeared because yields of clonal rubber are 3 times higher. In the landscapes under study, oil palm had the effect of a steamroller. Most local Dayak farmers exchanged land in a way that benefited the oil palm estates (the farmers gave up 5 ha of land in exchange for 2 ha planted with oil palm trees provided by the estate).

---

76. Published in Penot E, 2021. Rubber Agroforestry systems (RAS) in West Kalimantan, Indonesia: an historical perspective. *E3S Web of Conferences* (Vol. 305, p. 02001). EDP Sciences.

In 2024, most farmers cultivate an average of 2 ha of oil palm, 2 ha of rubber (partly clonal and some remaining jungle rubber) and have a small area for food crops or other crops. These farmers can no longer count on land being available, as they did some 25 years ago. We do not know the exact proportion of clonal rubber currently cultivated as agroforestry, but it could be more than 30%.

It is important to grasp the "pros" and "cons" of oil palm and how oil palm has significantly influenced land use, farmers' strategies and cropping patterns. The pros of oil palm are (i) limited labour requirements: 8 days a month/ha compared to 14 for rubber, (ii) secure income up to now, despite fluctuations, (iii) access to homes and to some social benefits, (iv) new roads and access to markets. The cons of oil palm are (i) loss of land (5.5 ha) according to concession regulations, (ii) the risk implicit in a monoculture: less resilience, (iii) one hectare of oil palm requires 700/1,000 kg of fertilisers/year so the farmers must have the necessary capital, and (iv) a recent decrease in the price of fresh fruit bunches.

Because of its advantages, oil palm is now the number one crop for local smallholders, jungle rubber has almost completely disappeared but clonal rubber is still being cultivated, partly as rubber agroforestry. Some local Dayak farmers also kept some jungle rubber as a land reserve while preserving *tembawang* (man-made agroforests with fruits and timber trees, which is possible under *adat* common law). In 2020, in our study area, we were able to estimate that in the 4 villages where the SRAP (Smallholder Rubber Agroforestry Project) was implemented, 70% of available land was under oil palm, 20% under clonal rubber (either monoculture or agroforestry) and 10% remained as old jungle rubber and *tembawang,* according to the farmers. In transmigration areas, the situation was different, as most farmers owned only 2 ha (sometimes 3 ha) mainly planted with clonal rubber. Oil palm companies did not intend to penetrate areas that enjoyed a special status. The Dayak farmers do not have the possibility to cultivate oil palm on new land on their own initiative (Penot and Chambon, 2003).

Deforestation

The very first responsibility for deforestation belongs to Indonesian logging companies. Theoretically, the terms and conditions of the forest exploitation contracts guaranteed sustainable logging (so-called "productive" forest status). It was the failure of forestry companies to respect these terms, leading to considerable over-logging, that made them largely responsible for the first deforestation (Gouyon, 1993), not slash-and-burn agriculture, which long served as an ideal scapegoat[77]. In fact, the TPI law of 1972 and the 1989 TPTI law on management systems[78] laid down in the land ownership legislation were far from being respected by the entire private sector (Cossalter, 1992; Durand, 1999). Indonesia passed legislation classifying 75% of its land as forest production area. In reality, the forestry potential of Indonesia in 1998 was 66 million hectares (Durand, 1999), i.e., 35% of the total area. For Dayak farmers in the Sanggau district of West Kalimantan Province (Borneo), the forest had long been a major resource for

---

77. In June 1998, the new Minister for the Environment in the first post-Suharto government (the Habibie government) officially recognised that the forest management situation was similar to that in the American Far West in the 19th century, i.e. lawless.

78. TPI (*Tebang Pilih* Indonesia) and TPTI (*Tebang Pilih Tanam* Indonesia) are laws defining felling methods and the duration of concessions.

hunting, gathering, collecting supplies of wood, medicinal plants, etc. However, the resource diminished following the introduction of tree crops: rubber in 1911, and oil palm and commercial forest crops, primarily *Acacia mangium*, in the 1990s.

Around the Indonesian State plantations (PTP), smallholder plantings (in practice they were controlled by the PTP) have developed and are referred to as NES (Nucleus Estate Smallholder Scheme). The very first tools of government action were smallholder development projects. Two types of projects were designed: sectorial development projects that targeted local farmers (rubber, oil palm, coconut) and transmigration projects that targeted external Javanese populations (rubber, oil palm, coconut and food crops). The State also acted through its policy of issuing forestry concessions, industrial concessions (HTI) and concessions for tree crop plantations to private companies, and through its transmigration programme. Two types of concessions exist: concessions for perennial crop estates (the majority decided to plant oil palm) and concessions for industrial tree crops (most planted *Acacia mangium* for pulp).

The second factor that without doubt intensified deforestation was the official concession policy, particularly that for oil palm. Deforestation was exacerbated by new planting companies (Potter, 1999), particularly in 1997, an El Niño year with a severe drought, when most of the fires that occurred in the area were associated with the planting operations undertaken by these companies (Laumonier and Legg, 1998). Oil palm boomed from 1997 to 2010. The concession policy ended officially in 2015. In the 1990s, many Indonesian companies obtained big concessions (mainly for oil palm) to profit from the oil palm boom. The government considered oil palm to be a "modern" development pathway as well as a valuable source of income and employment for the local population. In 1998, the total area under oil palm area in Indonesia was estimated at 2,634 million ha, up from 500,000 ha in less than 15 years. The different forest status classes are listed in Table S11 (Appendices), and the different types of forest and concession are listed in Table S12 (Appendices) along with their respective actors.

From *adat* to concessions, or the legal wresting of land control from local populations

Common law (*adat*) and land occupation in 1980

Until rubber was introduced at the turn of the 20th century, common land was abundant, it was managed by the community and had no value because there was no market for it. The extension of rubber plantations was accompanied by a gradual shift to land ownership but still based on *adat* local law (Michon et al., 1986). The traditional shared land ownership law grants individual tenure of any plot that is really farmed, and for as long as it continues to be farmed. Planting trees, including rubber, was thus a direct way of acquiring land, and, under the rules of usufruct, ultimately, this is comparable to the private ownership system (Roman law).

West Kalimantan Province as an illustration: the situation in 1995

The case of West Kalimantan Province is typical of the Indonesian situation. In the 1990s, there were 463,000 ha of rubber smallholdings in the province, of which 97.2% used jungle rubber systems (DGE, 1998). Smallholders were also responsible for deforestation, but the damage they caused was gradual, spread out over a whole century, and above all, was relatively limited. Jungle rubber systems are complex agroforests whose end-of-cycle biodiversity is similar to that of secondary forests (de Foresta, 1997). It has even been suggested that most of the remaining forest biodiversity could be found in jungle rubber systems in the 1990s in the central plains of Sumatra and in Kalimantan (de Foresta, 1992a) in the 1990s. Officially, in the 1980s, 74% of the land area of Indonesia was classified as "forest", and hence as under direct State control. The 1960 agrarian law recognised the common law used in the outer islands on one condition: *"in agrarian matters, common law applies provided it does not run contrary to the interests of the Nation and the State"* (Levang, 1997). The State therefore recognised *adat* in these areas until it decided how the land should be used. This enabled the government to recover land under its "royal prerogative" (in the sense that it can do what it likes with the land by virtue of a prerogative that is nevertheless not strictly legal) and redistribute it according to the policies it chooses. In practical terms, there is no way of opposing this, and the State thus behaves as if it were the "owner", in practice, if not entirely legitimately.

This kind of land allocation system can lead to conflict with local communities when land that does not appear to have an owner (particularly on maps), but in fact belongs to a smallholder community, is "given away" for projects (e.g., for transmigration programmes) or for concessions for private plantations. It is worth noting that the local populations are generally not informed about the changed status of the land. This gives rise to two worlds incapable of understanding one another, since they do not perceive land in the same way[79]; one perception is ancestral, based on *adat*, i.e., tradition, while the other is based on "legal" logic.

In West Kalimantan Province, a landscape previously dominated by a mosaic of fallows and jungle rubber was mostly replaced by oil palm plantations. In 2000, most forest had already disappeared. Data from the Ministry of Agriculture showed that Kapuas district became one of the districts with the largest oil palm plantations (Rahayu et al., 2021). Official data from Balai Penelitian Statistik (PBS) for Sanggau showed that between 1994 and 1996, the quantity of oil palm in that area was negligible, whereas in 2019, the land-use distribution was: (i) *Hutan lindung*/watershed protection forest: 100,221 ha, (ii) *Hutan produksi*/production forest that could be converted: 453,300 ha, (iii) plantations: 723,000 ha, (iv) smallholder rubber: 107,000 ha (52,300 families) = 28% of total tree crops, and (v) oil palm (including estates and smallholders): 283,500 ha (58,900 families) = 72% of total tree crops.

In 2024, oil palm represents almost 75% of land planted with tree crops. There was significant replacement of old jungle rubber gardens by smallholder oil palm. This was not the case in Rantau Pandan, where clonal rubber and upland rice mixed with fallows replaced natural forest.

1985-2000: very rapid changes in land occupation resulting from a government policy of granting concessions to planting companies

In view of the still high forestry potential of the area, as early as 1985, the government was planning significant extension of both forest and commercial plantations, and a major policy of concessions through the REPPROT[80] programme launched by

---

79. In 1980, the local communities were almost the only players, and they controlled the major part of Sanggau district, which was still mainly covered in secondary forest, logged forest (degraded primary forest), smallholder plantings (mainly jungle rubber or rubber agroforests) and *Imperata* savannah.
80. RePPProT: Regional Physical Planning Programme for Transmigration.

the Ministry for Transmigration. To this end, the State planned to grant "forest for conversion" in these areas. Under conventional logging, had these forests retained their "productive forest" status, they would have been left as they were for 30 years to enable tree regeneration, but as it was, they often burned, either accidentally or fires were set deliberately, with a view to requesting their classification as "forest for conversion", which opened the way for logging in the form of plantations.

Such practices reccurred each time there was a major drought (in 1983, 1987, 1991, 1994 and above all in 1997[81]), with often uncontrolled fires affecting several hundred thousand hectares. There was subsequently a marked increase, particularly in the 1990s, in the conversion of existing land to oil palm and *Acacia mangium*, which eventually jeopardised land availability to local communities.

The government policy of redistributing land to planting companies was linked to the introduction of new crops (oil palm and *Acacia mangium*), which proved to be extremely profitable alternatives for the agricultural sector, not only in Indonesia (high land availability, low labour costs) but worldwide (attractive prices and fast-growing markets). For smallholders, rubber continued to be worthwhile. However, despite the limited area in Kalimantan (less than 13,000 ha), existing rubber estates have not been extended. The size of agricultural concessions varies between 10,000 ha and 300,000 ha, while the area planted on each concession is generally between 3,000 and 20,000 ha.

For the last 20 years, the oil palm industry has played a leading role on the international oils and fats market thereby explaining the recent oil palm planting boom, which resembles the one in Malaysia in the 1980s. Moreover, as mentioned above, the Indonesian government considered oil palm to be a "modern means" of development, and as well as a source of income and employment, direct or indirect, for local populations in the outer provinces. Palm oil also became the leading non-petroleum export commodity in terms of value starting in 1988, ahead of rubber, and continues to be a precious source of foreign currency.

In 15 years, the total area under oil palm and production increased considerably (Figure 4.2), from 500,000 ha in 1984 to 2,634,000 ha in 1998 (DGE 1998), almost 2/3 of which were estates (either State owned or private). The development of private oil palm projects in Sanggau district has given some farmers a new way of diversifying their cropping systems, by benefiting from the loans offered by the planting companies. Palm oil became a serious rival for rubber, which was also confronted by major replanting problems after the original switch from traditional extensive jungle rubber systems to intensive clonal plantings (monoculture or agroforestry systems) in the 1990s.

Preliminary comparative analyses showed that the income per hectare from "clonal rubber" resembled that of oil palm under the conditions in the province (Penot, 2001). However, labour productivity is higher for oil palm except when a reduced-frequency tapping system with stimulation is used for clonal rubber (which in 2023 is not yet the case in Indonesia). Rubber still has a role to play in the rural economy, but the full credit policy and loans provided by planting companies and the relatively short three-year immature period of oil palm are a significant advantage for smallholders, who generally do not have sufficient capital to replant with clonal rubber.

---

81. Over 5 million hectares of land, including a small proportion of natural forest, burned in 1997, a figure that was exacerbated by "El Niño" (Laumonier, 1998).

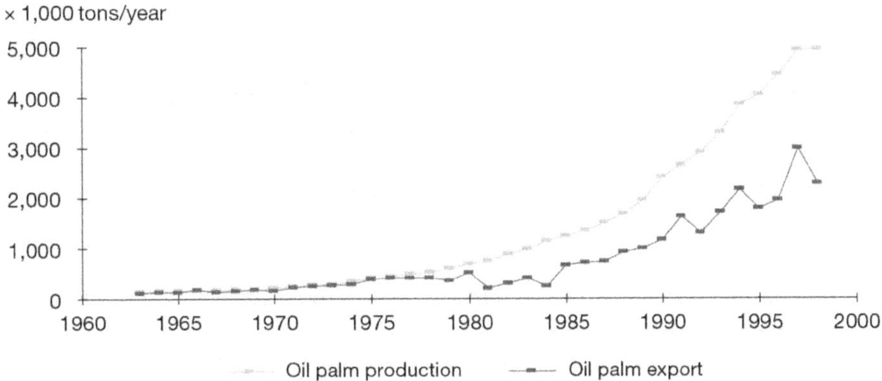

Figure 4.2. Palm oil produced in and exported from Indonesia (1960-1998)

▸▸ Conclusion

The communities concerned by the change in the status of their land continue to receive no information about the real legal threat their land is under. The map we were able to obtain for our 1998 study was published, but the local communities were not aware of it. There is one possible way of ensuring harmonious development in the province and that is by accounting for the rights and needs of farming communities on the one hand, and creating an economic environment that will favour the development of an industry centered on plantations, either smallholdings or estates, on the other. However, indicators that could be used to design an appropriate, rational development policy aimed at preventing conflict over land management and occupation, remain to be identified.

The eight points proposed and discussed by Durand in 1999 seem to us to be more relevant than ever in 2023. In fact, they have already been more or less incorporated in forestry policy, the problem is, the policy has never really been applied. We therefore consider it essential to follow the following eight pointers before making any changes to land management practices: 1) officially acknowledge deforestation has been and continues to be due to badly managed logging (and failure to enforce current legislation); 2) conduct a serious inventory of forest cover; 3) reactivat the traditional rights of local populations, not only over forest areas but also over other areas; 4) oversee the practical application of the results of community forestry research and development of agroforestry practices; 5) design and implement viable forest management systems; 6) reduce the size and extent of the concessions; 7) undertake realistic and precise planning of forest use. We would add rational management is required of the switch from "productive forest" to "forest for conversion" status, which opens the door to planting companies.

The main problem with rational and equitable use of land designated a forest area (whether or not it is actually covered by forest) in Indonesia does not stem from the lack of legislation, but rather from the failure to apply it and the lack of means or determination on the part of the State to control forestry and planting companies' activities. The State's wholesale encouragement of planting companies has made it largely responsible for deforestation and for redistributing land to the detriment of local populations. Instead, the State should act as a regulator to reconcile the issues involved in encouraging the private sector to develop a plantation economy which generates foreign currency, the different agricultural development projects, and meet the aspirations of the local community. It is worth noting that the areas conceded are already being reduced, as is the conversion of forested land into agricultural land, which have been achieved by encouraging optimum use of existing concessions before granting new ones.

In 2016, the Indonesian government recognised the need to respond to the interest shown by local populations in agroforestry systems, particularly those that can generate substantial income while maintaining a forest environment and a degree of biodiversity (as is the case with improved RAS. The merits of rubber and oil palm monocultures are also recognised as local development alternatives. It is crucial that local populations are able to manage their own land according to *adat*, which in practice, varies considerably from one region to another. It is consequently important to acknowledge – which was previously the case – that local (Dayak) or transmigrant (Javanese) populations are able to manage their own development without needing to rely on the development of estates and private companies.

In terms of land occupation, the development rationale of private planting companies is diametrically opposed to that of local populations. The State should therefore act as a "distributor", ensuring the right balance between the various development players and overseeing the rights and duties of each and every stakeholder.

Conclusion and perspectives for the future

Éric Penot, Bénédicte Chambon, Stéphane Boulakia

▸▸ Markets trigger agroforestry: the importance of understanding how income is generated in agroforestry

Flexible crop and tree production in agroforests are linked to the mature and immature periods of the crops or trees concerned. It is consequently indispensable to account for the life cycle of plants when conducting a long-term economic analysis. For instance, timber can only be harvested at the end of the agroforest's life span. Therefore, if detailed data are available for a reliable assessment of real income (including self-consumption), comparing systems will be very valuable (Penot, 2001, 2016). A multi-criteria analysis at both farm and community level is far more powerful than simple conventional cost-benefit analysis at cropping system level.

Taking into account externalities is still very difficult due to the lack of accurate dates. Re-internalizing RAS externalities to attribute a value to environmental and sustainability factors is a real challenge.

If the benefits of agroforests can be analysed using the market values of their products and services, then neo-classical environmental economics can be used and externalities can be included (or re-internalised) in the process of income generation. Growth or cost of pollution and delay can be taken into account as negative externalities or constraints to further development. Environmental services (for example, carbon sequestration potential; Albrecht and Kandji, 2003; Montagnini and Nair, 2004) can be valued according to a "system of values" which is perceived locally as being relevant at a higher, community or provincial level. The real problem is therefore understanding whether farmers can potentially or do benefit from the externalities and positive advantages of agroforestry.

Considering "commercially oriented agroforests" or "subsistence-oriented home-gardens" from a long-term perspective must be part of farmers' strategies. However, there is obviously a biased debate between the short term (economics) vs. the long term (ecology). In both cases, farmers have developed long-term farming practices through a long-haul innovation process that in the end, accounts for economics thanks to the risk buffering capacity of agroforests. In most cases, social organisation is closely linked with technical production constraints, reliance on food, securing an income and, possibly land control. There is a strong coherence between technical systems (technical pathways) and social systems (Penot, 2003a).

Economic analysis methods which use farming system modelling and integrates the outputs of mixtures of plants with different life cycles and enables the smoothing of long-term and patrimonial strategies required to accurately explain what farmers do and why they do so. Despite their positive externalities and advantages, agroforests are not a "magic bullet" but rather an ideal compromise between sustainability and risk spreading. Prospective analysis linked with value chains and existing markets make it possible to forecast future scenarios according to new emerging risks, i.e., climate change, market uncertainties, etc.

⇉ Rubber production and sustainability

As a commodity, rubber has a really secure future thanks to the link between natural rubber and the transport industry (road and air). The gradual rise in prices until the price surge in 2011, encouraged a massive increase in new plantations in some countries, particularly in Laos, Cambodia, China, Vietnam, Cambodia, and Côte d'Ivoire. Almost all the recent increase in plantations has been in monoculture, either by smallholders or estates. While global rubber production and consumption are concentrated in Asia, there is also a strong dynamics in West Africa, in particular in Côte d'Ivoire which became the 4th world producer in 2022.

The natural rubber market is influenced by many factors that cause price volatility and are linked to global growth, the oil market, inventories and rubber stocks, public policies, company standards, etc. Adoption of agroforestry systems is still limited and has no real impact on the rubber value chain as a whole, but does have a significant impact for the farmers who have adopted these systems. The low rubber prices since 2013 in fact were a good opportunity to boost agroforestry to both diversify and increase income to improve the sustainability of the rubber value chain, a key issue in the 2020s.

While the focus of "sustainability" issues is often on industrial plantations, it is important to remember that most growth is based on – and will continue to be – based on village plantations (family or employer). Important challenges include climate change, the risk of the spread of *Microcyclus*, environmental issues, the need for diversification to cope with rubber price volatility, optimisation of existing reservoirs of productivity (e.g. low tapping frequency with stimulation, upward tapping), but agroforestry could be one of the solutions if access to markets and diversification alternatives are sufficient where the biophysical conditions are favourable. In the 2020s, various initiatives have been launched and continue to evolve towards certification and different ways of achieving sustainability (IRSG, GPSNR, etc.).

We can present again our feeling expressed in 2004[82]. In the past, rubber farmers in Indonesia and Thailand developed a series of innovations to integrate rubber in their extensive agroforestry practices (jungle rubber) and, later, in the "estate" monoculture model, by associating rubber with annual or perennial crops. But, by the end of the 1980s, they had reached a point where further innovation was limited and any additional increase in productivity could only be obtained by using rubber clones and other external technologies that required a different management strategy. After passing through two intermediary stages, first between shifting cultivation and improved fallow, and second between improved fallow and a complex agroforestry

---

82. See Penot (2004), *Beyond tropical deforestation* in Babin D (ed), 554 p.

system (jungle rubber), they faced in the 1990s the challenge of how to significantly improve the productivity of their system though rubber clone adoption.

In 1996, Levang wrote: *"Classical Complex Agroforestry Systems such as jungle rubber can no longer compete with other agricultural systems which may be more risky but are more profitable in the short term"* (Levang, 1997). Agroforestry systems based on improved clonal rubber meet this challenge with reduced risk and an increase in environmental benefits. Farmers have shown their ability to develop remarkable innovations, endogenously or through participatory experimentation, for example, with the SRAP project in the 1990s. Jungle rubber covered more than 2.5 million ha in Indonesia in 2002 and probably less than 1 million ha in 2022. Most of it has been replaced by clonal rubber plantations (1/3 roughly) or oil palm (2/3). The challenge in 2024 is to help rubber farmers continue to acquire suitable innovations and to adopt RAS on a large scale.

Indonesia is still going through a stage of "late agricultural transformation", which began in the 1970s, observed by Barlow (1996) and continues in 20024 in the case of rubber. Political instability up to the 1960s and the subsequent priority given to a policy for self-sufficiency in rice production (achieved in 1984) prevented farmers from acquiring improved technologies for rubber on a large scale as was happening in Thailand and Malaysia. Jungle rubber was the most widely used system in Indonesia in the 1990s and still probably accounted for between 0.5 and 1 million ha in 2024, while sustained economic growth and new crop opportunities, in particular oil palm, invite farmers to increase the productivity of their rubber systems by shifting from jungle rubber to clonal rubber. The move from jungle rubber with unselected rubber seedlings to clonal rubber was a real revolution that was possible due to the increased availability of clonal planting material of different rubber development projects implemented from 1975 to 2000 (SRDP, TCSDP, etc., see chapter 2). By 2024, most producing rubber plots were clonal while jungle rubber was mostly no longer being tapped due to the poor prices since 2013 as well as low productivity. It is considered more as a land reserve for future plantations (oil palm, clonal rubber or any other opportunity that may arise).

The introduction of external technical innovations (low tapping frequency using stimulation, upward tapping, etc.) that take indigenous knowledge on agroforestry practices into account, the availability of micro-credits and relevant technical information on markets and farmers' organisations are key factors for the future of the rubber sector in the coming years.

Another major challenge is ensuring that all the different types of farmers have access to improved technologies suited to their particular strategies as well as to local resources; in other words, promoting equity as well as sustainability whether through agroforestry or monoculture. In a country such as Indonesia that has been able to develop millions of hectares of different types of sustainable complex agroforests, agroforestry still has great potential as long as environmental concerns are considered as a priority. This is also the case in Thailand, Sri Lanka and India.

As early as 1993, Michael Dove asked three important questions that "highlight the challenges of future development of the rubber sector" and are still relevant for Indonesia for agroforestry adoption:
– *Is it possible to promote exploitation of rubber, in the absence of a hierarchical political economic structure?* This raises the question of "producers' organisations" and their ability to control future changes in the commodity system themselves.

Up to now, the answer has been yes, as most farmers started growing rubber without help of any kind. But the use of external components (such as fruit and timber species in agroforestry) and the need for capital (investment) may change this situation. For instance, cooperatives and producers' associations are flourishing in Indonesia these days (Penot et al., 2023). Of course, this situation needs to be secured by appropriate policies on agroforestry.

– Is *it possible to attain goals of both ecological sustainability and socio-economic equity within a hierarchical structure?* The answer is probably yes if improved systems such as RAS, partially based on proven existing systems, are adopted by the farmers; and this seems to be the case, since signs of a move in this direction are becoming apparent. This question was particularly astute in 1993 while most of the world only began to think about it in the 2010s.

– *If both preceding solutions are not possible, what then?* The organisation of rubber farmers, and the availability of a wide range of rubber cropping patterns from semi-intensive rubber-based agroforests (RAS 1) to intensive RAS (RAS 2 and 3, for instance) and monoculture systems, are the main preconditions in terms of policy and technology development that will give environmentally friendly systems a chance to continue and to maintain the equilibrium of regional development with other crops.

The questions raised by Dove in 1993 remain relevant in 2024, as most countries have adopted more environmentally oriented policies since the 2010s. Rubber agroforestry systems as a mean of diversification within one plot may be one option amongst others (diversification at farm level with oil palm for instance), and these systems do not involve risks like crop failure, or uncertainties concerning the rubber market and outputs, as there is a steady and reliable demand for natural rubber.

As Barlow stated as early as 1989, "*It is assuredly appropriate to look seriously at policies which basically aim to help people to transform themselves, in an evolutionary approach where steady improvements are made from within the beginning framework of traditional agriculture*". Indeed, this is exactly what farmers have been doing with their agroforests since the beginning of the last century.

Some countries, including Thailand, Sri Lanka, Malaysia, Vietnam, have had real long-term rubber planting development programmes since the 1960s that continue to produce rubber sectors that perform well. Unfortunately, other countries, including Indonesia, Laos, Cambodia, and Côte d'Ivoire have more or less abandoned all governmental projects or extension services targeting rubber. In Indonesia, as well as in Thailand, the situation in 2023 resembles that in the 1990s except that most farmers already rely on clonal rubber plantations. In Côte d'Ivoire, a real boom linked with the cocoa situation, resulted in a highly performant sector where farmers obtain excellent rubber yields and master techniques like low frequency tapping with stimulation. However, this is a very specific situation based on the fact that planting cocoa after cocoa is problematic due to diseases, soil structure and fertilisation. In Côte d'Ivoire, planting rubber was the best way to break the cocoa/cocoa cycle and to have a forest like plantation (rubber) in which cocoa replanting is far easier.

In countries like Cambodia, Côte d'Ivoire, Cameroon, the sector is developing on its own. In all cases, the very low price of rubber is creating a long-term situation of depreciated prices and a context that is killing any incentive to plant rubber in the future. In some countries including Indonesia, the total area under rubber and rubber production are decreasing to the benefit of oil palm. On the bright side, from the point of view of rubber, such unfavourable conditions also create favourable conditions for income diversification and agroforestry in some countries (Thailand, Sri Lanka, India), but which are not favourable in other countries, for example in Indonesia where oil palm is a serious competitor for rubber.

All rubber producing countries have now adopted – to varying extents – global agricultural development policies that favour the environment, biodiversity conservation and agro-ecological practices including agroforestry in response to international demand. In other words, the political context is very favourable for rubber even if the economic context is not.

New organisations have appeared recently, for example, GPSNR (Global Platform for Sustainable Natural Rubber), which groups producers, cooperatives, processors, traders, tyre companies, civil society (NGOs), and a research institute (Cirad), was created in 2020 to promote sustainable natural rubber and explore ways to globally improve the rubber sector. It was originally initiated by WBCSD's Tire Industry Project members. GPSNR has initiated several activities including tracability feasibility studies, a digital Smallholders Knowledge Sharing Platform, capacity building, and insurance, including agroforestry workshops with champion farmers in Indonesia, Cambodia, Côte d'Ivoire and Liberia.

▶▶ A new political environment that accounts for environmental concerns

If an economic perspective with emphasis on local and regional levels is applied to integrate positive externalities such as agrobiodiversity management, improved nutrient cycling, integrated pest management, ecological sustainability and services, decision makers may be convinced that home gardens and agroforests are highly profitable ventures. If an "agroforest rent" approach is applied, policy makers and development agents will see that agroforests are a profitable long-term investment. Hopefully, this will give agroforests a better reputation in research and development programmes worldwide. If agroforests are still a success story for many farmers, it is obviously not for the sake of biodiversity conservation. Other values including social values, security (in terms of risk management and sustainability), diversity (and diversification), land control and land reserve ("rights" as a whole on land and trees with emphasis on tree tenure), are integral parts of the perception of agroforests by most farmers as one cropping pattern among others.

Most farmers who cultivate agroforests also include some monocrops in their farming system, depending on the local situation. If farmers maintain agroforests in some regions, e.g. in India (Kerala), Indonesia (jungle rubber, Pekarangan, Damar systems), Sri Lanka (Kandy agroforests), and West Africa (traditional oil palm based agroforests), it is probably because they have internalised the advantages of agroforests in their systems. A micro-economic analysis at farming system level including all sources of income, cost-benefit per activity and return to labour can explain such long-term strategies, provided it considers the time dynamics of perennial crops in home-gardens and other types of agroforests.

In addition to environmental concerns, rubber sustainability is becoming a real challenge for all the actors involved in the rubber value chain, including governments that have to account for the loss of biodiversity that accompanies the disappearance of forests and the carbon challenge, There is a need to improve the long-term sustainability of cropping systems, which, in the case of rubber, are already in their 3rd or even 4th cycle in some areas.

In 2024, several countries have public policies to support and/or control rubber production and expansion. However, few public policies seem to exist concerning the sustainability of rubber production, whereas at the same time, the biggest companies in the rubber sector are adopting new policies for sustainable supply chain management.

For the benefit of states which wish to be involved in the current process of improving the rubber supply chain, Gitz (2019) identified four possible levers: (i) limiting the negative impacts of land-use change, (ii) regulating land concessions and contract farming, (iii) supporting smallholders and farmers' groups and, (iv) promoting and improving diversified systems.

More generally, as mentioned by Costenbader et al. (2015) in the Mekong subregion of Vietnam, rubber plantation management has to be tackled through inter-sectorial coordination at the landscape, individual country and regional levels. New approaches offer opportunities for such coordination to take place in practice (e.g., landscape level planning, integrated watershed management, integrated and participatory land-use planning, and decentralisation). In order to have a real impact, political will is a prerequisite for the success of these approaches. Governments need to enhance their roles as facilitators in encouraging all sectors and stakeholders to proactively participate in broader natural resources management.

The long period when rubber prices were low not only caused many farmers to temporarily leave their rubber plantation to get work off the farm in another sector, but also in some countries (particularly in Indonesia) to shift to oil palm. In Malaysia and India, rubber trees compete with other crops. Competition is high with oil palm in Malaysia and Indonesia. In India, rubber do compete with high population pressure. Another factor may have contributed to the decline of rubber was the COVID 19 pandemic.

Large rubber estate plantation companies may have converted part of their rubber plantation into oil palm as did many smallholders. On the other hand, some may also have become interested in agroforestry because it is easy to manage like for instance timber based RAS in which timber is harvested at the end of rubber lifespan thereby covering all replanting costs. In Indonesia, mainly in Sumatra and Kalimantan, since the 1990s, rubber has faced fierce competition from oil palm in land allocation and productivity, in terms of both yields and return to labour. "Oil palm/rubber complexes" appeared in many areas in which the two crops competed or complemented each other.

In Vietnam, Cambodia, Côte d'Ivoire and Myanmar, competition with other perennial crops can be far lower, which may increase farmers' interest in agroforestry as a source of income diversification to cope with rubber price volatility, on the condition that there are local markets for associated products in RAS, which is a very important pre-requisite for its further development. In Côte d'Ivoire, there is another explanation for farmers' interest in the cocoa and rubber sectors, because including a rubber cycle makes it possible to interrupt the cacao disease cycles. After the rubber cycle, cacao can be planted again without destroying the naturel forest, which was previously the case.

Local governments need to seriously consider a comprehensive long-term programme for improvement of the rubber industry to support the sector, and to increase its productivity (compared to that of oil palm for instance which is the major challenger in Indonesia) mainly by providing training in tapping practices and RAS as well as ensuring the availability of good quality clonal planting material at an affordable price.

If in a capitalist world, nothing can be done about rubber prices, any and all activities that enable the use of good quality planting material, better tapping practices (including low frequency tapping with stimulation and upward tapping), improved return to labour and the development of RAS as a source of income diversification will significantly improve farmers' incomes and more globally, the long-term resilience of the rubber sector. A balance has to be found that allows all farmers to have balanced farming systems based on both oil palm and clonal rubber (including RAS) in order to be more economically resilient using environmentally friendly practices, in addition to finding a balance between on-farm and off-farm activities.

Agroforestry does have a future but cannot be considered as a "one size fits all" strategy. It needs to be adapted to local socio-economic conditions, to local soil and climate conditions and local markets. RAS offers a real opportunity to strengthen the situation of rubber farmers and to work towards more sustainable rubber production. However, the development of RAS requires both the creation of value chains for associated products, farmers need access to information and probably, for increased efficiency, the establishment of innovation platforms to provide farmers with information about the agroforestry practices that need to be adapted to local conditions, in particular to local climate conditions. The advantage of adopting RAS is to profit from local market opportunities for timber, fruits, gaharu, spices, medicinal plants, etc.

▸▸ Some innovative systems for the future

Here we suggest possible innovation pathways during and following the rubber cycle to maintain a certain level of biodiversity while promoting a landscape that is no longer dominated by monoculture.

For more productive adapted RAS, the challenge is to adapt and optimize what already exists

The easiest way to overcome this challenge is to observe and record the types of agroforestry patterns currently being developed by farmers in Indonesia, Thailand, Sri Lanka, China, India, Colombia and Brazil to adapt cropping systems to local conditions. GPSNR recently (2022/2023) boosted this trend by organising agroforestry training and discussion workshops in several countries. The creation of RAS innovation platforms was also suggested to Thai authorities to profit from the considerable reliable know-how of Thai farmers (Penot et al., 2022). There is tremendous scope for valorisation what already exists for the benefit of farmers who are still engaged in monoculture, but such policies require not only organisation, implementation and funding, but most of all, the adaptation of agroforestry patterns to local conditions and markets, including forecasting future climate conditions for the next 30 years. Technical information and know-how is there. Dissemination requires extension services and the willingness of public authorities to develop agroforestry as a possible solution among others.

Towards Indonesian *tembawang*: moving from rubber plantations to productive fruit/timber forests (based on durian, for instance)

The move from rubber plantations to productive fruit/timber forests (based on durian, for instance) is already underway in Sumatra and Kalimantan in Indonesia. After RAS, some farmers decided to change from a rubber-based plot to fruit/trees-based plots called "tembawang" in Kalimantan by Dayak farmers. Once they acquire fruit trees that can produce yields for more than 50 years along with timber species that require up to 60 years before being felled for sale (such as very high quality meranti), these farmers prefer to maintain the fruit/timber-based agroforestry cropping systems without rubber. In this case, rubber is planted on another plot. This system generally prevails where durian trees are growing in the plot as durian produces a good yield and a high-priced fruit.

This trend can also be observed in old jungle rubber plots where rubber is progressively disappearing while fruit/timber trees are preserved. Of the 60 SRAP research project plots dating from the 1990s (see chapter 2), 10% were preserved as *tembawang*, evidence that farmers still have a certain interest in this type of change to their plot. This option remains when farmers have still some land available for future plantation, possibly to replace their old jungle rubber.

Islands of agroforestry in a monoculture dominated landscape

Riparian/rubber corridors: towards the creation of biodiversity corridors using a landscape approach

The idea is to develop a landscape approach based on productive tree-crop plots such as rubber as monoculture or as RAS, and oil palm, with corridors containing local riparian varieties (39 different species), implemented in RAS in order to prepare the future and to have riparian corridors that function as such by the end of the RAS lifespan. The concept was designed and applied to oil palm in a big local private estate located in Sabah province in Malaysia as part of the Trails project. Trails is a Cirad project implemented with University Putra Malaysia, University Malaya, "Hutan" a French NGO and the private estate (Melangking Oil Palm plantations-MOPP). A total of 22 hectares have been planted with 3,000 associated local trees belonging to 15 different species in 3 blocks within the oil palm plantation. In this particular case, the RAS includes not only fruit/timber trees but also local riparian species with no particular productive function aside from biodiversity enrichment. The same system could easily be used for rubber, in particular by local private or government estates with the aim of creating biodiversity corridors and ending the 100% monoculture landscape that currently prevails in mainland Malaysia and in central Kalimantan, Indonesia.

Zemp et al. (2023) described a system based on tree islands located not far away, in Jambi, Indonesia, originally based on oil palm that could also be applied in rubber estates. This project is based on a large-scale, 5-year ecosystem restoration experiment in an oil palm landscape enriched with 52 tree islands, with assessments of 10 biodiversity indicators and 19 indicators of ecosystem functioning in order to compare multi-diversity and ecosystem multifunctionality in tree island systems and conventionally managed oil palm. Enriching oil palm-dominated landscapes with tree islands is a promising ecological restoration strategy, allthough it cannot replace protection of remaining forests.

Such systems are based on the fact that a small part of the plantation will not be replanted with rubber or oil palm, but will instead be devoted to scattered forest-like plots forming corridors or islands within the estate to create a landscape that is more suitable for wildlife.

The double nested cycles system: towards long-term productive forests with a high level of biodiversity

This idea, which was developed by Boulakia (Cirad) in 2002 is a particular type of agroforestry pattern designed to restore the complex and age-old forest cover with "nested" rubber cycles (Boulakia et al., 2010). Rubber plantations are reported to be drivers of deforestation in South East Asia. According to FAOSTAT, between 2000 and 2021, productive rubber area soared from 2.2 to 5.5 million ha in Cambodia, China, Myanmar, Thailand and Vietnam combined. In many regions, rubber expansion is being carried out to the detriment of the forest (To and Tran, 2014; Grogan et al., 2019; Sarathchandra et al., 2021; Bhagwat et al., 2017) with negative consequences for biodiversity and for the carbon balance (Min et al., 2019). In some of these regions, changes in the rubber-driven land use and land cover (LULC) are underway with contrasted types of beneficiaries, with a high percentage of smallholders in Thailand, while in Cambodia and Myanmar, estates drive the dynamics (Fox and Castella, 2013).

Considering these changes in LULC, Warren-Thomas et al. (2018) assessed the threshold value of tCO_2 to reduce the incentive to convert forests into rubber plantations, considering the type of forest that existed previously and the state of degradation. The resulting estimated US$30-51 per ton of CO_2 are far above contemporary market price (US$5-13 per tCO_2). Commenting on the payment for ecosystem services (PES) approach, Dove (2018) pointed out that to be able to reverse or at least limit current forest conversion processes, an economic valuation of PES would have to deal with diverse, complex and often conflictual contexts in terms of resources and land use rights. Based on the analysis of a nation-wide Chinese reforestation programme, Hua et al. (2016) call for a shift from mono or oligo-cultures to more complex planting designs to have a chance to restore biodiversity on similar levels to those of native forests.

Here, we propose some innovative and disruptive planting patterns to reforest or afforest with multi species, using rubber as a relay product in at least two successive rounds of production. The basic principles, presented in figure C.1, are quite simple:
– First, it consists of associating rubber trees planted in hedgerows, e.g. (13 m + 3 m) × 2.25 m, at a normal density of +/- 550 trees/ha, with various ligneous forest species, planted in the double inter-hedgerow space, for multiple production goals like high quality timber in the long term, or non-timber forest products (NTFPs) and ecosystem services as sources of income in the shorter term;
– Second, to escape from the usual around 30-year plantation cycle (a 6-year immature stage followed by 24 years of tapping), 20-25% of the rubber trees would not be opened at 5-6 years old, i.e., when they reach 50 cm-girth, but instead reserved for a second and relay tapping round; this round would exploit 100 to 120 trees/ha opened at a minimum of 31 years old.

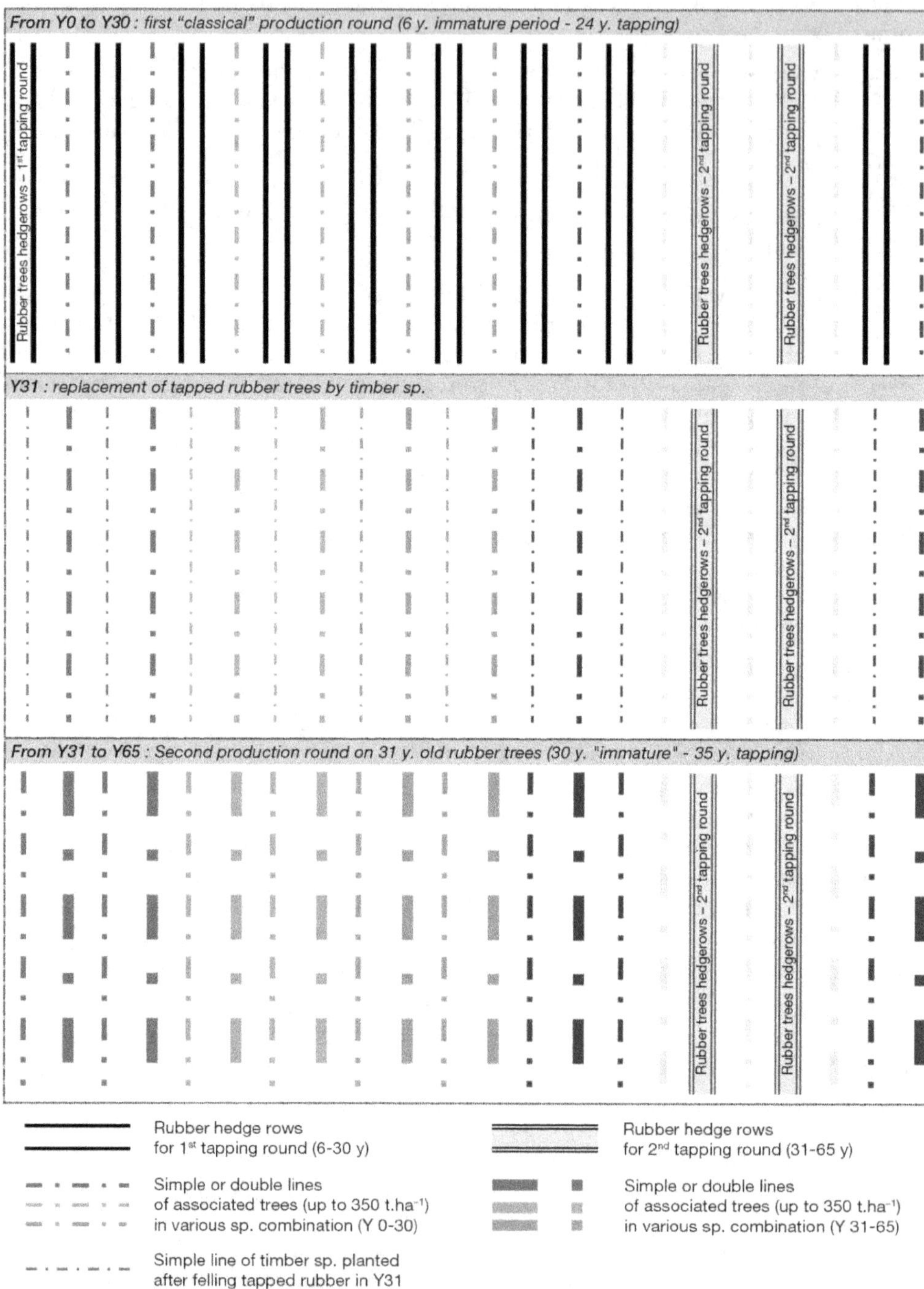

Figure C.1. Sketch of the first two stages of a nested cycle-based rubber agroforestry plantation

This stage makes it possible to imagine a third tapping round involving some of the trees that were tapped during the first round, then reopened about 30 years later (after the second round), thereby enabling regeneration of the bark and of the trees as a whole. This plantation design would produce less rubber per hectare than monocropping, even if the 6 year-long immature period is followed by almost 60 years of continuous tapping but such agroforests would supply complementary products and provide ecosystems services that should be acknowledged and paid for.

Labour is the main production cost of natural rubber and the cost of labour will continue to rise in the future. The design proposed above could be paired with improved labour productivity; after the first, *a priori* standard tapping round (1 tapping every 3 or 4 days, 6 days out of 7 with stimulation adapted to the physiology of the clone concerned), the second period will exploit very large rubber trees, prone to very low tapping frequency (D7 or every 7 days, or D14 or 1 tapping every 14 days) with a high level of stimulation. Rubber trees to be used in the 2nd tapping round should be concentrated in productive patches located on the margins of the planted area in order to limit the need for labour in the innermost part, which was tapped during the first round.

Complementary types of production can be set up in parallel with the progressive and long-term regeneration of complex ecosystems. The different kinds of associated products including NTFPs like fruit, leaves, bark and seeds used for multiple purposes, e.g., food, spices, medicinal plants, gene banks and tree nurseries, are possible sources of income. The most threatened species listed by UICN should be planted in patterns inspired by the structure and kinetics of natural ecosystems. In the long to very long term, timber trees that produce luxury wood with a zero-deforestation certificate will produce high incomes and before being felled, this "growing capital" might also enable access to long term credit secured by the high-quality wood "warrantee".

Throughout their life span, such plantations provide multiple ecosystems services through carbon sequestration and biodiversity recovery.

Carbon will accumulate in plant biomass, notably in the heartwood of luxury and first-class timber species. The sequestration in the biomass of future harvestable wood products (HWP) present two major advantages for the PES: first, it is relatively easy to monitor and report growth of the trees, and second, the sequestered carbon is highly unlikely to revert to CO_2 once the trees reach marketable size due to the extremely high value of their wood. Chayaporn et al. (2021) estimated that teak trees capture between 20 to 30 kg of carbon per year in their aerial biomass between 17 and 35 years old; in the absence of any references, we consider the lowest growth rate of high value timber species, the carbon biomass increment in the sole HWP of about 100 associated timbers could average 1.5 to 2 Mg CO_2e/year/ha, a significant source of income if carbon reaches the expected EU corridor price of between US\$60 and US\$120 per tCO_2e by 2030. Compared with a monocropping system based on a 30-year cycle from planting to logging and wood export, this long-term regenerative approach should lead to more intense and diversified inputs of fresh organic matter both above and below ground; this continuous supply of litter combined with reduced soil disturbance, should encourage soil organic carbon storage, enhance topsoil diversity and restore function (Panklang et al., 2022a,b), thereby allowing restauration of soil degraded by successive cycles of monocropping in traditional rubber producing areas.

This agroforestry design could also play a role in the conservation of endangered plant species. Beyond these possible conservation functions, Warren-Thomas et al. (2020) reported limited benefits for biodiversity at plot level, of rubber agroforestry systems compared to monoculture. Nevertheless, these authors underlined the positive influence of plant richness, multi-storey arrangements and the presence of neighbouring forest fragments on animal biodiversity assessed through birds, fruit-feeding butterflies and reptiles. Induced and emerging effects on biodiversity enhancement will depend on the scale of application and on connectivity with natural ecosystems. This regenerative agroforestry sequence could be conceived and established at different scales ranging from individual plots to community-managed agroforests; it can serve in conservation programmes designed to restore connections between forest patches or to develop long-term activities with communities in buffer zone programmes. Industrial estates could also use this type of pattern in marginal areas (unsuitable soil type, slope, remoteness, etc.) or to interrupt monocropping schemes after one or two cycles. Public policies could enforce such designs on allocated public land in anticipation of the development of amenities, offering local climate regulation through reforestation or afforestation as outlined by IPCC (2019), in rural zones that are planned to become urban or sub-urban areas in the coming century.

▸▸ The final word

Agroforestry systems have been widely applied during the immature period of the rubber trees using different combinations of intercrops, mainly food crops. These temporary agroforestry practices can be found almost everywhere in the world. But agroforestry practices during the rubber mature period combined with fruit and timber trees, resins, spices, food crops and other plants depend on local environmental conditions and on the planting design.

The appropriate degree of shade

Some plants can grow in deep shade but only a few species. According to farmers' experience and our own observations, to enable correct growth of fruit and timber trees or any other plants, the shade provided by rubber trees in a normal planting design should not exceed 70%.

In Indonesia, with a classical planting design, the development of leaf diseases (*Colletotrichum*, *Corynespora* and *Oidium*) to varying extents limited the rubber canopy. More recently the spread of *Pestalotiopsis* sp. has dramatically reduced the rubber canopy and hence the shade it provides. In Thailand, the widespread use of the clone RRIM 600, which naturally has a limited canopy (i.e. approximate 70% shade) created excellent conditions for agroforestry.

This is not the case in countries like Cambodia, Vietnam, Côte d'Ivoire where deep shade (around 90%) linked with the well-developed rubber canopy prevents any plants from growing with rubber. In such cases, the only option is to change to a different planting design with double or triple rubber rows and wide spacing between the rows of rubber trees.

In designs with double spacing and sufficiently large inter-rows (12 m up to maximum 25 m), it is possible to associate other tree species and plants as, depending on the spacing, they will be in full sun for 10 to 20 years. The shade provided by the rubber trees will only play a role in a limited part of the inter-row. In this example, it is important to design the plot with a minimum of 400 rubber trees to ensure a sufficient yield of rubber. Trials have shown that with 400 trees/ha, the reduction in the yield of rubber is usually limited to 10%, which is considered reasonable and can be largely offset by the value of the associated crops.

Market Opportunity

Agroforestry has expanded in countries where there is a local market, e.g. for fruits in Thailand, Indonesia and Columbia, and more recently for timber as a result of significant deforestation in Southeast Asia, for spices in India, medicinal plants in China, for tea in China and Sri Lanka, for coffee and sugar palm in North Sumatra, etc. Market opportunities clearly drive the development of agroforestry and are a pre-requisite for any further development.

The challenge posed by other crop opportunities

In some countries where rubber is the most widely grown local perennial crop, except for improving rubber growing practices, there are no other ways to obtain the highest possible yield and the best quality, the case in 2024 in Thailand and Côte d'Ivoire. But in other countries, associating another crop may be complementary but is more often competition. Côte d'Ivoire is a good example of complementarity between cocoa and rubber as rubber enables cocoa to be replanted after a rubber cycle of 25/35 years in good conditions close to those of traditional forests which have now almost completely disappeared.

Indonesia, on the other hand, is an example of extreme competition between rubber and oil palm. Other examples of competing crops are coffee and sugar palm (North Sumatra), cassava in Northeast Thailand, and tea in China.

The long periods with uninterrupted low rubber prices (the 12 years since rubber prices fell in 2012) created very unfavourable conditions for most rubber smallholders, resulting in the choice of other crops, particularly, oil palm, which provides double the gross margin/ha and a fourfold return to labour of rubber, and has led to rubber being replaced by oil palm almost everywhere in Indonesia. This long period of depreciated rubber price is the worst enemy of rubber and of any potential improvement in farmers' income from agroforestry practices. Planting oil palm in old rubber plots to eliminate the "no income effect" of immature impact is now very common.

In 2024, North Sumatra is characterised by the expansion of agroforestry practices based on sugar palm, coffee and lemon grass as a transition from a rubber-based system to a system in which the canopy is limited to 30/40% due to significant impact of leaf diseases.

New opportunities

Following the disappearance of local forests in many places in Africa and Southeast Asia, the market for good quality timber is focussed in teak, mahogany, *Dipterocarpaces* and local good value timber such as nyatoh and tembesu in Indonesia, and *cedro odorata* in Central America. Even if income from timber species generally only becomes available at the end of rubber lifespan, timber is already a valuable potential associated crop as it produces sufficient capital to renew a rubber plantation in good conditions with appropriate high quality planting material and the appropriate level of fertilisation. In 2024, tree tenure in Côte d'Ivoire and Indonesia is now favourable in that farmers are owners of their timber trees, meaning they can accumulate patrimonial capital that can be passed on to future generations.

Evolution of rubber systems

Under agroforestry, rubber plantations can also evolve into another system, for example, from rubber to a durian agroforestry system, to the Dayak people's *tembawang* system in West Kalimantan or to enriched forests. This kind of development can help create a new more balanced landscape with a mosaic of different perennial crops and a variety of forests and agroforests.

Rubber Agroforestry Systems remain an interesting alternative

Aside from the adoption of appropriate cultivation practices to improve rubber production and reach a yield of 1,700 kg/ha/year of rubber, the only possible way to increase the gross margin/ha of rubber plots is to adopt agroforestry practices to diversify both production and sources of income. For instance in Thailand, income can be increased by an average of 40% in this way, enabling a better economic result from the plot and helping farmers resist other opportunities.

Agroforestry would probably be more economically effective using double spacing to increase the profitability of associated crops and timber. Double spacing is a new paradigm for the majority of rubber smallholders and continues to represent a real challenge to the adoption of agroforestry. However, such a transition and the adoption of agroforestry is possible with help and support from local estates, as has been the case in Côte d'Ivoire, or from the government like in Thailand or Sri Lanka. Double spacing systems are probably the main challenge for smallholders as well as for countries which intend to maintain their current rubber production despite the presence of newcomers who own very large rubber plantations in Côte d'Ivoire, Vietnam, China, and Cambodia.

The multiplication of sources of income in the medium to very long term, either in the form of products or multiple ecosystem services, opens the way for a vast range of possible agreements between stakeholders, while simultaneously strengthening the resilience of the system. This nested cycle approach offers solutions to some of the limits of rubber monocropping including soil degradation, labour productivity and attractivity or social acceptability of the model. It also makes it possible to couple a response to the growing demand for natural rubber with large scale and self-financed reforestation/afforestation programmes. It creates pathways to establish rubber agroforests that enable the emergence of complex forested ecosystems with large rubber trees; it is a flexible approach that can be fairly adapted to the multiple social contexts encountered under the wet tropical regions of Asia, Africa and Latin America.

To conclude, in addition to the valorisation of existing agroforestry systems and practices, which are already well adapted to local contexts and offer a real economic advantage at plot level, we also perceive the potential for the creation of innovative, more sustainable landscape systems in the long term, landscapes that are more suitable for wildlife and biodiversity conservation. Many farmers who are involved in RAS have knowledge and know-how that could be promoted and disseminated through innovation platforms and at larger scales. Policies on both RAS dissemination and landscape approaches finally depend on governmental willingness to efficiently tackle with environmental concerns and economic sustainability of the rubber sector in the very near future.

Bibliography

Abraham J, Joseph P, 2015. A new weed management approach to improve soil health in a tropical plantation crop, rubber (H*evea brasiliensis*). *Exp Agric* 52:36–50.

Agarwood. 2023. https://dataintelo.com/report/agarwood-essential-oil-market/?utm_campaign=copy

Ahrends A, Hollingsworth PM, Ziegler AD, Fox JM, Chen H, Su Y, Xu J, 2015. Current trends of rubber plantation expansion may threaten biodiversity and livelihoods. *Global Environ Change* 34:48–58.

Ajayi OC, Catacutan D, 2012. Role of externality in the adoption of smallholder agroforestry: case studies from Southern Africa and Southeast Asia. *Externality: Economics, Management and Outcomes*, 167-188.

Albrecht A., Kandji ST, 2003. Carbon sequestration in tropical agroforestry systems. *Agric. Ecosyst. Environ.* 99, 15–27. https://doi.org/10.1016/S0167-8809(03)00138-5

Alliance R, and UTZ. (n.d.). Programme De Certification Rainforest Alliance 2020. *Rainforest Alliance Pour Les Entreprises*. https://www.rainforest-alliance.org/business/fr/solutions/certification/agriculture/2020-certification-program/

Alliot M, 2003. *Le droit et le service public au miroir de l'anthropologie*. Karthala éditions.

Alvim R, Nair PKR, 1986. Combination of cacao with other plantation crops: an agroforestry system in Southeast Bahia, Brazil. *Agrofor Syst* 4:3–15.

Anderson JR, Dillon JL, Hardaker JB, 1977. Agricultural decision analysis. Iowa State University Press, Ames.

Andriesse E, Tanwattana P, 2018. Coping with the end of the commodities boom: rubber smallholders in Southern Thailand oscillating between near-poverty and middle-class status. *J Dev Soc* 34(1):77–102. https://doi.org/10.1177/0169796X17752420

Aratrakorn S, Thunhikorn S, Donald PF, 2006. Changes in bird communities following conversion of lowland forest to oil palm and rubber plantations in southern Thailand. *Bird Conserv Int* 16:71–82.

Archibald SO, Capalbo SM, Antle JM, 1988. Incorporating externalities into agricultural productivity analysis. Agricultural Productivity: Measurement and Explanation. Washington DC: Resources for the Future, 366-394.

Arimalala N, Penot E, Michels T, Rakotoarimanana V, Michel I, Ravaomanalina H, et al., 2019. Clove based cropping systems on the east coast of Madagascar: how history leaves its mark on the landscape. *Agroforestry Systems* 93,1577-1592.

A/S C, 2018. *Feasibility study on options to step up EU action against deforestation*. Retrieved https://policycommons.net/artifacts/3339088/feasibility-study-on-options-to-step-up-eu-action-against-deforestation/4137934/

Assembe-Mvondo S, Putzel L, Atyi RE, 2015. Socioecological responsibility and Chinese overseas investments: The case of rubber plantation expansion in Cameroon. CIFOR. https://doi.org/10.17528/cifor/005474

Aznar O, Brettières G, Herviou S, 2005. Agriculture de service et services environnementaux : lien avec les politiques agricoles et rurales. In Guérin M, *Politiques de développement rural : Enjeux, modalités et stratégies*, Versailles, éditions Quæ.

Aznar O, Guérin M, Perrier-Cornet P, 2007. Agriculture de services, services environnementaux et politiques publiques : éléments d'analyse économique. *Revue d'économie régionale & urbaine*, (4), 573-587.

Babin D (ed), 2004. *Beyond tropical deforestation. From tropical deforestation to forest cover dynamics and forest development*, UNESCO/Cirad, 488 p.

Bagnall-Oakeley H, Conroy C, Faiz A, Gunawan A, Gouyon A, Penot E, et al., 1996. *Imperata* management strategies used in smallholder rubber-based farming systems. *Agroforestry Systems* 36:83–104.

Bardgett RD, van der Putten WH, 2014. Belowground biodiversity and ecosystem functioning. *Nature* 515, 505–511. https://doi.org/10.1038/nature13855

Barlow C, 1978. *The natural rubber industry, its development, technology, and economy in Malaysia.* New York, Oxford University Press.

Barlow, C., Jayasuriya, S. K., Price, E., Maranan, C., & Roxas, N. (1986). Improving the economic impact of farming systems research. *Agricultural systems, 22*(2), 109-125.

Barlow C, Shearing C, Derenda R, 1988. Alternatives approaches to smallholders rubber development. CPIS report. Jakarta.

Barlow C, Muharminto, 1988. Smallholders rubber in South Sumatra, towards economic improvement. Balai penelitian Perkebunan, Bogor and Australian national University.

Barlow C, 1989. Developments in plantation agriculture and smallholder cash crop production. Conference on indonesia's new order, Australian National University, Canbera.

Barlow C, 1993. Towards a planting material policy for Indonesia rubber smallholding: lessons from past projects. Unpublished paper.

Barlow C, 1996. Growth, structural changes and plantation tree crops: the case of rubber. Working papers in trade and development serie from the Australian National University. Department of economics, Paper 94/4.

Barlow C, 1997. Growth, structural change and plantation tree crops: the case of rubber. *World Dev* 25:1589–1607.

Barrett CB, Reardon T, Webb P, 2001. Nonfarm income diversification and household livelihood strategies in rural Africa: concepts dynamics, and policy implications. *Food Policy* 26:315–331. https://doi.org/10.1016/S0306-9192(01)00014-8

Baulkwill WJ, 1989. The history of natural rubber production. In Webster CC, Baulkwill WJ (eds), *Rubber.* Longman Scientific & Technical, Harlow, 1-56.

Baumol WJ, 1988. *The Theory of Environmental Policy.* Cambridge University Press.

Beer J, Muschler R, Kass D, Somarriba E, 1998. Shade management in coffee and cacao plantations. *Agrofor Syst* 38:139-164.

Belieres JF, Bonnal P, Bosc PM, Losch B, Marzin J, Sourisseau JM, 2015. *Family farming around the world: definitions, contributions and public policies.* http://publications.cirad.fr/une_notice.php?dk=576161

Benoît M, Deffontaines JP, Gras F, Bienaimé E, Riela-Cosserat R, 1997. Agriculture et qualité de l'eau. Une approche interdisciplinaire de la pollution par les nitrates d'un bassin d'alimentation. *Cahiers Agricultures* 6: 97-105.

Benoist A, Leconte A, 2020. Filière Hévéa en Côte d'Ivoire : Analyse fonctionnelle et diagnostic agronomique, Rapport d'étude Cirad, réalisé pour le FIRCA et l'APROMAC, 48p.

Beukema H, van Noordwijk M, 2004. Terrestrial pteridophytes as indicators of a forest-like environment in rubber production systems in the lowlands of Jambi, Sumatra. *Agric Ecosyst Environ* 104:63–73.

Beukema H, Danielsen F, Vincent G, Hardiwinoto S, van Ande J, 2007. Plant and bird diversity in rubber agroforests in the lowlands of Sumatra, Indonesia. *Agroforest Syst* 70:217–242. https://doi.org/10.1007/s10457-007-9037-X

Bhagwat T, Hess A, Horning N, Khaing T, Thein ZM, Aung KM, et al., 2017. Losing a jewel—Rapid declines in Myanmar's intact forests from 2002-2014. *PLoS ONE* 12(5): e0176364. https://doi.org/10.1371/journal.pone.0176364

Biatry N, 2021. Caractérisation agroéconomique des systèmes d'activités à hévéa en Côte d'Ivoire. Mémoire de stage. AgroParistech Ingénieur «Gestion environnementale des écosystèmes et forêts tropicales». Montpellier.

Biret C, Buttard C, Farny M, Lisbona D, Janekarnkij P, Barbier JM, Chambon B, 2019. Assessing sustainability of different forms of farm organization: Adaptation of IDEA method to rubber family farms in Thailand. *Biotechnologie, Agronomie, Société et Environnement* 23 (2), 74-87.

Bissonnette JF, 2009. Développement et palmier à huile : les enjeux de la gestion des territoires coutumiers ibans du Sarawak, Malaysia. *VertigO*, 8(3). https://doi.org/10.4000/vertigo.7082

Blagodatsky S, Xu J, Cadisch G, 2016. Carbon balance of rubber (*Hevea brasiliensis*) plantations: A review of uncertainties at plot, landscape and production level. *Agric. Ecosyst. Environ.* 221, 8–19. https://doi.org/10.1016/j.agee.2016.01.025

Blencowe JW, 1989. Organization and improvement of smallholder production. In Webster CC, Baulkwill WJ (eds), *Rubber.* Longman Scientific & Technical, Harlow, 499–538.

Booth A., 1988. Living standards and the distribution of income in colonial Indonesia: A review of the evidence. *Journal of Southeast Asian Studies, 19*(2), 310-334.

Boonkird SA, Fernandes ECM, Nair PKR, 1984. Forest villages: an agroforestry approach to rehabilitating forest land degraded by shifting cultivation in Thailand. *Agroforestry Systems* 2:87-102.

Boulakia S, Sona S, Vira L, Kimchhorn C, 2010. Five years of adaptive research for upland DMC based cropping systems creation in Cambodia. Ministry of Agriculture and Forestry [Laos].

Boutin D, Penot E, Wibawa G, 2001. Replantation des agroforêts à hévéas en Indonésie : les Systèmes Agroforestiers à base hévéa (RAS) : une alternative à la monoculture?, Cirad-TERA.

Boutin D, Penot E, Lubis R, Kramer E, 2000. Major agronomic results of rubber agroforestry systems on-farm experimentation network in East Pasaman (West-Sumatra, Indonesia). Cirad-TERA.

Brafman N, 2008. Le Brésil, les Français et l'arbre qui pleure. https://www.lemonde.fr/planete/article/2008/07/31/le-bresil-les-francais-et-l-arbre-qui-pleure_1079017_3244.html

Brauman A, Thoumazeau A, 2020. Biofunctool® : un outil de terrain pour évaluer la santé des sols, basé sur la mesure de fonctions issues de l'activité des organismes du sol. *Etude Gest. des Sols,* 27, 289–304.

Broughton WJ, 1977. Effect of various covers on soil fertility under *Hevea brasiliensis* Muell. Arg. and on growth of the tree. *Agro Ecosyst* 3:147–170.

Budiman AFS, Penot E, De Foresta H, Tomish T, 1994. Integrated rubber agroforestry for the future of smallholder rubber in Indonesia. National Rubber Conference, Medan.

Buranatham W, Chugamnerd S, Kongsripun S, 1997. Agroforestry under rubber plantation in Thailand. Thailand/Malaysia tehcnical seminar on rubber, Phuket, Thailand.

Byerlee D, 2014. The fall and rise again of plantations in tropical Asia: History Repeated? *Land* 3(3):574-597. https://doi.org/10.3390/land3030574

Cacho O, Hean R, 2001. Dynamic optimization for evaluating externalities in agroforestry systems: An example from Australia. 139-163. Chapter 8. In *An analysis of externalities in agroforestry systems in the presence of land degradation.* Ecological economics, Elsevier.

Campoy R, 2018. Slowing Chinese demand, supplies to weigh on rubber prices in 2019. Global Rubber Markets. https://globalrubbermarkets.com/141365/slowing-chinese-demand-supplies-to-weigh-on-rubber-prices-in-2019-assoc-official.html

Carr MKV, Stephens W, 1992. Climate, weather and the yield of tea. In Willson KC, Clifford MN (eds), *Tea—cultivation to consumption.* Champman & Hall, London, 87–135.

Cahyo AN, et al., 2024. Rubber-based agroforestry systems associated with food crops: A solution for sustainable rubber and food production? *Agriculture,* 14.7: 1038.

Ceccolini L, 2002. The homegardens of Soqotra island, Yemen: an example of agroforestry approach to multiple land-use in an isolated location, *Agroforestry Systems* 56(2): 107-115.

Challen G, 2014. Hevea brasiliensis *(rubber tree).* Kew Royal Botanic Gardens. http://www.kew.org/science-conservation/plants-fungi/hevea-brasiliensis-rubber-tree.

Chambon B, Penot E, Geissler C, 2002. L'Etat est-il un acteur indispensable à la modernisation de l'hévéaculture paysanne en Indonésie ? Cirad-Tera/Thi. 15 p.

Chambon B, 2001. Transfert et appropriation des techniques : le cas de la monoculture clonale d'hévéa dans la province de Ouest Kalimantan. Thèse de doctorat. Facultés des Sciences Economiques de Montpellier 1. Montpellier, France.

Chambon B, Ruf F, Kongmanee C, Angthong S, 2016. Can the cocoa cycle model explain the continuous growth of the rubber (*Hevea brasiliensis*) sector for more than a century in Thailand? *Journal of Rural Studies.* http://dx.doi.org/10.1016/j.jrurstud.2016.02.003

Chambon B, Bosc PM, Promkhambut A, Duangta K, 2018. Entrepreunarial and family business farms in Thaïland: Who took advantage of the rubber boom? *Journal of Asian Rural Studies.* http://agritrop.cirad.fr/588406/

Chambon B, Bosc PM, Gaillard C, Tongkaemkaew U, 2021. Using Labour to Characterise Forms of Agriculture: A Thai Family Rubber Farming Case Study, *Journal of Contemporary Asia*, https://doi.org/10.1080/00472336.2021.1901958

Chaudhuri PS, Bhattacharjee S, Dey A, Chattopadhyay S, Bhattacharya D, 2013. Impact of age of rubber (*Hevea brasiliensis*) plantation on earthworm communities of West Tripura (India). *J Environ Biol*, 34, 59–65.

Chauveau J, Cormier-Salem MC, Mollard E, 1999. L'innovation en Agriculture. IRD, collection « à travers champs ».

Choengthong S, Kongrithi W, Choengthong S, Chuaymuang B, 2014. The Current Status and Future Trends of Farm Production and Marketing of Para-Rubber in the Upper East Coast Provinces of Southern Thailand, *Advanced Materials Research*, 844, 30-33.

Charernjiratragul S, 1991. Le système productif agricole à base d'hévéaculture dans l'économie thaïlandaise du caoutchouc naturel. Thèse de doctorat. Montpellier 1.

Chayaporn P, Sasaki N, Venkatappa M, Abe I, 2021. Assessment of the overall carbon storage in a teak plantation in Kanchanaburi province, Thailand – Implications for carbon-based incentives. *Cleaner Environmental Systems*, (2), 2021, https://doi.org/10.1016/j.cesys.2021.100023

Chee YK, Faiz A,1991. Forage resources in Malaysian rubber estates. In Shelton HM, Stür WW (eds), *Forages for Plantation Crops*, Bali, Indonesia, 27–29 June 1990. ACIAR Proceedings, vol 32. Australian Centre for International Agricultural Research, 32–35.

Chen C, Liu W, Jiang X, Wu J, 2017. Effects of rubber-based agroforestry systems on soil aggregation and associated soil organic carbon: Implications for land use. *Geoderma* 299, 13–24. https://doi.org/10.1016/j.geoderma.2017.03.021

Chen C, Liu W, Wu J, Jiang X, Zhu X, 2019. Can intercropping with the cash crop help improve the soil physico-chemical properties of rubber plantations? *Geoderma* 335, 149–160. https://doi.org/10.1016/j.geoderma.2018.08.023

Chong DT, Tajuddin I, Abd. Samat MS, 1991. Productivity of Cover Crops and Natural Vegetation under Rubber in Malaysia. In: Shelton HM, Stür WW (eds), *Forages for Plantation Crops*, Bali, Indonesia, 27–29 June 1990. ACIAR Proceedings, vol 32. Australian Centre for International Agricultural Research, 36–37.

Chotiphan R, Vaysse L, Lacote R, Gohet E, Thaler P, Sajjaphan K, et al., 2019. Can fertilization be a driver of rubber plantation intensification? *Ind. Crops Prod.* 141, 111813. https://doi.org/10.1016/j.indcrop.2019.111813

Clarence-Smith W.G., Ruf F., 1996. Cocoa pioneer fronts: the historical determinants. *In Cocoa pioneer fronts since 1800: The role of smallholders, planters and merchants*, London, Palgrave Macmillan UK, 1-22.

Clarence-Smith WG, 2013. Rubber cultivation in Indonesia and the Congo from the 1910s to the 1950s: divergent paths. In Frankema E, Buelens F (eds), *Colonial exploitation and economic development: The Belgian Congo and the Netherlands Indies compared*. Routledge, 193–210.

Clément-Demange A, Priyadarshan PM, Tran Thi Thuy Hoa, Venkatachalam P. 2007. Hevea rubber breeding and genetics. In Janick J (ed.), *Plant breeding reviews*, Wiley, (29): 177-283. ISBN 978-0-470-05241-9.

Clermont-Dauphin C, Dissataporn C, Suvannang N, Pong-wichian P, Maeght JL, Hammecker C, Jourdan C, 2018. Intercrops improve the drought resistance of young rubber trees. *Agron Sustain Dev* 38:56. https://doi.org/10.1007/s13593-018-0537-z

Clough Y, Krishna V, Corre M, Darras K, Denmead L, Mei-jide A, Moser S, et al., 2016. Land-use choices follow profitability at the expense of ecological functions in Indonesian smallholder landscapes. *Nature Commun* 7(1):13137. https://doi.org/10.1038/ncomms13137.

Cohen PT, 2009. The post-opium scenario and rubber in northern Laos: alternative Western and Chinese models of development. *Int J Drug Policy* 20:424–430.

Compagnon P, 1986. *Le caoutchouc naturel.* Ed. Maisonneuve et Larose, Paris, 1986.

Continental, 2015. L'avenir du pneumatique, le caoutchouc à partir du pissenlit. https://www.continental-tires.com/be/fr/newsroom/newsroom-car-and-van/sustainable-materials/

Conway G, 1993. Sustainable agriculture: the trade-offs with productivity, stability and equitability. In Barbier EB (ed), *Economy and Ecology: New frontiers and sustainable development.* London, Chapman & Hall.

Cooter R, 1982. The cost of Coase. *The Journal of Legal Studies, 11*(1), 1-33.

Cossalter C, 1992. Les ressources forestières en Indonésie et les conditions économiques, sociales et politiques de leur exploitation.

Costenbader J, Broadhead J, Yasmi Y, Durst PB, 2015. Drivers Affecting Forest Change in the Greater Mekong Subregion (GMS): An Overview. *FAO, USAID and LEAF.* http://www.climatefocus.com/sites/default/files/GMS-Drivers-PolicyBrief.pdf

Courbet P, Penot E, Iwiang I, 1997. Farming systems characterization and innovations adoption process in West Kalimantan. Icraf, Srap workshop on Ras (Rubber Agroforestry Systems), Bogor, 29-30.

Cowie AL, Barron JO, Castillo Sanchez VM, et al., 2018. Land in balance: The scientific conceptual framework for Land Degradation Neutrality, *Environmental Science & Policy*, 79: 25-35.

CTIS, 2013. Cambodia Trade Integration Strategy 2014-2018. https://www.undp.org/cambodia/publications/cambodia-trade-integration-strategy-2014-2018

Danthu P, Penot E, Ranoarison KM, Rakotondravelo JC, Michel-Dounias I, Tiollier M, et al., 2014. The Clove tree of Madagascar a success story with an unpredictable future. *Bois & forêts des tropiques.* 320(2). http://publications.cirad.fr/une_notice.php?dk=574435

Danthu P, Razakamanarivo H, Razafy Fara L, Montagne P, Deville-Danthu B, Penot E, 2016. When Madagascar produced natural rubber: a brief, forgotten yet informative history. *Bois & forêts des tropiques.* 328 (2), 27-43. http://hal.cirad.fr/cirad-00771066/

Dao N, 2015. Rubber plantations in the Northwest: rethinking the concept of land grabs in Vietnam. *The Journal of Peasant Studies, 42*(2), 347–369. https://doi.org/10.1080/03066150.2014.990445

Dean W, 2002. *Brazil and the struggle for rubber: a study in environmental history.* Cambridge University Press.

De Blécourt M, Brumme R, Xu J, Corre MD, Veldkamp E, 2013. Soil Carbon Stocks Decrease following Conversion of Secondary Forests to Rubber (*Hevea brasiliensis*) Plantations. *PLOS One 8,* e69357. https://doi.org/10.1371/journal.pone.0069357

De Foresta H, 1992a. Botany contribution to the understanding of smallholder rubber plantations in Indonesia: an example from South Sumatra. Symposium Sumatra Lingkungan dan Pembangunan. Bogor, Indonesia: Biotrop.

De Foresta H, 1992b. Complex agroforestry systems and conservation of biological diversity: for a larger use of traditional agroforestry trees as timber in Indonesia, a link between environmental conservation and economic development. Proceedings of an international conference on the conservation of tropical biodiversity. Malayan Nature Journal.

De Foresta H, Michon G, 1996. The agroforest alternative to *Imperata* grasslands: when smallholder agriculture and forestry reach sustainability. *Agroforest Syst* 36, 105–120. https://doi.org/10.1007/BF00142869

De Foresta H, 1997. Smallholder rubber plantations viewed through forest ecologist glassess. An example from South Sumatra. ICRAF/SRAP workshop on RAS (Rubber Agroforestry Systems), Bogor.

De Foresta H, 2013. Advancing agroforestry on the policy agenda – a guide for decision-makers, *Forests, Trees and Livelihoods*, 22, 213-215. http://dx.doi.org/10.1080/14728028.2013.806162

Delarue J, 2009. Developing smallholder rubber production: lessons from AFD's experience. *Evaluation and capitalization series*, no 26. https://www.oecd.org/countries/vietnam/44662138.pdf

Delarue J, Chambon B, 2012. Thailand: first exporter of natural rubber thanks to its family farmers. *Écon Rurale* 330–331:191–213.

Denyer D, Tranfield D, 2009. Producing a systematic review. In Buchanan DA, Bryman A (eds), *The Sage handbook of organizational research methods.* Sage Publication, UK, 671–689.

Descroix F, Snoeck J, 2009. Environmental factors suitable for coffee cultivation. In Wintgens JN (ed), *Coffee: growing, processing, sustainable production.* Wiley-VCH, Wein- heim, 168–181.

Deyasi A, Sarkar A, 2018. Analytical computation of electrical parameters in GAAQWT and CNTFET with identical configuration using NEGF method. *International Journal of Electronics*, 105(12), 2144–2159. https://doi.org/10.1080/00207217.2018.1494339

DGE (Directorate General of Estates), 1995. *Statistik karet.* Jakarta, Indonesia: Ministry of Agriculture.

Diaz-Novellon S, Boutin D, 2001. Flora of rubber agroforests in the West Kalimantan Province. CD rom, SRAP, ICRAF-Cirad, Bogor.

Diaz-Novelllon S, Penot E, Arnaud M, 2002. Characterisation of biodiversity in improved rubber agro-forests in West-Kalimantan, Indonesia. Real and Potential uses for spontaneous plants. International Symposium Land-use, nature conservation and the stability of rainforest margins in Southeast Asia.

Diaz-Novellon S, Penot E, Arnaud M, 2004. Characterisation of biodiversity in improved rubber agro-forests in West-Kalimantan, Indonesia: real and potential uses for spontaneous plants. *In* Gerold G, Fremerey M, Guhardja E (eds), *Land use, nature conservation and the stability of rainforest margins in Southeast Asia.* Springer, Berlin, 426–444.

Diepart JC, Dupuis D, 2012. La colonisation agricole des territoires ruraux du nord-ouest cambodgien : Pouvoirs, politiques paysannes et différenciations spatiales.

Dijkman J, 1951. *30 years of rubber research.* University of Miami Press, USA.

Dove M, 1993. Smallholder rubber and swidden agriculture in Borneo: a sustainable adaptation to the ecology and economy of tropical forest. *Economic botany*, 47/2.

Dove M, 1995. Political vs techno-economic factors in the development of non timber forest products: lessons from a comparison of natural and cultivated rubbers in Southeast Asia and South America. *Society and natural ressources*, vol 8.

Dove MR, 2018. Rubber versus forest on contested Asian land. *Nature Plants* 4, 321–322. https://doi.org/10.1038/s41477-018-0169-y

Drescher J, Rembold K, Allen K, Beckschäfer P, Buchori D, Clough Y, et al., 2016. Ecological and socio-economic functions acrosstropical land use systems after rainforest conversion. *Philos. Trans. R. Soc.* B 371:20150275. https://doi.org/10.1098/rstb.2015.0275

Durand F, 1999. La gestion des forêts en Indonésie. Trois décennies d'expérimentation hasardeuse (1967-1998). *Bois & forêts des tropiques*, 262, 45-59.

Ekanayake PB, 2003. Crop diversification and intercropping in tea lands. *Trop Agric Res Ext*, 6, 66-70.

Esekhade TU, Orimoloye JR, Ugwa IK, Idoko SO, 2003. Potentials of Multiple Cropping Systems in Young Rubber Plantations, *Journal of Sustainable Agriculture*, 22:4, 79-94. https://doi.org/10.1300/J064v22n04_07

Esekhade TU, Idoko SO, 2010. Effect of intercropping in the development of rubber saplings in acid sand in southern Nigeria. *J. Sustain. Trop. Agric. Res.* 31:33-40.

Esekhade TU, Okore IK, 2012. Impact of Different Spacings of Cooking Banana Intercropped with Rubber on Soil Fertility Attributes and Maturity Rate of the Trees in a Humid Forest Area of South Eastern Nigeria. *Open J. For.* 2, 65–70. https://doi.org/10.4236/ojf.2012.22009

Fageria NK, Baligar VC, Moreira A, Moraes LAC, 2013. Soil phosphorous influence on growth and nutrition of tropical legume cover crops in acidic soil. *Commun Soil Sci Plant Anal* 44:3340–3364.

FAO, 2001. The economics of conservation agriculture. *Land and Water Division, FAO, Rome*, 2001.

FAO, 2019. *FAOSTAT.* http://www.fao.org/faostat/en/#data

FAOSTAT, 2013. FAO. https://www.fao.org/faostat/en/#home

FAOSTAT, 2021. FAOSTAT Data http://www.fao.org/faostat/en/#data/QC

FCBA/ONF, 2021. Carbone Forêt-Bois : Des faits et des chiffres. Document de travail. 15 p.

Feike T, Doluschitz R, Chen Q, Graeff-Hönninger S, Claupein W, 2012. How to overcome the slow death of intercropping in the North China Plain. *Sustainability* 4:2550–2565

Feintrenie L, Levang P. 2009. Sumatra's rubber agroforests: advent, rise and fall of a sustainable cropping system. Small Scale For 8:323-335. http://www.cifor.org/publications/pdf_files/articles/AFeintrenie0901.pdf

Feintrenie L, Chong WK, Levang P, 2010. Why do Farmers Prefer Oil Palm? Lessons Learnt from Bungo District, Indonesia. *Small-Scale Forestry*, 9, 379–396. https://doi.org/10.1007/s11842-010-9122-2

Feintrenie L, Schwarze S, Levang P, 2010. Are local people conservationists? Analysis of transition dynamics from agroforests to monoculture plantations in Indonesia. Indonesia. *Ecology and Society*, 2010, vol. 15, no 4.

Fenske J, 2013. "Rubber will not keep in this country": failed development in Benin, 1897–1921. *Explor Econ Hist* 50:316–333.

Fern, 2018. Agricultural commodity consumption in the EU - Policy Brief. (October), 1–4. https://www.fern.org/

Finlay M, 2013. *Guayule et autres plantes à caoutchouc. De la saga d'hier à l'industrie de demain ?* Versailles, éditions Quæ, 256 p. https://www.quae.com/produit/1176/9782759220908/guayule-et-autres-plantes-a-caoutchouc

Fourcin V, 2014. Contribution du giroflier à la sécurité alimentaire des ménages agricoles dans la région de Fénérive-Est, Madagascar. Modélisation économique et analyse prospective. Mémoire M2. IRC/SUPAGRO Montpellier.

Fox J, 2009. Crossing borders, changing landscapes: land-use dynamics in the Golden Triangle. *Asia Pacific Issues* 92:8.

Fox J, Castella JC, 2013. Expansion of rubber (*Hevea brasiliensis*) in Mainland Southeast Asia: what are the prospects for smallholders? *The Journal of Peasant Studies*, 40(1), 155–170. https://doi.org/10.1080/03066150.2012.750605

Fox J, Castella JC, Ziegler AD, 2014. Swidden, rubber and carbon: Can REDD+ work for people and the environment in Montane Mainland Southeast Asia? *Glob Environ Chang.* 29:318–326.

Fritts R, 2019. Huge rubber plantation in Cameroon halts deforestation following rebuke. https://news.mongabay.com/2019/05/huge-rubber-plantation-in-cameroon-halts-deforestation-following-rebuke/

Gajaseni J, Gajaseni N, 1999. Ecological rationalities of the traditional homegarden system in the Chao Phraya Basin, Thailand, *Agroforestry Systems* 46: 3-23.

Gapkindo, 2022. Tentang Kami. http://gapkindo.org/tentang-kami

Geissler C, Penot E, 2000. La déforestation et après ? Mon palmier à huile contre ta forêt : Déforestation et politiques de concessions chez les Dayaks, Ouest-Kalimantan, Indonésie. *Bois & forêts des tropiques*, 266 (4), 7-22.

George S, Suresh PR, Wahid PA, Nair RB, Punnoose KI, 2009. Active root distribution pattern of *Hevea brasiliensis* determined by radioassay of latex serum. *Agrofor Syst* 76:275–281.

Ghornabi A, Langenberger G, Feng L, Saubern L, 2011. Ethnobotanical study of medicinal plants utilised by Hani ethnicity in Naban River Watershed National Nature Reserve, Yunnan, China. *Journal of Ethnopharmacology* 13(3), 651-667.

Gillison AN, Liswanti N, 2000. An intensive biodiversity baseline study in Jambi Province, Sumatra. Impact of different land uses on biodiversity. Alternatives to Slash-and Burn report. International Centre for Research in Agroforestry, ICRAF, Nairobi, Kenya, 18-160.

Giroh DY, Adebayo EF, Jongur AAU, 2013. Analysis of labour productivity and constraints of rubber latex exploitation among smallholder rubber farmers in the Niger delta region of Nigeria. *Glob J Agr Res* 1(3):16–26.

Gitz V, 2019. How can rubber contribute to sustainable development in a context of climate change? CGIAR. http://www.foreststreesagroforestry.org/how-can-rubber-contribute-to-sustainable-development-in-a-context-of-climate-change/

Goh HH, Tan KL, Khor CY, Ng SL, 2016. Volatility and market risk of rubber price in Malaysia: pre- and post-global financial crisis. *J Quant Econ* 14(2):323–344. https://doi.org/10.1007/s40953-016-0037-4

Gohet E, Chantuma P, Lacote R, Obouayeba S, Dian K, Clément-Demange A, et al., 2003. Latex clonal typology of *Hevea brasiliensis*: physiological modelling of yield potential and clonal response to ethephon stimulation. *In* IRRDB Workshop on Exploitation Technology, 15 -18 December 2003, Kottayam, India.

Gomiero T, Paoleti M, 2011a. Is There a Need for a More Sustainable Agriculture? *Critical Reviews in Plant Sciences* 30(1-2):6-23.

Gomiero T, Pimentel D, Paoletti MG, 2011b. Environmental Impact of Different Agricultural Management Practices: Conventional vs. Organic Agriculture. Book chapter *Critical Reviews in Plant Sciences*, 30:95–124. doi:10.1080/07352689.2011.554355

Gosling E, Reith E, Knoke T, Gerique A, Paul C, 2020. Exploring farmer perceptions of agroforestry via multi-objective optimisation: a test application in Eastern Panama. *Agroforest Syst* 94(5):2003–2020. https://doi.org/10.1007/s10457-020-00519-0

Gouyon A, 1989a. Increasing the productivity of rubber smallholders in Indonesia: a study of agro-economic constraints and proposals. Rubber Growers Conference. RIMM, Malacca.

Gouyon A, Nancy C, Anwar C, Negri M, 1989b. Perspectives d'amélioration de la filière et comportement des agents. Séminaire «Économie des filières en région chaude». Montpellier, France, MESRU/Cirad.

Gouyon A, de Foresta H, Levang P, 1993. Does 'jungle rubber' deserve its name? An analysis of rubber agroforestry systems in southeast Sumatra. *Agroforestry Systems* 22(3):181–206.

Gouyon A, 1993. Les Plaines de Sumatra-Sud : de la forêt aux hévéas. *Revue Tiers monde* XXXVI (135).

Gouyon A, 1995. Paysannerie et hévéaculture : dans les plaines orientales de Sumatra : quel avenir pour les systèmes agroforestiers ? Thèse INA-PG. INA-PG, Paris.

Grass I, Kubitza C, Krishna VV, Corre MD, Musshoff O, Putz P, et al., 2020. Trade-offs between multifunctionality and profit in tropical smallholder landscapes. *Nat Commun* 11(1):1186. https://doi.org/10.1038/s41467-020-15013-5

Grogan K, Pflugmacher D, Hostert P, et al., 2019. Unravelling the link between global rubber price and tropical deforestation in Cambodia. *Nature Plants* 5, 47–53. https://doi.org/10.1038/s41477-018-0325-4

Guardiola-Claramonte M, Troch PA, Ziegler AD, Giambelluca TW, Durcik M, Vogler JB, Nullet MA, 2010. Hydrologic effects of the expansion of rubber (*Hevea brasiliensis*) in tropical catchment. *Ecohydrology* 3:306–314.

Guo Z, Zhang Y, Deegen P, Uibrig H, 2006. Economic analyses of rubber and tea plantations and rubber-tea intercropping in Hainan, China. *Agrofor Syst*, 66:117-127. https://doi.org/10.1007/s10457-005-4676-2

Haddaway NR, Macura B, Whaley P, Pullin AS, 2017. ROSES flow diagram for systematic maps. *Version*, 2017, vol. 1, no 10.6084, p. m9. https://doi.org/10.6084/m9.figshare.6085940

Haines WB, 1934. The uses and control of natural undergrowth on rubber estates. Rubber Research Institute of Malaya, Kuala Lumpur.

Haishui Z, Kejun H, 1991. Intercropping in rubber plantation and its economic benefit. Zhaohua Z, Mantang C, Shiyi W, 204-206.

Hajra NG, Kumar R, 1999. Seasonal variation in photosynthesis and productivity of young tea. *Experimental Agriculture*, 35(1), 71-85.

Hamel O, Eschbach JM, 2001. Impact potentiel du mécanisme de développement propre dans l'avenir des cultures pérennes : état des négociations internationales et analyse prospective à travers l'exemple de la filière de production du caoutchouc naturel. *Oléagineux, Corps gras, Lipides*, 8(6), 599-610.

Hauser I, Martin K, Germer J, He P, Blagodatskiy S, Liu H, et al., 2015. Environmental and socio-economic impacts of rubber cultivation in the Mekong region: Challenges for sustainable land use. *CAB Reviews: Perspectives in Agriculture, Veterinary Science, Nutrition and Natural Resources.* https://doi.org/10.1079/PAVSNNR201510027

Hendrickson CT, Lave LB, Matthews HS, 2006. *Environmental Life Cycle Assessment of Goods and Services: An Input-Output Approach* (1st ed.). Routledge. https://doi.org/10.4324/9781936331383.

Herath PHMU, Takeya H, 2003. Factors determining intercropping by rubber smallholders in Sri Lanka: a logit analysis, *Agricultural Economics*, 29(2): 159-168.

Herlinawati E, Montoro P, Ismawanto S, Syafaah A, Aji M, Giner M, et al., 2022. Dynamic analysis of Tapping Panel Dryness in Hevea brasiliensis reveals new insights on this physiological syndrome affecting latex production. *Heliyon*, 8(10).

Higonnet E, Oram J, Kamkuimo-piam P, Weiss H, Kran-annexstein M, 2019. Restoration & reparations - Reforming the world's largest rubber company.

Hougni DGJ, Chambon B, Penot E, Promkhambut A, 2018. The household economics of rubber intercropping during the immature period in Northeast Thailand. *Journal of Sustainable Forestry*, 37:787–803. https://doi.org/10.1080/10549811.2018.1486716

Hughes AC, Satasook C, Bates PJ, Bumrungsri S, Jones G, 2012. The projected effects of climatic and vegetation changes on the distribution and diversity of Southeast Asian bats, *Global Change Biology*, 18(6), 1854-1865.

Hu A, Huang D, Duan Q, Zhou Y, Liu G, Huan H, 2023. Cover legumes promote the growth of young rubber trees by increasing organic carbon and organic nitrogen content in the soil. *Ind. Crops Prod.* 197,116640. https://doi.org/10.1016/j.indcrop.2023.116640

Hua F, Wang X, Zheng X, et al., 2016. Opportunities for biodiversity gains under the world's largest reforestation programme. *Nat Commun* 7,12717. https://doi.org/10.1038/ncomms12717

Huang IY, James K, Thamthanakoon N, Pinitjitsamut P, Rattanamanee N, Pinitjitsamut M, et al., 2023. Economic outcomes of rubber-based agroforestry systems: a systematic review and narrative synthesis. *Agroforest Syst*, 97, 335–354. https://doi.org/10.1007/s10457-022-00734-x

Huang J, Pan J, Zhou L, Zheng D, Yuan S, Chen J, Li J, et al., 2020. An improved double-row rubber (*Hevea brasiliensis*) plantation system increases land use efficiency by allowing intercropping with yam bean, common bean, soybean, peanut, and coffee: A 17-year case study on Hainan Island, China. *Journal of Cleaner Production*, 263, 121493.

Hughes AC, 2017. Understanding the drivers of Southeast Asian biodiversity loss. *Ecosphere*, 8(1). https://doi.org/10.1002/ecs2.1624

ILO, 2023. Child labour in plantation. https://www.ilo.org/jakarta/areasofwork/WCMS_126206/lang--en/index.htm

Imbach C, 1995. Caractérisation de la diversité des systèmes de culture associant hévéas, cultures annuelles et cultures pérennes dans le bassin de Cotabato à Mindanao (Philippines). Montpellier, CNEARC, 200 p.

IPCC (Intergovernmental Panel on Climate Change), 2019. Summary for Policymakers. In: Climate Change and Land: an IPCC special report on climate change, desertification, land degradation, sustainable land management, food security, and greenhouse gas fluxes in terrestrial ecosystems. IPCC report https://doi.org/10.1017/9781009157988.001

Iqbal SMM, Ireland CR, Rodrigo VHL, 2006. A logistic analysis of the factors determining the decision of smallholder farmers to intercrop: a case study involving rubber–tea intercropping in Sri Lanka. *Agric Syst* 87:296–312.

Iqbal SMM, Rodrigo VHL, Munasinghe ES, Balasooriya BMDC, Kudaligama KVVS, et al., 2016. Early growth of rubber in the Dry Zone of Sri Lanka: an investigation in Vavuniya District. *Journal of the Rubber Research Institute of Sri Lanka*, 96, 12-23.

IRIN, 2006. Rubber plantation workers strike over conditions, pay, child labour. *The New Humanitarian*. http://www.thenewhumanitarian.org/report/58109/liberia-rubber-plantation-workers-strike-over-conditions-pay-child-labour

IRSG, 2018. International Rubber Study Group – Statistics. http://www.rubberstudy.com/statistics.aspx

IUCN, 2011. Assessment of Economic, Social and Environmental Costs and Benefits of Mitr Lao Sugar Plantation and Factory: Case Study in Savannakhet Province. (January). http://www.unpei.org/component/docman/doc_download/133-pei-lao-sugarcane-report-2011.html?Itemid=

Jaafar AA, Kasiran N, Muhammed S, Ismail WHW, 2000. Agroforestry practices in Malaysia – integrating plantation crops with timber species. Conférence at Faculty of Applied Science, Universiti Teknologi MARA (Pahang) Malaysia.

Jata SK, Nedunchezhiyan M, Maity SK, Mmallikarjun M, 2018. Cropping system and irrigation effects on Nutrient use efficiency and quality of elephant foot yam [*Amorphophallus paeoniifolius* (Dennst.) Nicolson]. *e-planet* 16(1), 26-33.

Jessy MD, Joseph P, George S, 2017. Possibilities of diverse rubber based agroforestry systems for smallholdings in India. *Agrofor. Syst.* 91, 515–526. https://doi.org/10.1007/s10457-016-9953-8

Jiang XJ, Liu W, Wu J, Wang P, Liu C, Yuan ZQ, 2017. Land Degradation Controlled and Mitigated by Rubber-based Agroforestry Systems through Optimizing Soil Physical Conditions and Water Supply Mechanisms: A Case Study in Xishuangbanna, China. *Land Degrad. Dev.* 28, 2277–2289. https://doi.org/10.1002/ldr.2757

Jiang XJ., Chen C, Zhu X, Zakari S, Singh AK, Zhang W, et al., 2019. Use of dye infiltration experiments and HYDRUS-3D to interpret preferential flow in soil in a rubber-based agroforestry systems in Xishuangbanna, China. *CATENA* 178, 120–131. https://doi.org/10.1016/j.catena.2019.03.015

Jiang XJ, Zakari S, Wu J, Singh AK, Chen C, Zhu X, et al., 2020. Can complementary preferential flow and non-preferential flow domains contribute to soil water supply for rubber plantation? *Forest Ecology and Management*, 461, 117948.

Jin S, Min S, Huang J, Waibel H, 2021. Falling price induced diversification strategies and rural inequality: Evidence of smallholder rubber farmers, *World Development*, 146: 105604.

Jongrungrot V, Thungwa S, Snoeck D, 2014a. Tree-crop diversification in rubber plantations to diversify sources of income for small-scale rubber farmers in Southern Thailand. *Bois & forêts des tropiques*, 68(321), 21–32.

Jongrungrot V, Thungwa S, 2014b. Resilience of rubber-based intercropping systems in southern Thailand. *Adv Mat Res* 844:24–29.

Jongrungrot V, Kheowvongsri P, 2021. Comparative study between rubber agroforestry systems and rubber monocultures and soil erosion decrease via rubber agroforestry profile and crown cover in lower southern Thailand, *Kasetsart Journal of Social Sciences*, 42(2): 255-261.

Joshi L, Wibawa G, Beukema H, Williams S, van Noordwijk M, 2000. Technological change and biodiversity in the rubber agroecosystem of Sumatra. *Tropical Agroecosystem*, 133-155.

Joshi L, Wibawa G, Sinclair FL, 2001. Local ecological knowledge and socio-economic factors influencing farmers' management decisions in jungle rubber agroforestry systems in Jambi, Indonesia. World Agroforestry Centre (ICRAF).

Joshi L, Wibawa G, Akiefnawati R, Mulyoutami E, Wulandari D, Penot E, 2006. Diversified rubber agroforestry for smallholder farmers–a better alternative to monoculture. Information booklet for workshop on rubber development in lao PDR: exploring improved systems for small-holder production, 9–11.

Junaidi U, Sultoni Arifin M, 1989. Cropping patterns for smallholder rubber up to 3 years old. RIEC, Sembawa, South Sumatra, RRIM rubber growers conférence, Kuala Lumpur, Malaysia.

Kaligis DA, Sumolang C, 1991. Forage species for rubber plantations in Malaysia. In Shelton HM, Stilr WW (eds), Forages for plantation crops, Sanur Beach, Bali, Indonesia. *ACIAR Proceedings*, vol 32, 49–53.

Kalkuhl M, von Braun J, Torero M, 2016. Volatile and extreme food prices, food security, and policy: an overview. In *Food price volatility and its implications for food security and policy*. Springer.

Kartasujana I, Martawijaya A, 1973. Commercial woods of Indonesia, their properties and uses, *Pengumuman* no.3. Lembaga Penelitian Hasil Hutan, Bogor.

Kaya Y, Piepel GF, Caniyilmaz E, 2013. Development of a rubber-based product using a mixture experiment: a challenging case study, *Progress in Rubber Plastics and Recycling Technology* 29(3): 123-150.

Keli JZ, Omont H, Assiri AA, Boko KA, Obouayeba S, Dea BG, Doumbia A, 2005. Associations culturales à base d'hévéa : bilan de 20 années d'expérimentations. *Agronomie africaine*, 17(1), 37-52. https://doi.org/10.4314/aga.v17i1.1656

Keli JZ, De La Serve M, 1988. Temporary association of hevea and food crops in lower Cote d'Ivoire. *Revue Générale des Caoutchoucs et Plastiques (France)*, 65(679).

Kheowvongsri P, 1990. Les jardins à hévéas des contreforts orientaux de Bukig Barisan, Sumatra, Indonésie.

Killick E, 2018. Rubber, Terra Preta, and Soy: A Study of Visible and Invisible Amazonian Modernities. *Journal of Anthropological Research*, 74(1), 32–53. https://doi.org/10.1086/696163

Killmann W, Hong LT, 2000. Rubber wood—the success of an agricultural by-product. *Unasylva* 51:66–72.

King RG, Plosser CI, Rebelo ST, 1988. Production, growth and business cycles: I. The basic neoclassical model. *Journal of monetary Economics*, 21(2-3), 195-232.

Kou W, Dong J, Xiao X, Hernandez AJ, Qin Y, Zhang G, et al., 2018. Expansion dynamics of deciduous rubber plantations in Xishuangbanna, China during 2000-2010. *GIScience & Remote Sensing*, 55(6), 905-925.

Kouadio YDM, Kouassi KH, Bahan FM, Keli ZJ, 2021. Impact de l'association Hevea-Cafeier sur la production sur la production des deux spéculations.

Kremen C, Miles A, 2012. Ecosystem services in biologically diversified versus conventional farming systems: Benefits, externalities, and trade-offs. *Journal Ecology and Society*, 17(4).

Kumar BM, Nair PKR, 2004. The enigma of tropical home gardens. *Agroforest Syst* 61:135-152.

Laosuwan P, Sripana P, Thongsomsri A, 1988. Research on mungbean as intercrop of young rubber and as sequential crop with rice, *Mungbean*, 392.

Langenberger G, Cadisch G, Martin K, 2015. Rubber plantations as repositories for endangered plant species? A case study from SW China. In Bredemeier M, Steckel J, Seele U (eds), *Ecology for a sustainable future*, University of Gottingen.

Langenberger G, Cadisch G, Martin K, Min S, Waibel H, 2017. Rubber intercropping: a viable concept for the 21st century? *Agroforest Syst* 91, 577–596. https://link.springer.com/article/10.1007/s10457-016-9961-8

Lass RA, 1985. Diseases. In Wood GAR, Lass RA (eds), *Cocoa*. Longman Group Limited, London, 265–365.

Laumonier Y, Legg C, 1998. Le suivi des feux de 1997 en Indonésie. Apport d'Internet et de la télédétection à une situation d'urgence. *Bois & forêts des tropiques*, 258, 5-18.

Lavigne-Delville P, Wybrecht B, 2009. Les diagnostics, outils pour le développement. In *Memento de l'Agronome*, Versailles, éditions Quæ, 1699 p.

Lawrence DC, 1996. Trade-offs between rubber production and maintenance of diversity: the structure of rubber gardens in West Kalimantan, Indonesia. *Agrofor Syst* 34:83–100.

Leakey R, 1996. Definition of agroforestry revisited. *Agroforestry today*, 8, 5-5.

Lehébel-Péron A, Feintrenie L, Levang P, 2011. Rubber agroforests' profitability, the importance of secondary products. *Forests, Trees Livelihoods* 20:69–84. https://doi.org/10.1080/14728028.2011.9756698

Lemmens RHMJ, 1989. *Plant resources of South-East Asia*. Pudoc.

Levang P, de Foresta H, 1991. Economic plants of Indonesia: a Latin, Indonesian, French and English dictionary of 728 species. *Economic plants of Indonesia*. Orstom/SEAMEO-BIOTROP, Bogor.

Levang P, Michon G, de Foresta H, 1997. Agriculture forestière ou agroforesterie. *Bois & forêts des tropiques*, 251 (1).

Li H, Aide TM, Ma Y, Liu W, Cao M, 2007. Demand for rubber is causing the loss of high diversity rain forest in SW China. *Biodivers Conserv* 16:1731–1745.

Li J, Zhou L, Lin W, 2019. Calla lily intercropping in rubber tree plantations changes the nutrient content, microbial abundance, and enzyme activity of both rhizosphere and non-rhizosphere soil and calla lily growth. *Industrial Crops and Products*, 132, 344-351.

Li W (ed), 2001. *Agro-ecological farming systems in China*, vol 26. Man and the biosphere series, Unesco, Paris.

Li XA, Ge TD, Chen Z, Wang SM, Ou XK, Wu Y, et al., 2020. Enhancement of soil carbon and nitrogen stocks by abiotic and microbial pathways in three rubber-based agroforestry systems in Southwest China. *Land Degrad. Dev.* 31, 2507–2515. https://doi.org/10.1002/ldr.3625

Li Z, Fox JM, 2012. Mapping rubber tree growth in mainland Southeast Asia using time-series MODIS 250 m NDVI and statistical data. *Appl Geogr* 32, 420–432.

Lin W, Zhou Z, Huang S, 1999. A review and prospect of intercropping in rubber plantation in China. *Chinese J. of Ecology.* 18(1), 43-52. (in Chinese).

Liu C, Jin Y, Hu Y, Tang J, Xiong Q, Xu M, et al., 2019. Drivers of soil bacterial community structure and diversity in tropical agroforestry systems. *Agric. Ecosyst. Environ.* 278, 24–34. https://doi.org/10.1016/j.agee.2019.03.015

Liu CA, Liang MY, Tang JW, Jin YQ, Guo ZB, Siddique KHM, 2021. Challenges of the establishment of rubber-based agroforestry systems: Decreases in the diversity and abundance of ground arthropods. *J. Environ. Manage.* 292, 112747. https://doi.org/10.1016/j.jenvman.2021.112747

Liu H, Blagodatsky S, Giese M, Liu F, Xu J, Cadisch G, 2016. Impact of herbicide application on soil erosion and induced carbon loss in a rubber plantation of Southwest China. *CATENA* 145, 180–192. https://doi.org/10.1016/j.catena.2016.06.007

Liu W, Luo Q, Lu H, Wu J, Duan W, 2017. The effect of litter layer on controlling surface runoff and erosion in rubber plantations on tropical mountain slopes, SW China. *CATENA* 149, 167–175. https://doi.org/10.1016/j.catena.2016.09.013

Liu W, Hu H, Ma Y, Li H, 2006. Environmental and socioeconomic impacts of increasing rubber plantations in Menglun township, southwest China. *Mountain Research and Development,* 26(3), 245-253.

Liu W, Zhu C, Wu J, Chen C, 2018. Are rubber-based agroforestry systems effective in controlling rain splash erosion? *CATENA,* 147, 16-24.

Majid NM, Ghani ANA, Hamid JA, 1990. Economic appraisal of sheep grazing under rubber plantation in Peninsular Malaysia. *Biotrop Special Publication,* (Indonesia).

Malaysian Rubber Board 2016. Malaysia – National rubber statistics. https://www.lgm.gov.my/webv2/home

Malaysian Rubber Board, 2019. *Malaysian Rubber Exchange - Monthly Average & Charts.* http://www3.lgm.gov.my/mre/YearlyAvg.aspx

Malaysian Rubber Board, 2023. *Rubber Board - Ministry of Commerce & Industry.* http://rubberboard.org.in/public

Malézieux E, Crozat Y, Dupraz C, Laurans M, Makowski D, Ozier-Lafontaine H, et al., 2009. Mixing plant species in cropping systems: concepts, tools and models: a review. *Sustainable agriculture,* 329-353.

Manivong V, Cramb RA, 2007. Economics of smallholder rubber production in Northern Laos, In: 51st Annual conference, Australian agricultural and resource econom-ics Society, Queenstown, New Zealand.

Mann CC, 2009. Addicted to rubber. *Science,* 325(5940), 564–566. https://doi.org/10.1126/science.325_564

Mann CC, Charles BR, 2016. Riding Rubber's Boom The rising global demand for car tires may pay off for Southeast Asia's poor, but at a cost to the planet. *National Geographic,* 2016.

Mariel J, Penot E, Herimandimby H, Danthu P, 2019. A combined landscape and farming system approach to analyze agroforestry heterogeneity in an agrarian territory in Madagascar. 4th World Agroforestry Congress (WAC 2019). Montpellier.

Markowitz H, 2008. *Harry Markowitz: selected works* vol 1. World Scientific Publishing Company, Hackensack NJ, USA.

Marsden C, Martin-Chave A, Cortet J, Hedde M, Capowiez Y, 2019. How agroforestry systems influence soil fauna and their functions - a review. *Plant Soil.* 453, 29-44. https://doi.org/10.1007/s11104-019-04322-4

Masae A, Cramb RA, 1995. Socio-economic aspects of rubber intercropping on acid soils in Southern Thailand. *In Plant-Soil Interactions at Low pH: Principles and Management:* Proceedings of the Third International Symposium on Plant-Soil Interactions at Low pH, Brisbane, Queensland, Australia, 12–16 September 1993, Springer Netherlands, 679-684.

McAdam JH, 2000. Tropical Agroforestry. By P. Huxley. Oxford: Blackwell Science (1999). *Experimental Agriculture,* 36(4), 519-519.

McAllister KE, 2015. Rubber, rights and resistance: the evolution of local struggles against a Chinese rubber concession in Northern Laos. *The Journal of Peasant Studies*, 42(3–4), 817–837. https://doi.org/10.1080/03066150.2015.1036418

Meade JE, 1952 External Economies and Diseconomies in a Competitive Situation, *The Economic Journal*, 62(245), 54-67.

Meng LZ, Martin K, Weigel A, Liu JX, 2012. Impact of rubber plantation on carabid beetle communities and species distribution in a changing tropical landscape (southern Yunnan, China). *J Insect Conserv* 16:423–432.

Menon PM, 2002. Natural rubber in India: past, present and future. In Jacob CK (ed), Global Competitiveness of Indian Rubber Plantation Industry. Rubber Planter's Conference, Rubber Research Institute of India, Kottaym, 17–32.

Mercer DE, 2004. Adoption of agroforestry innovations in the tropics: a review, *Agroforestry systems* 61: 311-328.

Mesike CS, Esekhade TU, Idoko SO, 2019. Determinants of smallholding rubber agroforestry systems (RAFS) adoption in Nigeria. *Proceedings of International Rubber Conference.*

Michelin, 2018. Michelin déploie Rubberway, une application visant à cartographier les pratiques RSE de sa chaîne d'approvisionnement en caoutchouc naturel. Michelin Corporate. Site Internet MICHELIN. https://www.michelin.com/communiques-presse/michelin-deploie-rubberway-une-application-visant-a-cartographier-les-pratiques-rse-de-sa-chaine-dapprovisionnement-en-caoutchouc-naturel/

Michon G, Mary F, Bompard J, 1986. Multistoried agroforestry garden system in West Sumatra, Indonesia. *Agroforestry systems*, 4, 315-338.

Michon G, de Foresta H, 1990. Complex agroforestry systems and conservation of biological diversity: Agroforests in Indonesia: the link between two worlds. Proceedings of an international conference on the conservation of tropical biodiversity. Malayan Nature Society. Kuala Lumpur. Malaysia.

Michon G, de Foresta H, 1991. Complex agroforestry systems and the conservation of biological diversity: Agroforests in Indonesia: the link between two worlds. *Malayan Nature Journal*, 45.

Michon G, de Foresta H, 1995. The Indonesian agroforest model. Forest resources management and biodiversity conservation. In Halladay P, Gilmour D, *Concerning biodiversity outside protected areas; the role of traditional agro-ecosystems.* IUCN/AMA.

Michon G, de Foresta H, Levang P, 1995. Stratégies agroforestières paysannes et développement durable : les agroforêts à damar de Sumatra. *Natures - Sciences - Sociétés* 3(3): 207-221.

Michon G, De Foresta H, 1997. Agroforest: predomestication of forest or true domestication of forest ecosystems? *Netherlands Journal of agricultural Science* 45: 451-462.

Michon GH, de Foresta H, Levang P, Verdeaux F, 2007. Domestic forests: a new paradigm for integrating local communities' forestry into tropical forest science. *Ecol Soc* 12. http://www.ecology-andsociety.org/vol12/iss2/art1/

Min S, Huang J, Bai J, Waibel H, 2015. Adoption of intercropping among smallholder rubber farmers in Xishuangbanna, China. 29th International conference of agricultural economists, Milan, Italy. http://purl.umn.edu/212463

Min S, Huang J, Bai J, Waibel H, 2017a. Adoption of intercropping among smallholder rubber farmers in Xishuangbanna China. *Int J Agric Sustain* 15(3):223–237. https://doi.org/10.1080/14735903.2017.1315234

Min S, Waibel H, Cadisch G, Langenberger G, Bai J, Huang J, 2017b. The Economics of Smallholder Rubber Farming in a Mountainous Region of Southwest China: Elevation, Ethnicity, and Risk. *Mountain Research and Development*, 37(3), 281–293. https://doi.org/10.1659/MRD-JOURNAL-D-16-00088.1

Min S, Huang J, Waibel H, Yang X, Cadisch G, 2019. Rubber Boom, Land Use Change and the Implications for Carbon Balances in Xishuangbanna, Southwest China. *Ecological Economics*, 156, 57-67, https://doi.org/10.1016/j.ecolecon.2018.09.009

Mollard E, 1993. Changement technique et environnement économique, relations et interrogations. In « Innovations et sociétés. Quelles agricultures ? Quelles innovations ? ». Actes du XIVème séminaire d'économie rurale, Montpellier, France.

Momberg F, 1993. Indigenous knowledge systems Resource management of Land-Dayaks in West Kalimantan. AGris report, FAO. Berlin, 125 p.

Montagnini F, Nair PKR, 2004. Carbon Sequestration: An Underexploited Environmental Benefit of Agroforestry Systems. *Agroforestry Systems*, 61-62, 281-295. https://doi.org/10.1023/B:AGFO.0000029005.92691.79

Muller RA, Berry D, Avelino J, Bieysse D, 2009. Coffee diseases. In Wintgens JN (ed), *Coffee: growing, processing, sustainable production*, 2nd ed. Wiley-VCH, Weinheim, 495–549.

Mulyoutami E, Ilahang, Joshi L, Wibawa G, Penot E, 2006. From degraded Imperata grassland to productive rubber agroforests in West Kalimantan. Hat Hai rubber seminar 2006. Thailand.

Murray MJ, 1992. 'White gold' or 'white blood'? The rubber plantations of colonial Indochina, 1910–40. *J Peasant Stud* 19:41–67.

Muschler RG, 2009. Shade management and its effect on coffee growth and quality. In: Wintgens JN (ed), *Coffee: growing, processing, sustainable production*, 2nd ed. Wiley-VCH, Weinheim, 395–422.

Myint H, 2013. Rubber Planting Industry in Myanmar: Current Situation and Potentials. ASEAN Plus Rubber Conference, Phuket.

Nair STG, 2018. The effects of financial crisis on Hedging efficiency of Indian rubber future markets. *Financial Stat J.* 4(1), 608. https://doi.org/10.24294/fsj.v1i3.608

Nair PKR, 1989. *Agroforestry systems in the tropics*, Forestry.

Nair PKR, 1993. State-of-the-art of agroforestry research and education. *Agroforestry systems*, 23, 95-119.

Newman SM, 1985. A survey of interculture practices and research in Sri Lanka. *Agrofor Syst* 3:25–36.

Neyret M, Robain H, De Rouw A, Soulileuth B, Trisophon K, Jumpa K, Valentin C, 2018. The transition from arable lands to rubber tree plantations in northern Thailand impacts weed assemblages and soil physical properties. *Soil use and management*. 34, 404–417. https://doi.org/10.1111/sum.12431

Ng KF, 1991. Forage species for rubber plantations in Malaysia. ACIAR Proceedings-Forages for Plantation Crops, Sanur Beach, Bali, Indonesia (ed. WW Stur), p. 1990.

Ng KF, Stür WW, Shelton HM, 1997. New forage species for integration of sheep in rubber plantations. *J Agric Sci* 128, 347–356.

N'Guessan JLL, Niamké BF, Yao NGJC, Amusant N, 2023. Wood extractives: Main families, functional properties, fields of application and interest of wood waste. *Forest Products Journal*, 73(3), 194-208.

Nouvelle L, 2018. Firestone réduit sa production de caoutchouc au Liberia - L'Usine Matières premières. https://www.usinenouvelle.com/article/firestone-reduit-sa-production-de-caoutchouc-au-liberia.N744779

Nugawela A, 1991. Review of the Plant Science Department The Rubber Research Institute of Sri Lanka. 11-28.

Nyhus P, Tilson R, 2004. Agroforestry, elephants, and tigers: balancing conservation theory and practice in human-dominated landscapes of Southeast Asia. *Agriculture, ecosystems & environment*, 104(1), 87-97.

Nyikahadzoi K, Pali P, Fatunbi AO, Olarinde LO, Njuki J, Adekunle AO, 2012. Stakeholder participation in innovation platform and implications for integrated agricultural research for development (IAR4D).

Oktavia F, Agustina DS, 2021. Progress of rubber breeding program to support agroforestry system in Indonesia. E3S Web of Conferences (Vol. 305, p. 03006). EDP Sciences.

Orozco AO., Salber M, 2019. Palmed off - an investigation into three industrial palm oil and rubber projects in Cameroon and the republic of Congo. Rainforest foundation UK. https://www.rainforestfoundationuk.org/media.ashx/palmoilreportenweb.pdf

Page S, Hewitt A, 2001. World commodity prices: still a problem for developing countries? Overseas Development Institute, London, UK.

Panklang P, Thoumazeau A, Chiarawipa R, Sdoodee S, Sebag D, Gay F, et al., 2022a. Rubber, rubber and rubber: How 75 years of successive rubber plantation rotations affect topsoil quality? *Land Degradation & Development*, 33(8), 1159–1169. https://doi.org/10.1002/ldr.4171

Panklang P, Thaler P, Thoumazeau A, Chiarawipa R, Sdoodee S, Brauman A, 2022b. How 75 years of rubber monocropping affects soil fauna and nematodes as the bioindicators for soil biodiversity quality index, *Acta Agriculturae Scandinavica, Section B — Soil & Plant Science*, 72:1, 612-622. doi:10.1080/09064710.2022.2034930

Pathiratna LSS, 2006a. Management of intercrops under rubber: implications of competition and possibilities for improvement. *Bull Rubber Res Inst Sri Lanka* 47:8–16.

Pathiratna LSS, 2006b. Cinnamon for intercropping under rubber. *Bull Rubber Res Inst Sri Lanka* 47:17–23.

Payne WJA, 1985. A review of the possibilities for integrating cattle and tree crop production systems in the tropics. *For Ecol Manage* 12:1–36.

Pearce D, 1998. Cost benefit analysis and environmental policy. *Oxford Review of Economic Policy*, 14(4) 84–100. https://doi.org/10.1093/oxrep/14.4.84

Peerawat M, Blaud A, Trap J, Chevallier T, Alonso P, Gay F, et al., 2018. Rubber plantation ageing controls soil biodiversity after land conversion from cassava. *Agriculture, Ecosystems and Environment*, 257, 92–102. https://doi.org/10.1016/j.agee.2018.01.034

Penot E, 1994. The non-project rubber smallholder sector in Indonesia: rubber agroforestry systems (RAS) as a challenge for the improvement of rubber productivity, rubber-based systems sustainability, biodiversity and environment. ICRAF Working Paper, Nairobi.

Penot E, 1995. Taking the jungle out of rubber. Improving rubber in Indonesian agroforestry systems. *Agroforestry Today*, 7, 3/4, 11-13.

Penot E, 1995. Rubber agroforestry systems, RAS, as sustainable alternatives to Imperata grasslands in West-Kalimantan, Indonesia. ICRAF Imperata workshop, Benjarmasin.

Penot E, Gouyon A, 1995. Agroforêt et plantations clonales : des choix pour l'avenir. Présenté au séminaire Cirad-MES.

Penot E, Wibawa G, 1996. Improved Rubber Agroforestry Systems in Indonesia: an alternative to low productivity of jungle rubber conserving agroforestry pratices and benefits. First results from on-farm experimentation in West-Kalimantan. IRRDB annual meeting, Colombo/Bentota, Sri Lanka.

Penot E, 1996a. Improving productivity in rubber based agroforestry systems (RAS) in Indonesia: a financial analysis of RAS systems. GAPKINDO seminar in Sipirok, North-Sumatra.

Penot E, 1996b. Sustainability through productivity improvement of Indonesian rubber-based agroforestry systems. 14th international symposium on sustainable farming systems. Colombo, Sri Lanka.

Penot E, Werner S, 1997. Prospects for the conservation of secondary forest biodiversity within productivite rubber agroforests. SRAP workshop, Bogor, Indonesia, 1997. SRAP/CIRAS-ICRAF CD rom.

Penot E, 1997. Introduction to SRAP methodology and concepts: summary of the preliminary results. The ICRAF/SRAP Workshop on RAS, ICRAF, Bogor.

Penot E, 1998. Jungle rubber improvement in Indonesia. IRRDB Inf Q 2:8–17.

Penot E, 1999. Rubber Agroforestry Systems (R.A.S.) methodology and main results: technical report. Cirad/ICRAF, project paper. Montpellier.

Penot E, Ruf F, 2001. Rubber cushions the smallholder: no windfall, no crisis. In Gerard F, Ruf F (eds), *Agriculture in crisis: people, commodities and natural ressources in indonesia, 1996-2000*, Cirad/CURZON. Richmond, UK, Curzon Press, 237-266.

Penot E, Chambon B, 2003. Processus d'innovation : dynamique agroforestières et changement technique : le cas de l'hévéaculture villageoise en Indonésie. Colloque International ; «Un produit, une filière, un terroir», Toulouse, mai 2001. Publié en ouvrage collectif en août 2003.

Penot E, 2004a. Cohérence entre systèmes techniques et systèmes sociaux et territoires. Evolution des systèmes de production hévéicoles et gestion de la ressource foncière : le cas de la province de Ouest-Kalimantan, Indonésie. *Cahiers Agricultures*, 13, 450-458. http://revues.cirad.fr/index.php/cahiers-agricultures/article/view/30465

Penot E, 2004. From shifting agriculture to sustainable rubber agroforestry systems (jungle rubber) in Indonesia: a history of innovations processes. *In*: Babin D (ed), *Beyond tropical deforestation*. UNESCO/Cirad, Paris, 221–250.

Penot E, Geissler C, 2004. Deforestation, agricultural concession policies and potential conflicts in Sanggau district, West Kalimantan province, Indonesia. *In* Babin D (ed), *Beyond tropical deforestation. From Tropical Deforestation to Forest Cover Dynamics and Forest Development*, Unesco-Cirad, p. 382.

Penot E, 2007. Simulation et modélisation du fonctionnement de l'exploitation agricole. *In :* Gafsi M, Brossier J, Dugué P, Jamin JY (eds), *Les exploitations familiales agricoles africaines : enjeux caractéristiques et élements de gestion*, Versailles, éditions Quæ, 556 p.

Penot E, Ollivier I, 2009. L'hévéa en association avec les cultures pérennes, fruitières ou forestières : quelques exemples en Asie, Afrique et Amérique latine. *Bois & forêts des tropiques*, 301(301), 67. https://doi.org/10.19182/bft2009.301.a20407

Penot É (ed), 2012. *Exploitations agricoles, stratégies paysannes et politiques publiques : les apports du modèle Olympe*, Versailles, éditions Quae, 336 p.

Penot E, Rivano F, Follin C, 2012. Stratégies de diversification et développement alternatif à la culture de la coca en Amazonie colombienne (Caquetá). Analyse du cas colombien dans la sous-région. *Économie Rurale*. 329, 64-82. http://publications.cirad.fr/une_notice.php?dk=564192

Penot E, Feintrenie L, 2014. L'agroforesterie sous climat tropical humide : Une diversité de pratiques pour répondre à des objectifs spécifiques et à des contraintes locales. *Bois & forêts des tropiques*, 68(321), 5–6.

Penot E (ed), 2016. *Processus d'innovation et résilience des exploitations agricoles à Madagascar.* Editions l'Harmattan, Paris.

Penot E, 2016. Rubber agroforestry systems in Indonesia and Thailand for a sustainable agriculture and income stability. Promoting environmental friendly and socially responsible rubber cultivation. Conference Xishuangbanna Tropical Botanical Garden by the Chinese Academy of Sciences, China.

Penot E, Fourcin C, Michel I, Danthu P, Jahiel M, 2017. Systèmes à base de giroflier, stratégies paysannes et sécurité alimentaire. Le cas de la région de Fénérive-Est à Madagascar. Colloque «Agricultures, ruralités et développement», Université libre de Bruxelles. http://agritrop.cirad.fr/587013/

Penot E, Chambon B, Nahum Benavidez López D, Tongkaemkaew U, 2018. Les plates formes d'innovation comme nouveau modèle de développement agricole inter-acteurs dans les pays en développement : regard croisé sur deux expériences en Thaïlande et au Nicaragua. Congrès RRI, VII forum de l'innovation, Nîmes. http://agritrop.cirad.fr/588185/

Penot E, Ilahang, Asgnari A, Dinas P, 2019. Rubber Agroforestry systems in Kalimantan, Indonesia. Which changes from 1994 to 2019? Report of the mission undertaken in October 2019 with support from the Forests, Trees and Agroforestry research program (FTA) of the CGIAR. Montpellier, Cirad. https://agritrop.cirad.fr/597296/

Penot E, Thaler P, Nouvellon Y, Chambon B, Sainte Beuve J, 2020. État des lieux sur la déforestation et les standards de durabilité. Rapport d'étude pour le CST/Forêt et l'AFD.

Penot E, 2021. Rubber Agroforestry systems (RAS) in West Kalimantan, Indonesia: an historical perspective. *E3S Web of Conferences*, 305, p. 02001, EDP Sciences.

Penot E, Thériez M, Michel I, Tongkaemkaew U, Chambon B, 2022. Agroforestry rubber networks and farmers groups in Phatthalung area in Southern Thailand and potential for an innovation platform. *Forest and society.* 6(2). https://doi.org/10.24259/fs.v6i2.12481

Penot E, Chambon B, Gitz V, Meybeck A, Ilahang, Wibawa G, et al., 2023. Rubber In Indonesia: a resilience analysis and policy recommendations. Rubber Resilience project/FTA/CIFOR final report.

Penot E, Le Guen A, Chevreux A, Gallier C, Ellis H, Laville J, et al., 2023. Agronomic and socio-economic options for rubber intercropping in Sri Lanka: a forward analysis in the Moneragala and Ampara regions. *Bois & forêts des tropiques*, 356: 43-65. https://doi.org/10.19182/bft2023.356

Penot E, Ilahang, Chambon B, Wibawa G, 2023. Rubber Agroforestry systems evolution from 1994 to 2021 in West Kalimantan province, Indonesia. Rubber works project/FTA/CIFOR final report.

Penot E, Chambon B, Sainte-Beuve J, 2023. An analysis and comparaison of the rubber smallholder sector in 5 countries: Cote d'Ivoire, Thailand, Indonesia, Vietnam and Cambodia, Montpellier, Cirad, 13 p.

Penot E, Wang Mei Hua M, Akichi JP, 2024. Rubber agroforestry systems (RAS) for a sustainable agriculture. Agroforestry systems based on rubber trees. Workshop GPSNR 2023 with rubber planters in Côte d'Ivoire.

Perron T, Kouakou A, Simon C, Mareschal L, Gay F, Soumahoro M, et al., 2022. Logging residues promote rapid restoration of soil health after clear-cutting of rubber plantations at two sites with contrasting soils in Africa. *Science of the Total Environment*, 816:151526. https://doi.org/10.1016/j.scitotenv.2021.151526

Persoon GA, 2007. Agarwood: the life of a wounded tree. *Int Inst Asian Stud* 45:24–25.

Phommexay P, Satasook C, Bates P, Pearch M, Bumrungsri S, 2011. The impact of rubber plantations on the diversity and activity of understorey insectivorous bats in southern Thailand. *Biodivers Conserv* 20:1441–1456.

Pierini SV, Warrell DA, De Paulo A, Theakston RDG, 1996. High incidence of bites and stings by snakes and other animals among rubber tappers and amazonian indians of the Jurua – Valley, Acre State, Brazil. *Toxicon* 34:225–236.

Pirard R, Petit H, Baral H, 2017. Local impacts of industrial tree plantations: An empirical analysis in Indonesia across plantation types. *Land Use Policy*, 60, 242-253.

Polthanee A, 2018. Cassava as an insurance crop in a changing climate: the changing role and potential applications of cassava for smallholder farmers in Northeastern Thailand, 121-137.

Popay J, Roberts H, Sowden A, Petticrew M, Arai L, Rodgers M, et al., 2006. Guidance on the conduct of narrative synthesis in systematic reviews: a product from the ESRC Methods Programme. ESRC. Methods Programm, 1:1–69. https://doi.org/10.13140/2.1.1018.4643

Potter CS, 1999. Terrestrial biomass and the effects of deforestation on the global carbon cycle: results from a model of primary production using satellite observations. *BioScience*, 49(10), 769-778.

Prance GT, 1979. Notes on the vegetation of Amazonia III. The terminology of Amazonian forest types subject to inundation. *Brittonia* 31:26–38.

Prasetyo LB. Kumazaki M, 1995. Land use changes and their causes in the tropics: a case study in South-Sumatra, Indonesia in 1969-1988. *Tropics*, 5(1/2).

Pretty JN, Brett C, Gee D, Hine RE, Mason CF, Morison JIL, et al., 2000. An assessment of the total external costs of UK agriculture. *Agricultural Systems* 65(2), 113-136.

Priyadarshan PM, 2011. *Biology of Hevea rubber*. CAB International, Wallingford.

Priyadarshan PM, Hoa TTT, Huasun H, de Gonçalves PS, 2005. Yielding potential of rubber (*Hevea brasiliensis*) in suboptimal environments. *J Crop Impr* 14:221–247.

Pullin AS, Frampton GK, Livoreil B, Petrokofsky G, 2018. Guidelines and standards for evidence synthesis in environmental management. Version 5.0.

Putri FA, Basukriadi A, Pradana DH, 2020. Study of the bird community in rubber plantations at Universitas Indonesia, Depok. *AIP Conference Proceedings* 2242, 050020 (2020). https://doi.org/10.1063/5.0012144

Putranto RA, Herlinawati E, Rio M, Leclercq J, Piyatrakul P, Gohet E, et al., 2015. Involvement of ethylene in the latex metabolism and tapping panel dryness of Hevea brasiliensis. *International journal of molecular sciences*, 16(8), 17885-17908.

Purnamasari R, Cacho O, Simmons P, 2002. Management strategies for Indonesian rubber production under yield and price uncertainty: a bioeconomic analysis. *Agroforestry systems*, 54: 121-135.

Qiu J, 2009. Where the rubber meets the garden. *Nature* 457:246–247.

Rahayu S, Santoso S, Wijayanti P, 2021. The analysis of palm oil plantation impact on the social geographic conditions in Kapuas sub-district, Sanggau district. IOP Conference Series: Earth and Environmental Science (Vol. 683, No. 1, p. 012136). IOP Publishing.

Rajasekharan P, Veeraputhran S, 2002. Adoption of intercropping in rubber smallholdings in Kerala, India: a tobit analysis. *Agrofor Syst* 56:1–11.

Rantala L, 2006. Rubber plantation performance in the Northeast and East of Thailand in relation to environmental conditions. Thesis, University of Helsinki.

Ratnasingam J, Ramasamy G, Ioras F, Kaner J, Wenming L, 2012. Production Potential of Rubberwood in Malaysia: Its Economic Challenges. *Notulae Botanicae Horti Agrobotanici Cluj-Napoca*, 40(2), 317–322. https://doi.org/10.15835/nbha4028006

Ren Y, Lin F, Jiang C, Tang J, Fan Z, Feng D, et al., 2023. Understory vegetation management regulates soil carbon and nitrogen storage in rubber plantations. *Nutr. Cycl. Agroecosystems*, 127, 209–224. https://doi.org/10.1007/s10705-023-10296-8

Rigg J, 2005. Poverty and livelihoods after full-time farming: a south-east Asian view. *Asia Pac Viewpoint* 46:2173–2184.

Roberts C, 2016. Don't Let Your Tires Destroy the World's Forests. Time. https://time.com/4391096/rubber-deforestation/

Robison L, Barry P, 1987. *The competitive firm's response to risk.* Macmillan, New York.

Rodrigo VHL, Nugawela A, Pathirathne LSS, Waidyanatha UPdeS, Samaranayake ACI, Kodikara PB, Weeralal JLK, 1995. Effect of planting density on growth, yield, yield related factors and profitability of rubber (*Hevea brasiliensis*). *J. Rubber Res. Inst. Sri Lanka.* 76: 55-71.

Rodrigo VHL, Stirling CM, Teklehaimanot Z, Nugawela A, 2001. Intercropping with banana to improve fractional interception and radiation-use efficiency of immature rubber plantations. *Field Crops Res* 69:237–249.

Rodrigo VHL, 2001. Rubber based intercropping systems. In Tillekeratne LMK, Nugawela A. (eds), *Handbook of Rubber Agronomy*, vol. 1. Rubber Research Institute of Sri Lanka, 139–155.

Rodrigo VHL, Silva TUK, Munasinghe ES, 2004. Improving the spatial arrangement of planting rubber (*Hevea brasiliensis* Muell. Arg.) for long-term intercropping. *Field Crops Res* 89:327–333.

Rodrigo VHL, Stirling CM, Silva TUK, Pathirana PD, 2005. The growth and yield of rubber at maturity is improved by intercropping with banana during the early stage of rubber cultivation. *Field Crops Res* 91:23–33.

Romyen A, Sausue P, Charenjiratragul S, 2018. Investigation of rubber-based intercropping system in Southern Thailand, *Kasetsart Journal of Social Sciences*, 39(1): 135-142.

RRIM Annual report, 1995. Kuala Lumpur, Malaysia.

Ruf F, 1994. La rente forêt, l'arbre et l'instabilité des marchés. In « la parabole du cacao. Quelques enseignements de l'Indonésie sur la durabilité de l'agriculture en régions tropicales humides ». Rapport de recherche sur l'appel d'offre : « l'arbre dans les systèmes de culture sédentaires dans les régions chaudes et humides ». Ministère de l'Enseignement Supérieur et de la Recherche, Paris, 32-56.

Ruf F, 1995. From forest rent to tree-capital: basic 'laws' of cocoa supply. *Cocoa cycles: the economics of cocoa supply*, 1-54.

Ruf F, 1995. Booms et crises du cacao, Karthala, Paris.

Ruf F, Penot E, Yoddang, 1999. After tropical forest, replantation of rubber trees and cocoa: a garden of Eden or of chemicals inputs? Jardin planétaire'99, Planetary garden'99: proceedings. INRA, Cirad, Savoie-Conseil général. Prospective 2100, 318-324. International symposium on sustainable ecosystem management.

Ruf F, 2011. The myth of complex cocoa agroforests: the case of Ghana. *Hum Ecol* 39:373–388.

Ruf F, 2019. Suite à des reproches, une immense plantation de caoutchouc camerounaise stoppe la déforestation. Mongabay. https://fr.mongabay.com/2019/07/suite-a-des-reproches-une-immense-plantation-de-caoutchouc-camerounaise-stoppe-la-deforestation/#

Russo M, Bouquet E, Chambon B, 2017. Sharing more than rubber: the economic and social lives of share-tapping contracts in Southern Thailand. 11èmes Journées de recherche en Sciences Sociales (SFER), Lyon.

Sahuri M, Rosyid J, and Agustina DS, 2016. Development of Wide Row Spacing to Increase Land Productivity of Rubber Plantation. Sembawa Research Centre Indonesian Rubber Research Institute CRRI & IRRDB International Rubber Conference, Cambodia.

Sainte-Beuve J, 2015. Caoutchouc. La déprime. In Chalmin P (ed), *Cyclope 2015 : les marchés mondiaux. « Pour qui sonne le glas ? »*, 430–434. http://publications.cirad.fr/une_notice.php?dk=580003

San N, Deaton BJ, 1999. Feasibility of integrating sheep and crops with smallholder rubber production systems in Indonesia. *J Agribusin* 17:105–122. https://doi.org/10.22004/ag.econ.14727

Sanchez MD, Ibrahim TH, 1991. Forage species for rubber plantations in Indonesia. Forages for Plantation Crops, Sanur Beach, Bali, Indonesia.

Sanial E, 2018. L'appropriation de l'arbre, un nouveau front pour la cacaoculture ivoirienne ? Contraintes techniques, environnementales et foncières.

Sankalpa JKS, Wijesuriya W, Ishani PGN, 2020. Do rubber-based agroforestry practices build resilience upon poverty incidence? A case study from Moneragala district in Sri Lanka. *Agroforestry Systems*, 94(5), 1795-1808.

Sarathchandra C, Abebe AY, Worthy FR, Wijerathne IL, Ma H, Yingfeng B, et al., 2021. Impact of land use and land cover changes on carbon storage in rubber dominated tropical Xishuangbanna, South West China, *Ecosystem Health and Sustainability*, 7:1, 1915183. https://doi.org/10.1080/20964129.2021.1915183

Scherr SJ, McNeely JA (eds.), 2012. *Farming with nature: the science and practice of ecoagriculture.* Island Press.

Steppler, Howard A., and PK Ramachandran Nair. "Agroforestry." *A Decade of Development* (1987).

Schrepfer L, Ulrich E, 2018. Recommandations pour une récolte durable de biomasse forestière pour l'énergie – Focus sur les menus bois et les souches. Paris: ECOFOR, Angers: ADEME, 50 p.

Schroth G, Coutinho P, Moraes VHF, Albernaz AL, 2003. Rubber agroforests at the Tapajós river, Brazilian Amazon—environmentally benign land use systems in an old forest frontier region. *Agric Ecosyst Environ* 97:151–165. https://doi.org/10.1016/S0167-8809(03)00116-6

Schroth G, da Fonseca G, Harvey C, Gascon C, Vasconcelos H, Izac AM, 2013. *Agroforestry and Biodiversity Conservation in Tropical Landscapes.* Island Press, Washington, 575 p.

Schroth G, Ruf F, 2014. Farmer strategies for tree crop diversification in the humid tropics a review. *Agron Sustain Dev* 34:139–154. https://doi.org/10.1007/s13593-013-0175-4

Schueller W, Penot E, Sunaryo I, 2003. The improved genetic rubber planting material (IGPM) availability and use by smallholders in the province of West Kalimantan, Borneo, Indonesia. ICRAF.

Schultes RE, 1992. Ethnobotany, biological diversity, and the Amazonian Indians. *Environ Conserv* 19:97–100.

Schwarze S, Euler M, Gatto M, Hein J, Hettig E, Holtkamp AM, et al., 2015. Rubber vs. oil palm: an analysis of factors influencing smallholders' crop choice in Jambi, Indonesia. *EFForTS Discussion Paper Series* 11:1–29. https://www.econstor.eu/handle/10419/117323

Schwinning S, 2010. The ecohydrology of roots in rocks. *Ecohydrology*, 3, 238–245. https://doi.org/10.1002/eco.134

Senevirathna AMWK, Stirling CM, Rodrigo VHL, Pathirana PD, Karunathilake PKW, 2010. High density banana/rubber intercrops have no negative effects on component crops under the smallholder conditions. *Journal of the Rubber Research Institute of Sri Lanka*, 90, 1-17.

Serrat O, 2017. The Sustainable Livelihoods Approach. In *Knowledge Solutions*. Springer, Singapore.

Serve MD La, Sainte-Beuve J, Gouyon A, 1991. L'heveaculture au Liberia : les planteurs face aux variations des prix du caoutchouc. *Revue Générale Du Caoutchouc et Des Plastiques*, 68(710), 181–190. http://publications.cirad.fr/une_notice.php?dk=404592

Sethuraj MR, 1996. Impact of natural rubber plantations on environment. International seminar on natural rubber as an environmentally friendy raw material and a renewable ressource. Trivandum, India.

Shapiro J, 2001. Mao's war against nature. Cambridge University Press, Cambridge.

Shelton HM, Stür WW (eds), 1991. Forages for Plantation Crops, ACIAR Proceedings No. 32, 168 p.

Shigematsu A, Mizoue N, Kajisa T, Yoshida S, 2011. Importance of rubberwood in wood export of Malaysia and Thailand. *New Forest* 41:179–189.

Shigematsu A, Mizoue N, Kakada K, Muthavy P, Kajisa T, Yoshida S, 2013. Financial potential of rubber plantations considering rubberwood production: wood and crop production nexus. *Biomass Bioenergy* 49:131–142.

Siju T, George KT, Lakshmanan R, 2012. Changing dimensions of intercropping in the immature phase of natural rubber cultivation: a case study of pineapple intercropping in central Kerala. *Rubber Science*, 25(2), 164-172.

Simanjutak R, Penot E, Danthu P, Michel I, Jahiel M, 2014. Document de travail AFS4FOOD n° 15. Analyse technico-économique de la filière amont de production d'essence de girofle à Fénérive-Est, Madagascar : de la feuille à l'alambic, 30 p. https://doi.org/10.13140/RG.2.1.4418.6725

Simien A, Penot E, 2011. Current evolution of smallholder rubber-based farming systems in southern Thailand. *J Sustain for* 30(3):247–260. https://doi.org/10.1080/10549811.2011.530936

Singh AK, 2018. Prospective of rubber-based agroforestry on swidden fallow for synergizing climate change mitigation. *Tropical Ecology* 59(4): 605–618.

Singh AK, Liu W, Zakari S, Wu J, Yang B, Jiang XJ, et al., 2021. A global review of rubber plantations: Impacts on ecosystem functions, mitigations, future directions, and policies for sustainable cultivation. *Science of the Total Environment*, 796, 148948.

Sivanadyan K, Moris N, 1992. Consequences of transforming tropical rainforest to Hevea plantations. FADINAP regional seminar on fertilization and environment, Chang mai, Thailand.

Snoeck D, Lacote R, Kéli J, Doumbia A, Chapuset T, Jagoret P, Gohet É, 2013. Association of hevea with other tree crops can be more profitable than hevea monoculture during first 12 years. *Ind Crops Prod* 43:578–586. https://doi.org/10.1016/j.indcrop.2012.07.053

Soemarwoto O, Conway GR, 1991. The Javanese homegarden, *J. Farming Systems Research-Extension* 2(3): 95-117.

Soerianegara I, Lemmens RHMJ, 1993. Plant resources of southeast Asia. No. 5 (1). *Timber Trees: Major Commercial Timbers*, 384-391.

Soerianegara I, Lemmens RHMJ, (eds.). 2002. *Sumber Daya Nabati Asia Tenggara 5(1): Pohon penghasil kayu perdagangan yang utama*. PROSEA – Balai Pustaka. Jakarta. ISBN 979-666-308-2. Hal. 7Secara internasional diperdagangkan sebagai resak. Sedangkan giam merujuk pada kayu-kayu berat marga *Hopea*. Lihat Soerianegara dan Lemmens, *op. cit.* hal. 141.

Somado EA, Becker M, Kuehne RF, Sahrawat KL, Vlek PLG, 2003. Combined effects of legumes with rock phosphorus on rice in west Africa. *Agron J* 95:1172–1178.

Somarriba, E. (1992). Revisiting the past: an essay on agroforestry definition. *Agroforestry systems*, *19*, 233-240.

Somboonsuke B, 2001a. Recent evolution of rubber-based farming systems in southern Thailand. *Kasetsart J* 22:61–74.

Somboonsuke B, Shirakoti DP, 2001b. Small holders of rubber-based farming systems in Songkhla Province Thailand: problems and potential solutions. *Kasetsart J* 22:79-97.

Somboonsuke B, Wetayaprasit P, Chernchom P, Pacheerat K, 2011. Diversification of smallholding rubber agroforestry system (SRAS) Thailand. *Kasetsart J* 32:327–339.

Sosef MSM, Hong LT, Prawirohatmodjo S, 1998. *Plant Resources of South-East Asia.*5, No. 3. SV Kerkwerve, The Netherlands: Backhuys.

Stanfield ME, 1998. *Red rubber, bleeding trees: Violence, slavery, and empire in northwest amazonia, 1850–1933*. University of New Mexico Press, Albuquerque.

Steppler HA, Nair PKR, 1987. *Agroforestry: A Decade of Development*, Agribookstore.

Stone R, 2008. Ecology. Showdown looms over a biological treasure trove. *Science* 319(5870):1604.

Strange N, Theilade I, Thea S, Sloth A, Helles F, 2007. Integration of species persistence, costs and conflicts: an evaluation of tree conservation strategies in Cambodia. *Biological Conservation*, 137(2), 223-236.

Stroesser L, Penot E, Michel I, Tongkaemkaew U, Chambon B, 2018. Income diversification for rubber farmers through agroforestry practices. How to overcome rubber price volatility in Phatthalung province, Thailand. *Revue Internationale du Développement/Éditions de la Sorbonne*, 235(3):117–145. https://doi.org/10.3917/ried.235.0117

Sun Y, Ma Y, Cao K, Li H, Shen J, Liu W, et al., 2017. Temporal Changes of Ecosystem Carbon Stocks in Rubber Plantations in Xishuangbanna, Southwest China. *Pedosphere* 27, 737–746. https://doi.org/10.1016/S1002-0160(17)60327-8

Sukviboon S, Meegangwal S, Suwangerd W, 1986. Study on some cover crops planted in rubber plantation.

Schueller, Wilfrid. *Production et utilisation du matériel végétal amélioré d'hévéa brasiliensis par les petits planteurs de la province de Kalimantan Ouest, Indonésie.* Diss. ENITAB, 1997.

Tajuddin I, 1986. Integration of animals in rubber plantations. *Agroforest Syst* 4(1):55–66. https://doi.org/10.1007/BF01834702

Takeya H, 2003. Factors determining intercropping by rubber smallholders in Sri Lanka: a logit analysis. *Agric Econ* 29:159–168.

Tenywa MM, Tukarhirwa KPC, Bruchara R, Adekunle AA, Mugabe J, Wanjiku C, 2011. Agricultural innovation platform as a tool for development oriented research: Lessons and challenges in the formation and operationalization.

Tetteh EN, Abunyewa AA, Tuffour HO, Berchie JN, Acheampong PP, Twum-Ampofo K, et al., 2019. Rubber and plantain intercropping: Effects of different planting densities on soil characteristics. *PLoS One*, 14(1), e0209260.

Tharian George K, Haridasan V, Sreekumar B, 1988. Role of government and structural changes in rubber plantation industry. *Econ Polit Weekly* 23:M158–M166.

Theriez M, Penot E, Tongkaemkaew U, Chambon B, 2017. De l'indifférence à la reconnaissance : potentiel de développement de plateformes d'innovation sur les systèmes agroforestiers hévéicoles du sud de la Thailande. JJSS, 11es Journées de Recherche en Sciences Sociales, Lyon. http://agritrop.cirad.fr/587012/

Therumthanam AC, Mathew J, Vazhacharickal J, 2014. Studies on Pineapple (*Ananas comosus*) cultivation as an intercrop in rubber replant: effects on soil microorganisms. https://api.semanticscholar.org/CorpusID:54087875

Thoumazeau A, Bessou C, Renevier MS, Panklang P, Puttaso P, Peerawat M, et al., 2018. Biofunctool®: A new framework to assess the impact of land management on soil quality. Part B: Investigating the impact of land management of rubber plantations on soil quality with the Biofunctool® index. *Ecological Indicators* 97, 429-437. https://doi.org/10.1016/j.ecolind.2018.10.028

Tiffen M, 1995. Population density, economic growth and societies in transition: Boserup reconsidered in a Kenyan case-study. *Development and Change*, 26(1), 31-66.

Tillekeratne LMK, 1996. Role of hevea rubber in protecting the environment in Sri Lanka. International seminar on natural rubber as an environmentally friendy raw material and a renewable ressource. Trivandum, India.

Tiong, G. L., Seng, C. J., & Yee, H. C. (1994). Economic evaluation of planting rubber as a dual purpose tree. In *International Planters Conference, Malaysia* (pp. 242-288)

Tittonell P, Scopel E, Andrieu N, Posthumus H, Mapfumo P, Corbeels M, et al., 2012. Agroecology-based aggradation-conservation agriculture (ABACO): Targeting innovations to combat soil degradation and food insecurity in semi-arid Africa. *Field Crops Research*, 132, 168-174.

To Xuan Phuc, Tran Huu Nghi, 2014. Rubber Expansion and Forest Protection in Vietnam. Tropenbos International, Viet Nam.

Tomish T, 1992. Smallholder Rubber development in Indonesia. In Perkins DH, Roemer M (eds), *Reforming economic systems in developing countries*, Harvard University.

Tomish TP, van Nordwick M, 1995. Agroforestry technologies for social forestry: tree crop interactions and forestry farmer conflicts. Social forestry and sustainable management., 168 -193. Jakarta.

Tongkaemkaew U, Sukkul J, Sumkhan N, Panklang P, Brauman A, Ismail R, 2018. Litterfall, litter decomposition, soil macrofauna, and nutrient contents in rubber monoculture and rubber-based agroforestry plantations. *For. Soc.* 2, 138–149. https://doi.org/10.24259/fs.v2i2.4431

Tongkhaenkhew U, Penot E, Chambon B, 2020. Rubber Agroforestry Systems in Mature Plantations in Phatthalung Province, Southern Thailand, *Asean Journal of Scientific and Technological Reports*, 23(1).

Tongkhaenkhew U, Penot E, Chambon B, 2020 Characterization of rubber agroforestry systems in mature plantations in Southern of Thailand. *Taksin journal*, 23, 1. https://ph02.tci-thaijo.org/index.php/tsujournal/article/view/240091

Torquebiau E, 1992. Are tropical agroforestry home gardens sustainable? *Agriculture, ecosystems & environment*, 41(2):189-207.

Torquebiau E, Mary F, Sibelet N, 2002. Les associations agroforestières et leurs multiples enjeux. *Bois & forêts des tropiques*, 271, 23-35.

Torquebiau E, Penot E, 2006. Ecology vs Economics in Tropical Agroforests. *In*: Kumar BM, Nair KM (eds), 2006. *Tropical Homegardens: A Time-Tested example of Sustainable Agroforestry*. Advances in Agroforestry. Springer Science, The Netherlands, 269-282. https://link.springer.com/book/10.1007/978-1-4020-4948-4

Toxopeus H, 1985. Botany, types and populations. In Wood GAR, Lass RA (eds), *Cocoa*, 4th edn. Longman Group Limited, London, 11–37.

Traoré KBS, 2018. Défis de politique, de développement et de durabilité du caoutchouc naturel en Côte d'ivoire. Abidjan.

Trouillard K, 2001. Les systèmes agroforestiers à base d'hévéa clonal : une solution pour la relance des plantations villageoises : un système de culture complémentaire du palmier à huile. CNEARC, Master of Science « Développement agricole tropical : option valorisation des productions ». Montpellier, 150 p.

Tu E, Idoko SO, Osazuwa E, Okore IK, 2014. Effect of intercropping on the gestation period of rubber. *Wudpecker J. of Agric. Res*, 3(8), 150-153.

Ule E, 1901. Ule's Expedition nach den Kautschuk-Gebieten des Amazonenstromes. Erster Bericht über den Verlauf der Kautschuk-Expedition bis zum Beginn des Jahres 1901. Notizblatt des Königl. botanischen Gartens und Museums zu Berlin, 3:111–118.

Ule E, 1901. Ule's Expedition nach den Kautschuk-Gebieten des Amazonenstromes. Zweiter Bericht über den Verlauf der Kautschuk-Expedition vom 1. Januar bis zum Mai des Jahres 1901. Notizblatt des Königl. botanischen Gartens und Museums zu Berlin 3:129–134.

Ule E, 1903. Ule's Expedition nach den Kautschuk-Gebieten des Amazonenstromes. Dritter Bericht über den Verlauf der Kautschuk-Expedition vom Mai bis zum November des Jahres 1901. Notizblatt des Königl. botanischen Gartens und Museums zu Berlin 3:224–237.

Ule E, 1903. Ule's Expedition nach den Kautschuk-Gebieten des Amazonenstromes. Vierter Bericht über den Verlauf der Kautschuk-Expedition vom November 1901 bis zum Marz 1902. Notizblatt des Königl. botanischen Gartens und Museums zu Berlin 4:92–98.

Umar HY, Ugwa IK, Abubakar M, Giroh DY, 2008. Analysis of bites and stings by snakes, insects and other animals among rubber (*Hevea brasiliensis*) tappers in Southern Nigeria. *J Agric Soc Sci* 4:174–176.

Vagneron I, Chambon B, Nay Myo Aung, Saw Min Aung, 2017. Rubber production in Tanintharyi Region. Yangon, Myanmar: WWF, 80 p.

van Noordwijk M, Tomich TP, de Foresta H, Michon G, 1997. To segregate –or to integrate? The question of balance between production and biodiversity conservation in complex agroforestry systems. *Agroforestry Today*, 9(1): 6-9.

Varkey LM, Kumar P, 2013. Price risk management and access to finance for rubber growers: the case of price stabilisation fund in Kerala Indian. *J Agric Econ* 68:67–88.

Viswanathan PK, Shivakoti GP, 2008. Adoption of rubber-integrated farm-livelihood systems: contrasting empirical evidence from the Indian context. *Journal of Forest Research*, 13(1), 1–14. https://doi.org/10.1007/s10310-007-0047-3

Vongkhamheng C, Zhou J, Beckline M, Phimmachanh S, 2016. Socioeconomic and Ecological Impact Analysis of Rubber Cultivation in Southeast Asia. *OALib*, 03(01), 1–11. https://doi.org/10.4236/oalib.1102339

Vrignon-Brenas S, Gay F, Ricard S, Snoeck D, Perron T, Mareschal L, et al., 2019. Nutrient management of immature rubber plantations. A review. *Agronomy for sustainable development*, 39, 1-21.

Waidyunatha UP de S, Wijesinghe DS, Stauss R, 1982. Zero-grazed pasture under immature Hevea rubber: improved productivity of *Panicum maximum* and *P. maximum* + *Pueraria phaseoloides* and their competition with Hevea. *J Rubber Res I Sri Lanka* 60:17–24.

Waliszewski WS, 2010. *Variation in* Thaumatococcus daniellii *(Benn.) Benth. and its potential as an intercrop with* Hevea brasiliensis *(Willd. Ex A. de Juss) Mueller-Argoviensis in West Africa*. Bangor University of Wales.

Wan AR, Wan Y, Jones K, 1996. Rubber as a green commodity. International seminar on natural rubber as an environmentally friendy raw material and a renewable ressource. Trivandum, India.

Wang J, Ren C, Cheng H, Zou Y, Bughio MA, Li Q, 2017. Conversion of rainforest into agroforestry and monoculture plantation in China: Consequences for soil phosphorus forms and microbial community. *Sci. Total Environ.* 595, 769–778. https://doi.org/10.1016/j.scitotenv.2017.04.012

Wang J, Zou Y, Di Gioia D, Singh BK, Li Q, 2020. Conversion to agroforestry and monoculture plantation is detrimental to the soil carbon and nitrogen cycles and microbial communities of a rainforest. *Soil Biol. Biochem.* 147, 107849. https://doi.org/10.1016/j.soilbio.2020.107849

Warren-Thomas E, Dolman PM, Edwards DP, 2015. Increasing Demand for Natural Rubber Necessitates a Robust Sustainability Initiative to Mitigate Impacts on Tropical Biodiversity. *Conservation Letters*, 8(4), 230–241. https://doi.org/10.1111/conl.12170

Warren-Thomas EM, Edwards DP, Bebber DP, Chhang P, Diment AN, Evans TD, et al., 2018. Protecting tropical forests from the rapid expansion of rubber using carbon payments. *Nat Commun* 9, 911. https://doi.org/10.1038/s41467-018-03287-9

Warren-Thomas E, Nelson L, Juthong W, Bumrungsri S, Brattström O, et al. 2020. Rubber agroforestry in Thailand provides some biodiversity benefits without reducing yields. *J Appl Ecol.* 57(1), 17– 30. https://doi.org/10.1111/1365-2664.13530

Watson GA, 1989. Field maintenance. In Webster CC, Baulkwill WJ (eds), *Rubber*. Longman Scientific & Technical, Harlow, 245–348.

Webster CC, Paardekooper EC, 1989. The botany of the rubber tree. In Webster CC, Baulkwill WJ (eds), *Rubber*. Longman Scientific & Technical, Harlow, 57–84.

Werner S, 1997. Biodiversity of jungle rubber in West-Kalimantan. ICRAF/SRAP workshop on RAS (Rubber Agoforestry Systems), Bogor.

Wessel M, 1985. Shade and nutrition. In Wood GAR, Lass RA (eds), *Cocoa*. Longman Group Limited, London, 166–194.

Wezel A, Bender S, 2003. Plant species diversity of homegardens of Cuba and its significance for household food supply, *Agroforestry systems* 57: 39-49.

Wibawa G, Joshi L, Van Noordwijk M, Penot E, 2006. Rubber based Agroforestry Systems (RAS) as Alternatives for Rubber Monoculture System. IRRDB annual conference, Ho-chi-minh city: Vietnam.

Williams SE, Van Noordwijk M, Penot E, Healey JR, Sinclair FL, Wibawa G, 2001. On-farm evaluation of the establishment of clonal rubber in multistrata agroforests in Jambi, Indonesia. *Agrofor Syst* 53:227–237.

Wilson JR, Ludlow MM, 1991. The environment and potential growth of herbage under plantations. Forages for Plantation Crops, Sanur Beach. Bali, Indonesia.

Win S, Kumazaki M, 1998. The History of Taungya Plantation Forestry and Its Rise and Fall in the Tharrawaddy Forest Division of Myanmar (1869-1994). *J. For plann.*, 4:17-26.

Winarni B, Lahjie AM, Simarangkir B, Yusuf S, Ruslim Y, 2018. Forest gardens management under traditional ecological knowledge in West Kalimantan, Indonesia. *Biodiv J Biolog Div* 19:77–84. https://doi.org/10.13057/biodiv/d190113

Wintgens JN, 2009. The coffee plant. *In* Wintgens JN (ed), *Coffee: growing, processing, sustainable production*, 2nd edn. Wiley-VCH, Weinheim, 3–24.

Wintgens JN, Descroix F, 2009. Establishing a coffee plantation. In Wintgens JN (ed), *Coffee: growing, processing, sustainable production*, 2nd edn. Wiley-VCH, Weinheim, 182–249.

Witness G, 2013. *Rubber Barons.* 52. https://www.globalwitness.org/fr/campaigns/land-deals/rubberbarons/

Wolff XY, Coltman RR, 1989. Productivity under shade in Hawaii of five crops grown as vegetables in the tropics. *J Am Soc Hort Sci* 115:175–181.

Wood GAR, 1985. Environment. In Wood GAR, Lass RA (eds), *Cocoa*, 4th edn. Longman Group Limited, London, 38–79.

Wood GAR. 1985. Establishment. In Wood GAR, Lass RA (eds), *Cacao*. Longman Group Limited, London and New York, 119–165.

Woods K, 2012. The Political Ecology of Rubber Production in Myanmar: An Overview. http://www.burmalibrary.org/docs20/The_Political_Ecology_of_Rubber_Production_in_Myanmar.pdf

Wu J, Liu W, Chen C, 2016. Below-ground interspecific competition for water in a rubber agroforestry system may enhance water utilization in plants. *Sci. Rep.* 6, 19502. https://doi.org/10.1038/srep19502

Wu J, Liu W, Chen C, 2017. How do plants share water sources in a rubber-tea agroforestry system during the pronounced dry season? *Agric. Ecosyst. Environ.* 236, 69–77. https://doi.org/10.1016/j.agee.2016.11.017

Wu J, Zeng H, Chena C, Liu W, 2019. Can intercropping with the Chinese medicinal herbs change the water use of the aged rubber trees? *Agricultural Water Management*, 226, 105803. https://doi.org/10.1016/j.agwat.2019.105803

Wu J, Zeng H, Zhao F, Chen C, Liu W, Yang B, Zhang W, 2020. Recognizing the role of plant species composition in the modification of soil nutrients and water in rubber agroforestry systems. *Sci. Total Environ.* 723, 138042. https://doi.org/10.1016/j.scitotenv.2020.138042

Wu ZL, Liu HM, Liu LY, 2001. Rubber cultivation and sustainable development in Xishuangbanna, China. *The International Journal of Sustainable Development & World Ecology*, 8(4), 337-345.

Wulan YC, Budidarsono S, Joshi L, 2006. Economic analysis of improved smallholder rubber agroforestry systems in West Kalimantan, Indonesia. Linking research to strengthen upland policies and practices, Sustainable sloping lands and watershed management conference.

WWF, (n.d.). Transforming the global rubber market. https://www.worldwildlife.org/projects/transforming-the-global-rubber-market

WWF, 2015. Sauvons les forêts en péril. Rapport Forêts Vivantes Du WWF: Chapitre 5, 2.

WWF, 2017. Le caoutchouc naturel, une ressource durable? ConsoGlobe website: https://www.consoglobe.com/caoutchouc-naturel-ressource-durable-cg

Xiao C, Li P, Feng Z, 2019. Monitoring annual dynamics of mature rubber plantations in Xishuangbanna during 1987-2018 using Landsat time series data: A multiple normalization approach. *International Journal of Applied Earth Observation and Geoinformation*, 77, 30-41.

Xie H, Wang P, Yao G, 2014. Exploring the dynamic mechanisms of farmland abandonment based on a spatially explicit economic model for environmental sustainability: a case study in Jiangxi Province, China. *Sustainability* 6:1260–1282.

Xinhua (agence de presse chinoise), 2012. Cameroun: 54.000 hectares de terres des forêts pour l'hévéaculture. *Journal Du Cameroun*. https://www.journalducameroun.com/fr/cameroun-54-000-hectares-de-terres-des-forets-pour-lheveaculture/

Xu J, 2006. The political, social and ecological transformation of a landscape: the case of rubber in Xishuangbanna, China. *Mt Res Dev* 26:254–262.

Xu J, Grumbine RE, Beckschäfer P, 2014. Landscape transformation through the use of ecological and socioeconomic indicators in Xishuangbanna, Southwest China, Mekong Region. *Ecol Indicators* 36:749–756. https://doi.org/10.1016/j.ecolind.2012.08.023

Yahya AK, 2006. Modelling the rubber tree system. Thesis. University of London.

Yahya MS, Atikah SN, Mukri I, Oon A, Hawa A, Sanusi R, et al., 2023. Potential of agroforestry orchards as a conservation set-aside initiative in industrial rubber tree and oil palm plantations for avian biodiversity. *Biodiversity and Conservation*, 32(6), 2101-2125.

Yang B, Meng X, Singh AK, Wang P, Song L, Zakari S, Liu W, 2020. Intercrops improve surface water availability in rubber-based agroforestry systems. *Agric. Ecosyst. Environ.* 298, 106937. https://doi.org/10.1016/j.agee.2020.106937

Yang B, Meng X, Zhu X, Zakari S, Singh AK, Bibi F, et al., 2021. Coffee performs better than amomum as a candidate in the rubber agroforestry system: Insights from water relations. *Agricultural Water Management*, 244, 106593.

Yi ZF, Cannon CH, Chen J, Ye CX, Swetnam RD, 2014. Developing indicators of economic value and biodiversity loss for rubber plantations in Xishuangbanna, southwest China: a case study from Menglun township. *Ecol Indicators* 36:788–797.

Yun T, Jiang K, Hou H, An F, Chen B, Jiang A, et al., 2019. Rubber tree crown segmentation and property retrieval using ground-based mobile LiDAR after natural disturbances. *Remote Sensing*, 11(8), 903.

Yung JM, Zaslavsky J, 1992. Pour une prise en compte des stratégies des producteurs. Cirad-SAR, 72 p.

Zaidon A, 2016. *Low density wood from poor to excellent.* Universiti Putra Malaysia Press. ISBN: 978-967-344.

Zaw ZN, Myint H, 2016. Common Agricultural Practices and Constraints of Natural Rubber Industry in Myanmar. *Songklanakarin Journal of Plant Science*, 3(3), 1-8.

Zaw ZN, Sdoodee S, Lacote R, 2017. Performances of low frequency rubber tapping system with rainguard in high rainfall area in Myanmar. *Australian Journal of Crop Science*, 11(11), 1451–1456. https://doi.org/10.21475/ajcs.17.11.11.pne593

Zemp DC, Guerrero-Ramirez N, Brambach F, Darras K, Grass I, Potapov A, et al., 2023. Tree islands enhance biodiversity and functioning in oil palm landscapes. *Nature* 618, 316–321. https://doi.org/10.1038/s41586-023-06086-5

Zeng X, Cai M, Lin W, 2012. Improving planting pattern for intercropping in the whole production span of rubber tree. *Afr J Biotechnol* 11, 8484–8490.

Zhang L, Kono Y, Kobayashi S, Hu H, Zhou R, Qin Y, 2015. The expansion of smallholder rubber farming in Xishuangbanna, China: A case study of two Dai villages. *Land Use Policy*, 42, 628–634. https://doi.org/10.1016/j.landusepol.2014.09.015

Zhao F, Yang B, Zhu X, Ma S, Xie E, Zeng H, Li C, Wu J, 2023. An increase in intercropped species richness improves plant water use but weakens the nutrient status of both intercropped plants and soil in rubber–tea agroforestry systems. *Agric. Water Manag.* 284, 108353. https://doi.org/10.1016/j.agwat.2023.108353

Zheng H, He K, 1991. Intercropping in Rubber Plantation and Its Economic benefit. In Zhu Z, Cai M, Wang S, Jiang Y (eds), *Agroforestry Systems in China.* International Development Research Centre (Canada), Chinese Academy of Forestry (P.R. China), 204-206.

Zhou S, 1993. Cultivation of *Amomum villosum* in tropical forests. *For Ecol Manage* 60:157–162.

Zhou Z, 2000. Landscape changes in a rural area in China. *Landscape Urban Plann* 47:33–38.

Zhu L, Wang X, Chen F, Li C, Wu L, 2019. Effects of the successive planting of *Eucalyptus urophylla* on soil bacterial and fungal community structure, diversity, microbial biomass, and enzyme activity. *Land Degrad. Dev.* 30, 636–646. https://doi.org/10.1002/ldr.3249

Zhu X, Chen C, Wu J, Yang J, Zhang W, Zou X, Liu W, Jiang X, 2019. Can intercrops improve soil water infiltrability and preferential flow in rubber-based agroforestry system? *Soil Tillage Res.* 191, 327–339. https://doi.org/10.1016/j.still.2019.04.017

Ziegler AD, Fox JM, Xu J, 2009. The rubber juggernaut. *Science* 324(5930):1024–1025. https://doi.org/10.1126/science.1173833

Various sources of information

Source for tropical timber

https://p2k.unkris.ac.id/id6/3065-2962/Daftar-kayu-di-Indonesia_64803_p2k-unkris.html*

Source for Porang

https://www.kompas.com/food/read/2021/08/12/180800375/sejarah-dan-perkembangan-porang-umbi-asli-indonesia-yang-mendunia#google_vignette

https://www.ipb.ac.id/news/index/2021/04/why-people-hunt-porang-many-this-is-the-explanation-of-ipb-university-experts/3ffb69c47d1732bad45be20b64f31a3a

https://www.courrierinternational.com/article/flore-en-indonesie-le-boom-du-phallus-amorphe

Source per country:

India

https://sites.google.com/site/zhuangfangnanayi/home/my-academic-presentations/rubberagroforestryorrubbermonoculture---theoralpresentationinworldcongressonagroforestrydelhiindia

https://www.cabdirect.org/cabdirect/abstract/20123010415

Thailand

https://www.mightyearth.org/2020/10/02/can-agroforestry-provide-a-lifeline-for-struggling-rubber-farmers-and-threatened-forests/

http://ocean.ait.ac.th/wp-content/uploads/sites/10/2018/07/Buncha_Rubber-based-Agroforestry-System-toward-Livelihood-of-Thai-Rubber-farmers.pdf

Myanmar

https://www.lift-fund.org/en/news/myanmar-looks-agroforestry-its-uplands

Gabon

Au Gabon, une plantation géante d'hévéas dérange. 2015. *Sciences et Avenir*. https://www.sciencesetavenir.fr/nature-environnement/au-gabon-une-plantation-geante-d-heveas-derange_16010

Photo gallery

https://agritrop.cirad.fr/611057/

List of abbreviations

AFD: Agence Française de Développement

AFS: Agroforestry Systems

AHER: Area Harvest Equivalent Ratio

ANRPC: Association of Natural Rubber Producing Countries

APROMAC: Association des Professionnels du Caoutchouc Naturel de Côte d'Ivoire

ARP: Assisted Replanting Project

BLIG: Polyclonal seedlings from Bal Lias Isolated Garden in North Sumatra

CAF: Complex Agroforestry Systems

CAP: Common Agricultural Policy

CDM: Clean Development Mechanism

CNRA: National Agronomic Research Centre (Côte d'Ivoire)

CPE: Cumulative Pan Evaporation

CRRI: Cambodian Rubber Research Institute

CSIR-CRI: Council for Scientific and Industrial Research-Crops Research Institute (Ghana)

DBH: Diameter at Breast Height

DFI: Drought Factor Index

DOAE: Department of Agricultural Extension

ELC: Economic Land Concessions

FCFA: Franc CFA

FGT: Fast-Growing Trees

FTA project: Forest, Trees and Agroforests project

GCC: Group Coagulating Centre

GPSNR: Global Platform for Natural Sustainable Rubber

HWP: Harvestable Wood Products

ICRAF: International Center for Research in Agroforestry

IGPM: Improved Genetic Planting Material

IP: Innovation Platform

IRRDB: International Rubber Research and Developemt Board

IRRI: Indonesian Rubber Research Institute

IRSG: International Rubber Study Group

ITTO: International Tropical Timber Organization

KKU: Khon Kaen University

LCA: Life Cycle Analysis

LCC: Legumes Cover Crops

LSCT: Large Scale Clone Trial

LULC: Land Use and Land Cover

NES: Nucleus Estate Smallholder Scheme

NGS: Next Generation Sequencing

NPV: Net Present Value

NSSDP: North Sumatra Smallholder Development project

NTFP: Non-Timber Forest Products

OECD: Organisation for Economic Co-operation and Development

ORRAF: Office of Rubber Replanting Aid Fund

PES: Payment for Ecosystem Services

PIR: Indonesian equivalent of NES (Perkebunan Inti Rakhyat)

PSU: Prince of Songkla University

RAOT: Rubber Authority of Thailand

RAS: Rubber Agroforestry Systems

RRIC: Rubber Research Insitute of Cambodia

RRIT: Rubber Research Institute of Thailand

RRIV: Rubber Research Institute of Vietnam

SAF: Simple Agroforestry Systems

SNR initiative: Sustainable Natural Rubber initiative

SRAP/SRDP: Smallholders Rubber Agroforestry Project

SSCT: Small-Scale Clone Trials

TCSDP: Tree Crop Smallholders Development Project

TIP: Tire Industry Project

TPD: Tapping Panel Dryness

TRAILS: climaTe Resilient lAndscapes for wIldLife conservation

TSU: Thaksin University

WBCSD: World Business Council for Sustainable Development

WSSDP: West Sumatra Smallholder Development project

Appendices

Table S1. Description of all agroforestry plot patterns

Plot	1	2	3	4	5	6	7	8	9	10	11	12	13	14	15	16	17	18	19
Area ha	2.4	4	0.32	0.8	0.16	0.96	0.96	0.32	0.72	1.6	0.92	0.8	2.24	1.12	0.64	1.72	4.8	0.64	1.12
Value																			
Ruber	3,511			768	11,556		555		14,597	6,175	5,078		7,924		10,033	8,471	3,325	2,809	10,805
Timber																		54	
Fruit trees	9					3,996		623				2,730	697	3,777		580			
Other plants		31					3,169		3,303	173	75		418	140	5,417		4,681	325	434
Total	3,520	31		786	11,556	3,396	4,524	623	17,900	6,348	5,135	2,750	9,039	3,917	15,450	9,051	8,006	3,188	11,239
Operational cost																			
Seedling												607	131	22					
Organic fertilizer	401	47		57	520			58				8	58	464			673		232
Chemical fertilizer							497	309	398	87	199		21	305	390	513	646		
Fuel		6		6	164	596	77	113	14	37	131	26	61	6	147	171	89	82	129
Small equipment						109									70		3		
EthyleneLabour									68						36		13		
Total	401	53		63	684	695	1,007	618	480	124	332	641	233	1,248	1,727	684	2,451	82	358
Margin/ha	3,119	22		705	10,872	2,701	3,517	15	17,420	6,224	4,821	2,089	5,766	2,669	13,273	8,367	5,555	3,106	10,881

Table S2. Species of associated plants found in rubber plots

Family	Southern name	Common name	Scientific name
Fruit trees			
1. Anacardiaceae	Ma-prang	Marian plum	1. *Bouea burmanica* Griff
	Ma-pring	Plum Mango	2. *Bouea oppositifolia* Meissn
	Ma-moung	Mango	3. *Mangifera indica* Linn
2. Bombacaceae	Turian	Durian	4. *Durio zibethinus* Linn
3. Guttiferae	Mangkod	Mangosteen	5. *Garcinia magostana* Linn
4. Leguminosae – Mimosodeae	Niang	-	6. *Archidendron jiringa* Jack
5. Meliaceae	Longkong	Longkong	7. *Lansium domesticum* Corr
	Langsat	Langsat	8. *Lansium domesticum* Corr
6. Minosaceae	Sator	-	9. *Parkia specioca* Hassk
7. Moraceae	Khanun	Jack fruit	10. *Artocarpus heterophyllus* Lamk
	Champedak	Champedak	11. *Artocarpus champenden* Spreng
8. Myrtaceae	Wa	-	12. *Syzygium cacuminis* (Craib) Chantar
9. Palmae	Maproaw	Coconut	13. *Cocos nucifera* Linn
	Rakam,	Salak	14. *Salacca rumphii* Wall
	Sala	Sala	15. *Salacca zalacca* Gaertn.
	Mak	Areca palm, Areca nut palm	16. *Areca catechu* Linn.
	Lumphi	Kelumi, Asam paya	17. *Eleiodoxa conferta* Griff
10. Phyllanthaceae	Mafai	Burmese grape	18. *Baccaurea ramiflora* Lour.
11. Sapindaceae	Ngoa	Rambutan	19. *Nephelium lappaceum* Linn
	Lumyai	Longan	20. *Dimocarpus longan* Lour
12. Stilaginaceae	Sommoa	Black Currant Tree	21. *Antidesma ghaesembills* Gaerth
Timber trees			
1. Dipterocarpaceae	Takian	-	1. *Hopea odorata* Roxb
	Takian Thong	Iron wood,	2. *Hopea odorata* Roxb
	Payom	Shorea	3. *Shorea roxburghii* G.Don
	Yang na	Yang , Gurjan	4. *Dipterocapus alatus* Roxb. ex G. Don
2. Labiatae	Sak,	Teak	5. *Tectona grandis* L.f.
3. Lauraceae	Tang	Litsea	6. *Litsea grandis* Hook.f.
4. Leguminosae – Minosoideae	Katintapa	Brown salwood	7. *Acacia mangium* Willd
5. Magnoliaceae	Champak		8. *Michelia champaca* Linn
6. Malvaceae	Porjong,	Large-Leaved Hau	9. *Talipariti macrophyllum* Fryxell
	Ujong	Small-Leaved Hua	10. *Talipariti macrophyllum* Fryxell

Family	Southern name	Common name	Scientific name
Timber trees			
7. Meliaceae	Ma-hokagany	Bay wood, Honduras Mahogany	11. *Wietenia macrophylla* King
	Sadoa	Siamese neem Tree	12. *Azardirachta excelsa* (Jack) Jacobs
	Yomhome,	Mahogani, Cigar box cedar, Moulmein cedar	13. *Toona ciliata* M.Roem
8. Rubiaceae	Taku		14. *Anthocephalus chinensis* (Rich. ex Walp.)
9. Barringtoniaceae	Bantan	Karuk, Putat, Tampalang, Tempalang.	15. *Barringtonia macrostachya* (Jack) Kurz,Rep.Pegu
Perennial vegetables			
1. Araceae	Phak nam	-	1. *Lasia spinosa* Linn
2. Brommeliaceae	Samparod	Pine apple	2. *Ananas comosus* Merr
3. Gnetaceae	Phak Miang	-	3. *Gnetum gnemon* Linn
4. Gramineae	Pai	Bamboo	4. *Bambusa multiplex* (Lour) Raeusch
5. Padanceae	Toei-home	Pandanus palm, Fragrant pandan, Pandom wangi	5. *Pandanus amaryllifolius* Roxb.
6. Palmae	Whaiy	Rattan palm, Rattan	6. *Calamus caesius* Blume

Table S3. List of species used in RAS in the Brazilian and Colombian Amazon

Fruit trees (2000/2002)

Brazilian Amazonia
- açaï/acai berry (*Euterpe olearacea*)
- pupuna/peach palm (*Bactris gasipaes*)
- cocona/ little ora *(Lulo amazonico)*
- castanera/Brazil nut (*Berthollieta excelsa*)
- anones/ Rollinia apple (*Rollinia* spp.)
- guarana (*Paullinia cupana*)
- araça-boi *(Eugenia stipitata),* different from Colombia
- copoaçu (*Theobroma grandiflorum*)
- banana (*Musa* spp.)
- Urucu *(Bixa orellana),* used as a food colourant
- Acerola (*Malpighia glabra*)
- Bacuri (*Platonia insignis*)
- Graviola/soursop (*Anona muricata*)
- jack fruit (*Artocarpus heterophyllus*)
- Avocado (*Persea americana*)
- mango (*Mangifera indica*)
- papaya (*Carica papaya* L.)
- maracuja/passion fruit *(Passiflora* spp.*)*
- Goyava guava (*Psidium guajava*)

- Ananas/pineapple (*Ananas comosus*)
- *Citrus* spp.
- other minor fruit trees (figueira – *Ficus carica* L, cajá/cajou – *Anacardium occidentale* L, biribá/ cerimoya/corrossol sauvage – *Annona mucosa* Jacq, Jambo – *Syzygium malaccense, jalbolanum, cumini*), Genipapo – *Genipa americana*)

Colombian Amazonia
- Chontaduro/peach palm (*Bactris gasipaes*), pupuna in Brazil
- cocona/little orange *(Lulo amazonico)*
- uva/tree grape (*Uva caimarona*)
- mangoustan (*Garcinia dulcis*)
- tamarillo/ tree tyomato *(Solanum betaceum)*
- anones/corrossol/Rollinia apple (*Rollinia* spp.)
- bacuri/genipap (*Enipa americana*)
- guarana (*Paulinia capona*)
- araza (*Eugenia stipitata*), arasa in Brazil
- copoaçu (*Theobroma grandiflorum*)
- borojo (*Borojoa patinoi*)
- banana (*Musa* spp.)
- caimo/star apple (*Pauteria caimilo*)
- lulo amazonico/potato tree *(Solanum grandiflorum)*

Cash crops
- coffee (*Robusta canephora*)
- cacao (*Theobroma cacao*)
- black pepper (*Piper nigerum*)
- coconut tree (*Cocos nucifera*)

Timber species

Brazilian Amazon
- mogno/ mahogany (*Swietenia macrophylla*)
- Teca/teak (*Tectonia grandis*)
- Castanheira/Brazil nut, (*Bertholletia excelsa* Humb. and Bonpl.)
- Andiroba/crabwood *(Carapa* spp.*)*
- acapu (*Wacapoua americana*)
- cedro/cedarwood (*Cedrela odorata* L.)
- ipês or ipe/yellow lapacho *(Tabebuia serratiolia)*
- Maçarandouba/ bulletwood (*Manilkara* sp.*)*
- Parica/Brazilian fern tree (*Schizolobium parahiba* var amazonicam)

Colombian Amazon
- nogal/Colombian walnut (*Cordia alliodora*)
- pelnemono (*Apeiba Asperii*)
- Abarco (*Cariniana piriformis*)
- Saman/monkeypod (*Samanea samar*)
- Teak (*Tectonia grandis*)
- Ahumado (*Minguartia guineensis*)
- caoba/large leaf mahogany (*Switenia macrophylla*)
- carrecillo (*Bombacopsis quinata*)
- flormorado/ rosy trumpet tree *(Tabebuia rosea)*
- Guayacan/golden triumpet tree *(Tabebuia chrysantha)*
- Camu camu (*Myrcinia dubia*)
- Cedro odorata
- Aceituno/tree of heaven (*Aceituno Simaruba glauca*)
- Chilli pepper (*Capsicum baccatum*)
- balsamo/balsa (*Ochroma pyramidale*)
- canelo/mountain pepper (*Drimys winteri*)

Other species
- neem tree (*Azadirachta excelsa*)

Table S4. Tropical timber in Indonesia in jungle rubber or RAS systems

No.	Vernacular name	Latin name	Local name
1	Damar	*Araucaria* spp. (mis. *A. cunninghamii, A. hunsteinii* K. Schum.)	Alloa, ningwik, pien (Pap.). Ingg.: *araucaria.*
2	Durian	Durio spp. (terutama Durio carinatus Mast.); Coelostegia spp.	Durian burung, lahong, layung, apun, begurah, punggai, durian hantu, enggang
3.	Medang	Cinnamomum spp.	Sintuk, sintok lancing, ki teja, ki tuha, ki sereh, selasihan
4.	Meranti kuning	Shorea spp. (di antaranya: S. acuminatissima Sym., S. balanocarpoides Sym., S. faguetiana Heim, S. gibbosa Brandis, Shorea scollaris V.Sl.;	Damar hitam, damar kalepek; Damar hitam katup; Bangkirai guruk, karamuku; Damar buah, mereng-kuyung; Damar tanduk, *yellow seraya.*
5.	Meranti merah	Shorea spp. (di antaranya: S. johorensis Foxw., S. lepidota Bl., S. leprosula Miq., S. ovalis Bl., S. palembanica Miq., S. platyclados V.Sl. ex Foxw., S. leptoclados Sym., dll.)	Majau, meranti merkuyung; Meranti ketrahan; Meranti tembaga, kontoi bayor; Meranti kelungkung; Tengkawang majau; Banio, ketir; Seraya merah, campaga, lempong, kumbang, meranti ketuko, cupang. Ingg.: *red seraya, red lauan.*
6.	Meranti putih	Shorea spp. (di antaranya: S. assamica Dyer, S. bracteolata Dyer, S. javanica K. et. Val., S. lamellata Foxw., S. ochracea Sym., S. retinodes V.SI., S. virescens Parijs, S. koordersi Brandis, dll.)	Damar mesegar; Bunyau, damar kedontang; Damar mata kucing, damar kaca, damar kucing; Damar tunam, damar pakit; Damar kebaong, baong, bayong, baung, belobungo, kontoi tembaga; Balamsarai, damar mansarai; Damar maja, kontoi sabang; Kikir, udang, udang ulang, damar hutan, anggelam tikus, maharam potong, pongin, awan punuk, mehing (Smt., Kal.); Damar lari-lari, lalari, temungku, tambia putih (Slw.), Damar tenang putih, hili, honi (Mlku.). Ingg.: *white meranti.*
7.	Merawan	*Hopea* spp. (mis. *H. dasyrrachis* V.Sl., *H. dyeri* Heim, *H. sangal* Korth., dll.)	Tekam, tekam rayap; Bangkirai tanduk, emang, amang besi; Cengal, merawan telor; Ngerawan, cengal balau
8.	Merbau	*Intsia* spp. (terutama *I. bijuga* O.K., *I. palembanica* Miq.)	Merbau asam, ipi (NT.), kayu besi (Papua); Ipil, anglai, maharan; Tanduk (Mlku.)
9.	Nyatoh	Palaquium spp., Payena spp., Madhuca spp.	Suntai, balam, jongkong, hangkang, katingan, mayang batu, bunut, kedang, bakalaung, ketiau, jengkot, kolan
10.	Pulai	Alstonia spp. (di antaranya A. pneumatophora Back., A. scholaris R.Br., A. spatulata Bl., A. macrophylla Wall., A. spectabilis R.Br.)	Kayu gabus, rita, gitoh, bintau, basung, pule, pulai miang. Ingg.: *white cheesewood, milkwood, milky pine.*

No.	Vernacular name	Latin name	Local name
11.	Rasamala	Altingia excelsa Noroña	Tulasan (Smt.), mandung (Min.), mala (Jw.)
12.	Resak	Vatica spp.; mis. V. maingayi Dyer, V. oblongifolia Hook.f., V. rassak Bl.	Damar along, resak putih
13	Berumbung	Adina minutiflora Val.); Pertusadina spp.	Kayu lobang, Barumbung, Kayu gatal
14.	Ketapang	Terminalia spp.	Kalumpit, Klumprit, Jelawai, Jaha
15.	Ketimunan	Timonius spp.	Seranai, Temirit, Kayu reen
16.	Lancat	Mastixiodendron spp.	Kundur, Modjiu, Raimagago
17	Lara	Metrosideros spp. dan Xanthostemon spp.	Lompopaito, Nani, Langera
18	Mahang	Macaranga spp.	Merkubung, Mara, Benua
19	Medang	Litsea firma Hook f.; Dehaasia spp.	Manggah, Huru kacang, Keleban, Wuru, Kunyit
20	Sengon	Paraserianthes falcataria (L) Nielsen	Jeungjing, Tawa kase, Sika (Maluku)
21	Surian	Toona sureni Merr.	Suren, kalantas
22	Tembesu	*Fagraea* spp.; mis. *F. fragrans* Roxb., *F. sororia* J.J. Sm.	Tomasu (Smt.), kulaki (Slw.), malbira, ki tandu
23	Terap	Artocarpus spp.	Cempedak, Kulur, Tara, Teureup
24	Eboni bergaris	Diospyros celebica Bakh.	Maitong, Kayu lotong, Sora, Amara
25	Eboni hitam	Diospyros rumphii Bakh.	Kayu hitam, Maitem, Kayu waled
26	E b o n i	Diospyros spp.; di antaranya D. areolata King et G., D. cauliflora Bl., D. ebenum Koen, D. ferrea Bakh., D. lolin Bakh., D. macrophylla Bl.	Baniak, Toli-toli, Kayu arang, Kanara, Gito-gito, Bengkoal, Malam
27	Mahoni	Swietenia spp.; mis. S. macrophylla King, S. mahagoni (L.) Jacq.	Mahoni
28	Ramin	Gonystylus bancanus Kurz	Gaharu buaya, Medang keladi, Keladi, Miang
29	Sungkai	Peronema canescens Jack	Jati seberang, Jati londo
30	Tertulis 1	Eugenia spp.,	marga yang tidak tepat
31	Tertulis 2	Cassia spp.,	nama yang usang

Sources: Soerianegara and Lemmens, 1993 ; Lemmens, 1989. https://p2k.unkris.ac.id/id6/3065-2962/Daftar-kayu-di-Indonesia_64803_p2k-unkris.html

Table S5. List of Indonesian commercial wood species (Kartasujana and Martawijaya, 1973)

Local name	Latin name
Berumbung	*Adina minutiflora*
Durian	*Durio carinatus, Durio oxleyanus; Durio zibethinus*
Ebony	*Diospyros celebica, Diospyros ebenum, Diospyros ferrea, Diospyros lolin, Diospyros macrophylla, Diospyros pilosanthera, Diospyros rumphii*
Jati	*Tectona grandis*
Jelutung	*Dyera costulata*
Albizia falcataria	
Mahoni	*Swietenia mahagoni Swietenia macrophylla*
Medang	*Litsea firma*
Meranti merah	*Shorea acuminate, Shorea spp; Shorea compressa, Shorea lepidota, Shorea leprosula, Shorea leptoclados, Shorea macroptera, Shorea ovalis, Shorea ovata, Shorea pachyphylla, Shorea palembanica, Shorea parvifolia, Shorea pauciflora, Shorea pinanga, Shorea platycarpa, Shorea platyclados, Shorea quadrinervis, Shorea sandakensis, Shorea selanica, Shorea smithiana, Shorea stenoptera, Shorea teysmanniana, Shorea uliginosa*
Meranti putih	Shorea bracteolate, Shorea gysbertsiana, Shorea javanica, Shorea koordersii, Shorea ochracea, Shorea retinodes, Shorea sororia, Shorea virescens
Meranti kuning	*Shorea acuminatissima, Shorea faguetiana, Shorea gibbosa, Shorea multiflora*
Merbau	*Intsia bijuga, Intsia palembanica*
Nyatoh	*Palaquium burcki, Palaquium ferox, Palaquium gutta, Palaquium hexandrum, Palaquium javense, Palaquium leiocarpum, Palaquium luzoniense, Palaquium microphyllum*
Surian	*Toona sureni*
Tembesu	*Fagraea fragrans, Fagraea sororia, Sloetia elongate*

Table S6. RAS importance reported in different countries

Region/ country	Species used in the immature period	Mature period			Proportional land use
		Fruits	Timber	Other	
Asia					
Indonesia	Rice, palawijas (secondary annual crop)	All fruits found in the original forests	All timber species found in the original forests		Occupies from 10-40% of the area under rubber depending on the province?
Thailand	Pineapple, banana, *Gnetum*	All fruits	Commercial timber species		15% in the southern area? southern part of the rubber growing area?

Region/ country	Species used in the immature period	Mature period			Proportional land use
		Fruits	**Timber**	**Other**	
Asia					
China				Tea, cacao	Partially developed but only on 14% of the rubber growing area
Vietnam	Vegetables	Durian, cashew nut		Coffee, pepper, livestock	Very few AFS
Cambodia	Cassava, mungbean, peanut	Durian, cashew nut		Pepper	Very few AFS
Laos	Monoculture				No AFS reported
Myanmar (AFS only in trials)	Maize, pineapple, banana, pigeon pea	Durian, footyam		Coffee	Very few AFS, only in trials
India	Very diversified			Spices	Extensive AFS (> 80%)
Sri lanka	Banana, pineapple	soursop		Cacao, timber, cinnamon	AFS moderately developed
Malaysia					no AFS
Africa					
Côte d'Ivoire Current	Pure monoculture				No AFS
Côte d'Ivoire Potential	Intercropping with annual crops	Orange, mango, avocado		Cola tree	
Ghana					no AFS
Liberia	No information				
Cameroun					No AFS
Nigeria					What remains from local jungle rubber Some AFS trials
South America					
Brazil		Castabho do brazil and pepper Palmito on Michelin estate		Coffee; cacao,	Very few AFS with rubber

Region/ country	Species used in the immature period	Mature period			Proportional land use
		Fruits	Timber	Other	
South America					
Columbia		Local fruits acai, pupuna, *Theobromas grandis*, palmito			AFS moderately developed
Central America					
Mexico	Pure monoculture				No AFS reported
Guatemala		Palmito/ heart of palm, pineapple		Cacao, cardamom, medicinal plants, local native food plants	Very few AFS

Table S7. Complete list of plant species used as intercrops in rubber agroforestry systems (Adapted from Landenberger and Penot, 2022)

Annual intercrops

Arachis hypogaea L. / groundnut

Capsicum annuum L. / chili pepper

Colocasia esculenta (L.) Scho/ taro

Dioscorea alata L. / purple yam

D. cayenensis Lam. /yellow yam

Glycine max (L.) Merr. /soybeanf

Ipomoea batatas L. (Lam.) / sweet-potato *Manihot esculenta* / cassava

Nicoana spp. / tobacco

Oryza sava L. / upland rice

Pisum savum L. / pea

Sorghum bicolor (L.) Moench / sorghum

Vigna radiata (L.) R. Wilczek / mung bean

Voandzeia subterranea (L.) Thouars / bambara groundnut

Zea maiys L. / maize

Vigna unguiculata /cowpea

Multi-annual crops

Annanas comosus (L.) Merr. / pineapple

banana Musa x paradisiaca L. / banana, plantain

Passiflora edulis Sims / passion fruit

Pogostemon cablin (Blanco) Benth. /patchouly *Cymbopogon citratus* (DC.) Stapf / lemon grass

Elettaria cardamomum) cardamom *Saccharum officinarum* L. / sugarcane

Cover crops

Calopogonium caeruleum (Benth.) C. Wright

Centrosema pubescens Benth.

Mucuna bracteata DC. ex Kurz

Mucuna cochinchinensis (Lour.) A. Chev.
Pueraria phaseoloides (Roxb.) Benth. Senna sp.
Stylosanthes gracilis Kunth *Cassia cobanensis* (Brion) Lundell
Crotalaria sp. Desmodium ovalifolium (Prain) Wall ex Ridley
Flemingia macrophylla (Willd.) Kuntze ex Merr
Flemingia congesta. Mimosa invisa Mart. ex Colla
Mimosa invisa var. inermis Adelb.
Psophocarpus palustris Desv.
Psophocarpus tetragonolobus (L.) DC.
Stenolobium brachycarpum var.
brachystachyum Benth. (Fabaceae)

Permanent intercrops: Standard species grown for export

Thea sinensis L. / tea *Theobroma cacao* L. / cacao
Coffea sp./coffee

Fruits trees

Anacardium occidentale L. / cashew nut
Annona reculata L. / custard-apple
Areca catechu / betel nut
Artocarpus sp. Carica papaya L. / papaya
Durio zibethinus Rumph. ex Murray/durian *Lansium domescum* Corrêa / langsat
Macadamia sp. / macadamia nut
Mangifera indica L. / mango
Morus sp. / fig
Nephelium lappaceum L. / rambutan
Garcinia mangostana L. / mangosteen
Parkia speciosa Hassk. / stink bean

Spice plants

Cinamomum verum J. Presl / cinnamon
Piper nigrum L. / pepper

Timber trees

Aquilaria sp. / eaglewood
Azadirachta indica A. Juss. / neem tree
Dalbergia sp. / i.a. rosewood
Dipterocarpus sp. Eucalyptus sp. / eucalyptus
Hopea sp. Pterocarpus sp. / padouk, narra
Parashorea sinensis H. Wang / wang an shu
Salacca zalacca (Gaertn.) Voss /snake fruit
Shorea macrophylla P.S. Ashton / light red meran
Swietenia mahagoni (L.) Jacq. / mahogany /t Taxus mairei
Tectona grandis L. f. / teak *Adina minutiflora* /Berumbung *Durio carinatus, Durio oxleyanus; Durio zibethinus,* /Durian *Diospyros celebica, Diospyros ebenum, Diospyros ferrea, Diospyros lolin, Diospyros macrophylla, Diospyros pilosanthera, Diospyros rumphii,* / Ebony *Dyera costulata,* /Jelutung *Swietenia mahagoni Swietenia macrophylla,* /Mahoni-Mahogany *Litsea firma,* /Medang
Shorea acuminate, /Meranti merah
Shorea spp; Shorea compressa Shorea lepidota Shorea leprosula, Shorea leptoclados Shorea macroptera Shorea ovalis Shorea ovata Shorea pachyphylla, Shorea palembanica Shorea parvifolia Shorea pauciflora Shorea pinanga Shorea platycarpa Shorea platyclados Shorea quadrinervis Shorea sandakensis Shorea selanica, Shorea smithiana Shorea stenoptera Shorea teysmanniana Shorea uliginosa

Shorea bracteolate Shorea gysbertsiana Shorea javanica Shorea koordersii, Shorea ochracea Shorea retinodes Shorea sororia Shorea virescens, Shorea acuminatissima Shorea faguetiana Shorea gibbosa Shorea multiflora, /Meranti putih
Palaquium burcki Palaquium ferox Palaquium gutta Palaquium hexandrum, /Meranti kuning
Palaquium javense Palaquium leiocarpum Palaquium luzoniense Palaquium microphyllum, /Nyatoh
Toona sureni, /Surian
Fagraea fragrans Fagraea sororia Sloetia elongate /Tembesu
+ Amazonian species

Firewood trees /fast growing trees
Acacia mangium Willd. / acacia
Gmelina arborea Roxb. ex Sm. / gmelina
Paraserianthes falcataria (L.) I.C. Nielsen /white albizia (before Albizia falcataria)

Miscellanous
Gnetum gnemon L. / gnemon
Thaumatococcus daniellii (Benn.) Benth
Afzelia sp. Amorphohallus konjac K. Koch / konjac
Archidendron pauciflorum I.C. Nielsen /dogfruit
Betula alnoides Buch.-Ham. ex D. Don /alder birch
Endospermum malaccense Benth. ex Müll.Arg.
Fagraea fragrans Roxb. ex Carey and Wall.
Nyssa yunnanensis W. Q. Yin ex H. N. Qin and
Phengklai/ protecon
Peronema canescens Jack
Rhus sp. / sumac /
Citrullus lanatus (Thunb.) Matsum. and Nakai, / water melon
Cucurbita spp. Gossypium spp. / coon
Morinda officinalis F.C. How / morinda
Osteospermum spp. / African Daisy
Pachyrrhizus tuberosus (Lam.) Spreng. / yam bean
Phallus indusiatus Vent. / bamboo fungus
Cannabis sava L. / hemp /
Erythroxylum coca, coca
Amorphophassus spp

Multi-purpose trees
Calliandra sp. / false mesquite
Gliricidia spp. / gliricidia

Plants tested, but not actually grown
Axonopus compressus (Sw.) P. Beauv.
Mikania micrantha Kunth (Asteraceae) O
Ochloa nodosa (Kunth) Dandy
Panicum maximum Jacq.
Paspalum conjugatum P.J. Bergius

Noxious weeds
Brachiaria brizantha (Hochst. ex A. Rich.) Stapf
Brachiaria muca (Forssk.) Stapf
Imperata cylindrica (L.) P. Beauv., alang alang

Medicinal plants
Alpinia oxiphylla Amomum longiligulare T.L. Wu / hai nan sharen
Ammomum villosum Lour. / sha ren
Ricinus communis L. / castor bean /

Sources: Landenbergen (2016), Penot (2001), Chee and Faiz (1991), Chong et al. (1991), Ekanayake (2003), Herath and Takeya (2003), Ng (1991), Ng et al. (1997), Pathiratna (2006a, b), Priyadarshan (2011), Rantala (2006), Sanchez and Ibrahim (1991), Shelton and Stur (1991), Shigematsu et al. (2013), Waliszewski (2010), Watson (1989), Wibawa et al. (2006), Williams et al. (2001), Zhou (2000), as well as information by F. Harich (pers. com.), own observations and trials.

Table S8. Technical and economic externalities of agroforestry systems: how to assess value

Positive	Negative	Value assessment
Efficient water use and maintenance of soil fertility	Erosion if there is no soil cover provided by weeds, forest regrowth or crops	Difficult
No weeds grow during the mature period	Competition with weeds during the immature period	Cost of herbicide
Maintenance or increase in soil fertility/limitation of land degradation	Limited effect on soil fertility, effect in the absence of a soil cover	Assessment is difficult when the cost cannot be compared with previous situation with no agroforest
More plant and animal biodiversity conservation in complex AFS	Limited in simple AFS	Medicinal plants No family expenses needed for fuel wood or other products
Better resilience to hurricanes		Possible through decrease in yield
More synergy between plants and less disease	More moisture increases the risk of diseases	Difficult
Shade provides a better environment enabling a longer life span for cocoa and coffee	Shade can limit plant yields	Easy by comparing plant life spans
Increases carbon storage (as a C sink)		Possible and valuable with C value on carbon credit market
Increases global effects of environmental and eco-systemic services		Difficult
Provides a refuge/sanctuary for bees, birds and other pollinators		Invaluable
Enables access to fuel wood (for cooking, distilleries): added value e.g. through the production of essential oil (Madagascar, Comores Islands)		Reduces family expenses

Table S9. Positive and negative economic externalities

Positive	Negative
Improves labour efficiency	Requires more labour than monoculture
Enables income diversification	Markets for associated trees may change and prices of products may drop
Sale of timber covers the cost of planting a new crop at the end of agroforestry system	Depends on tree tenure
Provides a forest-like environment	May reduce yields of some associated perennial crops depending on competition and shade

Table S10. Positive and negative externalities at specific landscape/land-use levels

Positive	Negative
Watershed protection on slopes and prevents silting (Krui, Maninjau lake in Indonesia)	Requires social organisation
Landscape value (tourism, cultural vision, international patrimony)	Limits land use
Increases the ratio of tree covered soil to bare soil	Competition with high value crops like oil palm

Table S11. TGHK forestry classification and its main characteristics

Class	Main Objectives	Uses
Conservation (HSA)	Conservation of natural resources	Extraction and agricultural activity are forbidden
Protection (HL)	Soil protection Water conservation	Extraction and agricultural activity are forbidden
Limited production (HPT)	Timber or wood production with erosion control	Selective cutting
Normal production (HPB)	Wood production	Selective or clear cutting
Conversion (HPK/HK)	Conversion of land for agricultural purposes	Clear cutting

TGHK: Tata Guna Hutan Kesepakatan/Consensus-based Forest Land-Use Planning; HAS, Hutan Suaka Alam dan Wisata; HL, Hutan Lindung; HPT, Hutan Produksi Terbatas; HPB, Hutan Produksi Biasa; HPK/HK, Hutan Produksi Konversi. Source: Cossalter, 1992.

Table S12. Forest, concessions, and actors

Type of forest and concession	Actors
Forest	
Production forest	Forestry concessions
Protected forest	The Indonesian State
Conversion forest	The State provides concessions to private or semi-public estates
Agricultural activities	Multiple actors
Transmigration projects	The State (Ministry of Transmigration)
Plantation concessions (estates)	
Recent concessions for perennial crops: actually planted	Private estates
Recent concessions for perennial crops: not yet planted before 2015	Private estates
Old concessions for perennial crops: actually planted	Private estates
Concessions for estates for industrial tree crop (*A mangium*)	Semi-public estates
Concessions for industrial tree crops in transmigration programmes	The State (Ministry of Transmigration) HTI/NES
Project PTP/NES	The State and local communities
Smallholder crops, fallow and uncropped non-forested land	Local communities (under *adat*)

List of authors

Joseph Adelegan, IRSG

Noé Biatry, ISTOM, Angers, France

Bénédicte Chambon, CIRAD UR Abcys, France

Phillipe Courbet, Nestlé, France

Stephanie Diaz-Novellon, CFA (Chambre des métiers et de l'artisanat de l'Hérault), France

Ilahang, SNV, Indonesie

Vichot Jongrungrot, PSU, Thailand

Laxman Joshi, ICRAF, Indonesie

Hugo Lehoux, Agrarian Systems Consulting, Montpellier, France

Isabelle Michel, Supagro, Montpellier, France

Pascal Montoro, Cirad, Montpellier, France

Elok Mulyoutami, ICRAF, Indonesia

Lekshmi Nair, retired

Éric Penot, Cirad, UMR Innovation, Montpellier, France

Adrien Perroches, Agrarian Systems Consulting, Montpellier, France

Lucie Polinen Agrarian Systems Consulting, Montpellier, France

Jerôme Sainte-Beuve, Cirad, retired

Wilfried Shueller, ENITA Bordeaux, France

Aude Simien, ISTOM, Angers, France

Didier Snoeck, Cirad, retired

Laetitia Stroesser, CIVAM, Poitou, France

Phillipe Thaler, Cirad, Abcys, Montpellier, France

Marion Theriez, Initiatives Paysannes - Territoires Hauts de France, France

Alexis Thoumazeau, CIRAD ABSys, Montpellier, France

Somiot Thungwa, PSU, Thailand

Uraiwan Tongkaemkaew, TSU, Thailand

Karine Trouillard, Bio Nouvelle-Aquitaine, GAB17, France

Gede Wibawa, IRRI, Indonesia, retired

Diah Wulandari, IRRI, Indonesia

Layout: Hélène Bonnet Studio 9

Imprimé pour vous par Libri Plureos GmbH (Allemagne)